D0145762

Pharmacokinetics

Processes, Mathematics, and Applications

Second Edition

Peter G. Welling
Institut de Recherche Jouveinal

PROPERTY OF
SENECA COLLEGE
LEARNING COMMONS
SENECA@YORK

WITHDRAWN

170101

ACS Professional Reference Book

American Chemical Society
Washington, DC

Library of Congress Cataloging-in-Publication Data

Welling, Peter G.
    Pharmacokinetics: processes, mathematics, and applications / Peter G. Welling—2nd ed.
        p.   cm.—(Professional reference book)
    Previously published in 1986 with subtitle: Processes and mathematics.
Includes bibliographical references and index.

ISBN 0–8412–3481–7

1. Pharmacokinetics. I. Title. II. Series: ACS professional reference book.

RM301.5.W45   1997
615'.7—dc21   97–12306
                                                        CIP

The paper used in this publication meets the minimum requirements of American National Standard for Information Sciences—Permanence of Paper for Printed Library Materials, ANSI Z39.48-1984.

Copyright © 1997 American Chemical Society

All Rights Reserved. Reprographic copying beyond that permitted by Sections 107 or 108 of the U.S. Copyright Act is allowed for internal use only, provided that the per-chapter fee of $17.00 base + $.25/page is paid to the Copyright Clearance Center, Inc., 222 Rosewood Drive, Danvers, MA 01923, USA. Republication or reproduction for sale of pages in this book is permitted only under license from ACS. Direct these and other permission requests to ACS Copyright Office, Publications Division, 1155 16th St., N.W., Washington, DC 20036.

The citation of trade names and/or names of manufacturers in this publication is not to be construed as an endorsement or as approval by ACS of the commercial products or services referenced herein; nor should the mere reference herein to any drawing, specification, chemical process, or other data be regarded as a license or as a conveyance of any right or permission to the holder, reader, or any other person or corporation, to manufacture, reproduce, use, or sell any patented invention or copyrighted work that may in any way be related thereto. Registered names, trademarks, etc., used in this publication, even without specific indication thereof, are not to be considered unprotected by law.

PRINTED IN THE UNITED STATES OF AMERICA

To Caroline, Christine, Graham, Luisa, and Stephen

About the Author

PETER G. WELLING, a native of London, obtained his B.Sc., M.Sc., and Ph.D. degrees from Sydney University in Australia. After completing postdoctoral studies at the University of Michigan and a brief period at Pfizer (U.K.), he joined the faculty of the School of Pharmacy at the University of Wisconsin in 1972, as Assistant Professor of Pharmaceutics. He was appointed to Associate Professor in 1975, and Full Professor in 1978. He served as Associate Dean for graduate studies during 1979–1983. In 1984, he joined Warner-Lambert/Parke Davis as Director of the Department of Pharmacokinetics and Drug Metabolism. He was appointed Vice President, Pharmacokinetics and Drug Metabolism in 1987. He held this position until 1996 when he joined the Institut de Recherche Jouveinal in Paris as Director of Pharmacokinetics and Metabolism.

Dr. Welling has published over 200 research articles and reviews and has authored or edited ten books on pharmacokinetics. He has served as a permanent member on the Pharmacology Study Section at the National Institutes of Health, and as a member of the Board of Visitors, University of Wisconsin. He is a Fellow of the American Association of Pharmaceutical Scientists. He was awarded the D.Sc. Degree from Sydney University in 1980.

Contents

Preface   xiii

**DRUG ABSORPTION, DISTRIBUTION, METABOLISM,
AND EXCRETION**   1

1. **Drug Absorption, Distribution, Metabolism, and Excretion**   3

    Different Approaches to Pharmacokinetics / 4
    References / 9

2. **Drug Transport**   11

    The Cell Membrane / 11
    Membrane Transport / 12
    Summary / 17
    References / 17

3. **Parenteral Routes of Drug Administration**   19

    Intraarterial Administration / 20
    Intrathecal Administration / 20
    Intravenous Administration / 21
    Intramuscular Administration / 23
    Buccal Administration / 24
    Intranasal Administration / 25
    Inhalation / 28
    Transdermal Administration / 31
    Vaginal Administration / 37
    Rectal Administration / 38
    Summary / 39
    References / 40

4. Enteral Routes of Drug Administration   **43**

Physiology of the GI Tract / 43
GI Structure and Motility Factors / 45
Gastrointestinal Blood Flow in Relation to Drug Absorption / 50
Special Transport Mechanisms / 51
Animal Models for Oral Drug Absorption / 52
Summary / 59
References / 60

**5. Factors Influencing Absorption and Bioavailability After Enteral
Administration**   **63**

Metabolism in the Gut / 64
Metabolism in the Lung / 65
Hepatic Metabolism and the First-Pass Effect / 66
Enterohepatic Circulation / 69
Factors Influencing the First-Pass Effect / 71
Summary / 73
Problems / 73
References / 73

6. Physicochemical and Formulation Factors Affecting Drug Absorption   **75**

Chemical Factors / 75
Physical Factors / 76
Formulation Factors / 78
In Vitro–In Vivo Correlations / 80
Excipients / 81
Coated Tablets / 82
Controlled-Release Formulations / 82
Absorption Enhancers / 89
Summary / 93
References / 93

7. Clinical Factors and Interactions Affecting Drug Absorption   **95**

Influence of GI Disease on Drug Absorption / 95
Drug–Drug Interactions Affecting Absorption / 100
Drug–Food Interactions Affecting Absorption / 105
Conclusions / 116
Summary / 116
References / 117

8. Drug Distribution   **119**

Distribution–Concentration Relationships / 120
Capillary Permeability / 121
Perfusion and Diffusion Effects / 122
Binding of Drugs to Tissues / 124

Intravascular and Extravascular Drug Binding / 128
Penetration of Drug into the Central Nervous System / 135
Methods of Determining Drug Distribution / 139
Summary / 141
Problems / 142
References / 142

9. Drug Metabolism   **145**

Sites of Drug Metabolism / 145
Mechanisms of Drug Metabolism / 147
Species, Sex, and Age Differences / 156
Steric Factors / 159
Enzyme Induction / 159
Enzyme Inhibition / 160
Summary / 160
References / 161

10. Effect of Liver Disease on Drug Metabolism and Pharmacokinetics   163

Function and Structure of the Liver / 164
Types and Severity of Liver Disease / 164
Effects on Pharmacodynamics and Pharmacokinetics / 165
Examples of Effects of Liver Disease on the Pharmacokinetics of Some Drug
      Classes / 169
Conclusions / 171
Summary / 172
References / 172

11. Renal Excretion   **175**

Structure and Function of the Kidney / 175
Clearance / 178
Renal Clearance and Plasma Clearance / 179
Relationship Among Clearance, Drug Elimination Rate, and Half-Life / 182
Summary / 184
Problems / 184
References / 185

12. Drug Elimination in Renal Impairment   **187**

Methods of Measuring Renal Function / 188
Use of Creatinine Clearance to Predict the Effect of Renal Impairment on Drug
      Elimination / 190
Methods of Dosage Adjustment / 191
Maintenance of Patients with End-Stage Kidney Disease / 194
Summary / 195
Problems / 196
References / 197

THE MATHEMATICS OF PHARMACOKINETICS   199

13. The One-Compartment Open Model with Intravenous Dosage   201

The One-Compartment Open Model with Bolus Intravenous
    Injection / 202
The Trapezoidal Rule / 207
Urinary Excretion Kinetics / 210
Construction of Sigma-Minus Plots / 211
Zero-Order Drug Input and First-Order Elimination / 213
Summary / 219
Problems / 219
References / 221

14. The One-Compartment Open Model with First-Order Absorption and Elimination   223

General Aspects of First-Order Absorption and Elimination / 223
Graphical Estimation of Parameters / 226
Other Parameters Associated with First-Order Absorption and
    Elimination / 230
Drug Absorption Plots / 232
Area Under the Drug-Concentration Curve (AUC) / 234
Absorption Lag Time / 236
Cumulative Urinary Excretion of Unchanged Drug / 237
Summary / 241
Problems / 242
References / 243

15. Multiple-Dose Kinetics   245

Drug Accumulation with Repeated Doses / 245
Drug Accumulation Rate / 250
The Degree of Drug Accumulation with Repeated Dosing and
    the Loading Dose Required To Instantaneously Achieve
    Steady-State Levels / 252
First-Order Absorption Case / 253
Summary / 256
Problems / 256
References / 257

16. Metabolite Pharmacokinetics   259

Pharmacokinetics of Metabolite Formation and Elimination / 259
Analysis of Metabolite Concentrations in Plasma, and Urinary
    Excretion / 262
Resolving the Pharmacokinetics of Metabolite and Parent Drug / 265
Summary / 268
Problems / 268
References / 269

17. The Two-Compartment Open Model with Intravenous or Oral Administration   271

The Two-Compartment Model with Bolus Intravenous Injection / 272
Interpretation of Drug-Concentration Profiles in Plasma To Obtain
   Parameter Estimates / 273
Graphical Estimation of Kinetic Parameters / 276
Derivation of $AUC^{0\to\infty}$, Plasma Clearance, and Renal Clearance / 278
Volume of Distribution at Equilibrium / 279
Kinetics of Tissue Distribution / 282
Obtaining Parameter Estimates from Urinary Excretion Data / 283
The Two-Compartment Model with First-Order Drug Input / 285
Drug Concentration in the First Compartment / 288
Some Model-Independent Parameters for a Drug that Obeys
   Two-Compartment Model Kinetics / 292
Summary / 294
Problems / 295
References / 295

18. Physiological Pharmacokinetic Models   297

Description of Physiological Pharmacokinetic Model / 297
Organ Clearance / 299
Blood Flow Rate-Limited Transport / 300
Membrane-Limited Transport / 305
Experimental Considerations / 306
Summary / 308
Problems / 309
References / 309

19. Nonlinear Pharmacokinetics   311

Saturable Processes / 311
Expressions Useful in Elimination Kinetics / 313
Obtaining Estimates of V_m and K_m from Plasma-Level Data / 316
Influence of Saturable Kinetics on Drug Elimination, Area Under the
   Drug-Concentration Curve, and First-Pass Effect / 318
Summary / 321
Problems / 321
References / 322

APPLICATIONS OF PHARMACOKINETICS IN DRUG DISCOVERY AND DEVELOPMENT   323

20. Pharmacokinetics and Toxicokinetics   325

Pharmacokinetics and Pharmacodynamics / 325
Toxicokinetics and Toxicodynamics / 326
Technical Differences Between Pharmacokinetics and Toxicokinetics / 326

Philosophical Differences Between Pharmacokinetics and
    Toxicokinetics / 328
Conclusions / 329
Summary / 329
References / 329

21. Role of Pharmacokinetics in Drug Discovery and Development   331

Regulatory Submissions / 331
Drug Discovery and Development / 332
Regulatory Submissions and Drug Labeling / 340
Postsubmission and Postmarketing Studies / 341
Conclusions / 341
Summary / 342
References / 342

22. Integration of Pharmacokinetics into Research and Development 343

The R&D Sequence / 343
Statement of the Problem / 344
The Alternatives / 344
Comparison of the Alternatives / 345
Conclusions / 350
Summary / 352
Reference / 353

APPENDIXES   355

Appendix I: Computer Methods and Software for Pharmacokinetic
    Data Analysis   357

Appendix II: Worked Answers to Problem Sets   359

Appendix III: Nomenclature   373

Appendix IV: Glossary   377

INDEX   381

Preface

ince the first edition of this book was published in 1986, there have been considerable changes in the scope and importance of pharmacokinetics and the closely related subject of drug metabolism. Pharmacokinetics has expanded to play a larger and more central role in drug discovery and development and in therapy. The use of pharmacokinetic principles in drug dosage design, treating special patient populations, and in increasing our understanding of factors controlling drug disposition has continued to evolve. The use of metabolism technology, promoted by development of in vitro systems and increasing availability of metabolism data banks, has revolutionized the role of this discipline in drug discovery and development.

To enable the reader to keep abreast of these dramatic and ongoing changes, this book has been extensively revised. The first section, on absorption, distribution, metabolism, and excretion, has been updated and expanded to include chapters on drug transport and on drug disposition in patients with hepatic or renal function impairment. The second section, on the mathematics of pharmacokinetics, has been extensively upgraded consistent with the evolutionary nature of this topic. As in the first edition, pharmacokinetic topics in this section are presented in order of increasing complexity, from the relatively simple one-compartment model to multicompartment and physiological models and nonlinear kinetics. Problems are included with all chapters dealing with the mathematics of pharmacokinetics, and worked answers are provided as an appendix.

A third section has been added describing the extensive applications of pharmacokinetics and drug metabolism in drug discovery and development. This section contains chapters dealing with a critical comparison of pharmacokinetics and toxicokinetics as contributing disciplines in drug discovery and development, the role of pharmacokinetics in pharmaceutical R&D, and the optimum structure for integration of pharmacokinetics and drug metabolism into the discovery and development environments.

While it is not the intent of this book to address computer applications in pharmacokinetics, a brief description of some current computer methods and associated software is provided in an appendix. The reader should find this, and the associated reference, to be useful as a source for further investigation of this topic.

By expanding the book and including an "applications" section, I have attempted to provide the reader with a broad spectrum of information that not only teaches the sciences of drug pharmacokinetics and metabolism but also puts these into perspective relative to the pharmaceutical R&D environment. I am indebted to my many colleagues and friends in academia, regulatory organizations, and industry who have continually tried to educate me in this fascinating subject. I am particularly indebted to my close colleague and friend Theresa Davis, who kept the whole thing, and me, together.

PETER G. WELLING
Institut de Recherche Jouveinal
July, 1997

Drug Absorption, Distribution, Metabolism, and Excretion

Drug Absorption, Distribution, Metabolism, and Excretion

Pharmacokinetics is one of the many disciplines that contribute to the discovery, development, and use of drugs. This was not always the case. Although the basic mathematics of pharmacokinetics was described almost 60 years ago (*1, 2*), it was not until 30 years later that pharmacokinetics, and the closely related discipline of drug metabolism, started to make an appreciable contribution to the understanding and management of drug action. Rapid advancement in these disciplines during the 1960s was the result of the combined influences of improved analytical and computer technology, and increased awareness by drug researchers and health care providers of the importance of pharmacokinetics in drug development and therapy. Rapid advancement was also due, in no small part, to the ingenuity, foresight, and entrepreneurial skills of a number of scientists, predominantly in the United States but also in Europe, including Beckett and Moffatt (*3*), Benet (*4*), Bischoff (*5, 6*), Dedrick (*5, 6*), Dost (*7*), Garrett (*8*), Gibaldi (*9, 12*), Gillette (*10*), Krüger-Thiemer (*11*), Levy (*12, 13*), Nelson (*13*), Resigno and Segre (*14*), Loo and Riegelman (*15*), Teorell (*16*), Wagner (*17*), Yamaoka, Nakagawa, and Uno (*18*), and others.

Prior to 1960, pharmacokinetic and metabolism data made only a minor contribution to regulatory marketing submissions. Not only was there little appreciation of the importance of this type of information, but also little means to provide it even if there were.

The situation has changed dramatically since then, and it continues to evolve as greater emphasis is placed on concentration–effect relationships, both in drug research and in clinical practice.

The discipline of pharmacokinetics, in all its various forms, now holds a unique position in medicine. As greater and greater emphasis is placed on concentration–effect relationships, the influence of pharmacokinetics has spread throughout the drug discovery and development processes and into the practice of medicine. Pharmacokinetics has become a highly interactive and critical discipline influencing a vast spectrum of activities ranging from molecular–structure decisions during early drug discovery to designing optimal dosage regimens in patients with various disease conditions, often complicated by single- or multiple-organ failure. The advance of pharmacokinetics has been so dramatic, and its influence in decision-making processes so great, that in the pharmaceutical industry pharmacokinetic resources are frequently exceeded by demands from other disciplines for its input. This change has given cause to reflect on what pharmacokinetics should and should not do in drug development in the current paradoxical environment of increasing regulatory demands (resulting in no small part from the spectacular advances in the science) and severe cost containment that is facing the pharmaceutical industry. So important is the role of pharmacokinetics in drug discovery, development, and therapy that a separate section is devoted in this book to this intriguing subject. However, before we address this topic and other practical, political, and economic aspects of pharmacokinetics, a longer journey is necessary to understand the principles and mathematics associated with pharmacokinetics and drug metabolism.

Different Approaches to Pharmacokinetics

In the first edition of this book, three different philosophical approaches to pharmacokinetics were identified. These were compartment modeling, physiological modeling, and model-independent pharmacokinetics.

In the compartment modeling approach, the body is assumed to be made up of one or more compartments. These compartments may be spatial or chemical in nature. For example, if a drug is converted in the body to a metabolite, then the metabolite may be considered a separate compartment to the parent drug. In most cases, however, the compartment is used to represent a body volume or group of similar tissues or fluids into which a drug is distributed. Typical compartment models of this type are illustrated in Figure 1.1.

In the physiological model approach, pharmacokinetic modeling and interpretation are based on known anatomical or physiological values. Unlike the compartment modeling approach, in which drug movement between compartments is based largely on reversible or irreversible first-order processes, drug movement using the physiological model approach is based on blood flow rates through particular organs or tissues and experimentally determined blood–tissue concentration ratios. The basic unit that describes the relationship between blood and a particular target tissue is shown in Figure 1.2, and a typical physiological model is shown in Figure 1.3.

The main advantages of the physiological model are that drug movement can be predicted in specific organs and tissues, and that changes in tissue perfusion due to pathological conditions, such as fever or congestive heart failure, can be taken into account when predicting tissue levels. The main disadvantage of the physiological model is that the associated mathematics can become complex and unwieldy. Models must also be developed in vitro or in experimental animals because it is difficult or impossible to validate a model in humans. The physiological model approach has been used extensively for anticancer compounds and for

ONE-COMPARTMENT OPEN MODEL, BOLUS INTRAVENOUS INJECTION

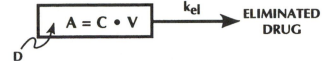

TWO-COMPARTMENT OPEN MODEL, ORAL ADMINISTRATION

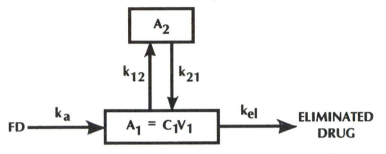

Typical compartment models. One-compartment: D is the dose, A is the amount of drug in the body, C is the concentration of drug in distribution volume V, and k_{el} is the first-order rate constant for drug elimination. Two-compartment: FD is the fraction of dose absorbed, k_{12} and k_{21} are the first-order rate constants for the transfer of drug between compartments, and k_a is the first-order rate constant for drug absorption. Subscripts denote first (central) or second (peripheral) compartments.

FIGURE 1.1

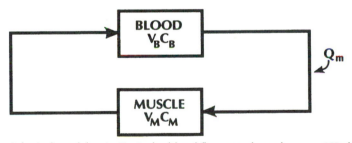

Basic physiological model unit. Q_m is the blood flow rate through organ, V is the organ volume, and C is the drug concentration. Subscript denotes the organ, in this case blood (B) or muscle (M).

FIGURE 1.2

other agents where drug or metabolite location at particular tissues or organs is important.

Model-independent pharmacokinetics represents a trend away from complex modeling systems toward a less complex approach based purely on mathematical description of blood or plasma profiles of drugs or metabolites, and calculation of useful pharmacokinetic values without invoking a particular model.

In many situations, particularly during drug development, and also when comparing plasma drug profiles from different oral formulations in clinical practice, it is sufficient to characterize plasma profiles in terms of maximum plasma

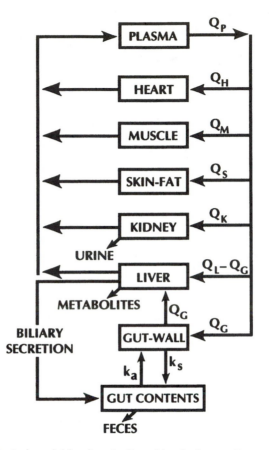

FIGURE 1.3 Complete physiological model for digoxin disposition in the rat. (Reproduced with permission from reference 19.)

levels, C_{max}, time of maximum levels, t_{max}, area under the plasma curve, AUC, and perhaps also elimination half-life, $t_{1/2}$. These parameters can generally be obtained with reasonable and sufficient accuracy by simple inspection of plasma profiles, without having to fit the data to a particular pharmacokinetic model. More complete characterization of plasma profiles in terms of specific models may become more important when (1) more specific pharmacokinetic characterization is required, as when the data are used to predict drug levels under different situations, and (2) accurate models are required to examine relationships between drug disposition and pharmacologic or therapeutic effects. Because of the elegant simplicity of the model-independent approach, it is commonly used throughout the pharmaceutical industry, in regulatory submissions, and in clinical practice.

In addition to these three approaches to pharmacokinetic data interpretation, a fourth approach has emerged that represents a combination of pharmacokinetic and statistical concepts. This approach has become known as population pharmacokinetics, or mixed-effect modeling. Whereas classical or traditional pharmacokinetics focuses on interpretation of data from particular individuals, for example in deriving parameter values from individual plasma drug profiles, the population approach focuses on the central tendency of diverse data that are accrued across a patient or subject population, and on the variability and source of variability of these data among individuals. Pharmacokinetic parameters obtained using traditional and population approaches are identical; they are simply arrived at in totally different ways.

Two examples of the population approach to pharmacokinetic data analysis are described here. The first is shown in Figure 1.4, in which the elimination rates of the nonsteroidal anti-inflammatory agent isoxicam are plotted against steady-state concentration using actual data from a variety of sources. Open circles in the figure represent steady-state isoxicam concentrations obtained during Phase 3 clinical trials in patients, and closed circles represent concentrations obtained from a dose proportionality study in a limited number of subjects. The dashed curve represents the estimated central tendency from these combined data.

Whereas the data from the closed circles alone may be interpreted to indicate an approximate linear relationship between elimination rate and concentration, that is, linear pharmacokinetics, the data from the larger population, which received a much broader range of doses, clearly show a nonlinear relationship

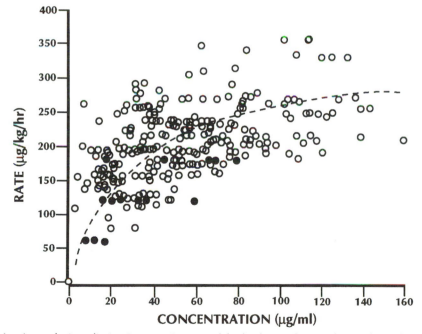

Isoxicam dosing elimination rate (corrected for body weight) vs. observed steady-state isoxicam plasma concentration. Open circles depict Phase 3 clinical trials, closed circles depict a dose proportionality study, and the dashed curve represents the NONMEM estimated central tendency. (Reproduced with permission from reference 20.)

FIGURE 1.4

between drug elimination rate and concentration, and the presence of nonlinear pharmacokinetics. These extended data obtained from a patient population thus provide information regarding isoxicam elimination kinetics and potential accumulation characteristics that would not have been obtained or anticipated from the traditional dose proportionality study.

The second example indicates the type of approach that may be used with sparse data obtained from several individuals to obtain pharmacokinetic characterization of a drug. The example, illustrated in Figure 1.5, is based on a simulation in which 25 individuals each yielded 3–4 data points at various times after oral drug administration. None of the individual data sets describes a complete plasma profile or even a major part of it. Nonetheless, by fitting individual data to a simple one compartment first-order kinetic model by means of the program NONMEM, a composite profile is obtained that is highly characterized both pharmacokinetically and statistically. Typically, the pharmacokinetic parameter values obtained as a result of NONMEM analysis of the combined simulated data were almost identical to the values used to generate the original data points following introduction of a deliberate error factor, giving rise to the observed individual data noise. Thus, population methods typically may not use intensive sampling from one individual to define a pharmacokinetic profile but rather a few samples from a larger number of patients to find the central tendency.

Because of the shift in perspective away from the individual that results from analysis of a population central tendency, intensive characterization of a single

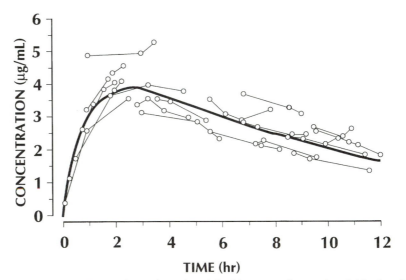

FIGURE 1.5 Example of NONMEM analysis of simulated, highly incomplete and variable data from 25 individuals to yield a pharmacokinetic profile of central tendency. Mean pharmacokinetic parameter values used for individual simulations: dose, 1000 mg; clearance, 20 L/h; distribution volume, 200 L; absorption rate constant, 1 h^{-1}; and a 5% assay error. Model mean estimated parameter values: clearance, 19.7 L/h; distribution volume, 213 L; and absorption rate constant, 1 h^{-1}. (Reproduced with permission from data in reference 21.)

individual is not critical to the method. Relaxation of restrictions concerning the number of samples required from each individual permits cost-effective sampling from a greater number of individuals. This in turn permits generation of a broad base of pharmacokinetic information in a large population at costs equivalent to those required to conduct formal, intensive, traditional pharmacokinetic studies in a small number of subjects.

However, the limited sampling approach associated with population pharmacokinetics has been perceived by some as a major disadvantage of this method. For example, reduced restrictions on sampling may result in loss of control over study conditions and sample documentation. For further discussions of the various aspects of population pharmacokinetics and mixed-effect modeling, the reader might find some recent references useful (*22, 23*). The reader is warned, however, that in order to obtain maximum benefit from these reviews it may be useful to first become familiar with other material in this book and with a substantial body of statistics before venturing into this new and intriguing but complex approach to pharmacokinetic data analysis.

References

1. Teorell, T. *Arch. Int. Pharmacodyn.* **1937,** *57*, 205–225.
2. Teorell, T. *Arch. Int. Pharmacodyn.* **1937,** *57*, 226–240.
3. Beckett, A. H.; Moffatt, A. C. *J. Pharm. Pharmacol.* **1970,** *22*, 15–19.
4. Benet, L. Z. *J. Pharm. Sci.* **1972,** *61*, 536–541.
5. Bischoff, K. B.; Dedrick, R. L. *J. Pharm. Sci.* **1968,** *87*, 1347–1357.
6. Dedrick, R. L.; Bischoff, K. B. *AIChE Symp. Ser.* **1968,** *64*, 32–44.
7. Dost, F. H. *Der Blutspiegel*; Georg Thieme: Liepzig, East Germany, 1953.
8. Garrett, E. R. *Acta Pharmacol. Toxicol.* **1971,** *29*, 1–29.
9. Gibaldi, M. *Pharmacokinetics and Clinical Pharmacokinetics*, 3rd ed.; Lea and Febiger: Philadephia, PA, 1984; p 330.
10. Gillette, J. R. *Ann. N. Y. Acad. Sci.* **1971,** *179*, 43–66.
11. Krüger-Thiemer, E. *Eur. J. Pharmacol.* **1968,** *4*, 317–324.
12. Levy, G.; Gibaldi, M. *Ann. Rev. Pharmacol.* **1972,** *12*, 85–98.
13. Levy, G.; Nelson, E. *J. Pharm. Sci.* **1965,** *54*, 812.
14. Resigno, A.; Segre, G. *Drug and Tracer Kinetics*; Blaisdell: Waltham, MA, 1966.
15. Loo, J. C. K.; Riegelman, S. *J. Pharm. Sci.* **1968,** *57*, 919–928.
16. Teorell, T. *Arch. Int. Pharmacodyn.* **1937,** *57*, 205–225.
17. Wagner, J. G. *Fundamentals of Clinical Pharmacokinetics*; Drug Intelligence Publications: Hamilton, IL, 1975; p 461.
18. Yamaoka, K.; Nakagawa, T.; Uno, T. *J. Pharmacokinet. Biopharm.* **1978,** *6*, 547–556
19. Harrison, L. I.; Gibaldi, M. *J. Pharm. Sci.* **1977,** *66*, 1138–1142.
20. Olson, S. C. *European Cooperation in the Field of Scientific Research. New Strategies in Drug Development and Clinical Evaluation: The Population Approach*; Office for Official Publications of the European Communities: Brussels, Belgium, 1992; pp 143–152.
21. Olson, S. C.; Kugler, A. R. *Pharm. Res.* **1990,** *7*, S262.
22. Aarons, L. *Br. J. Clin. Pharmacol.* **1991,** *32*, 669–670.
23. Beal, S. L.; Sheiner, L. B. *Crit. Rev. Biomed. Eng.* **1982,** *8*, 195–222.

Drug Transport

2

A drug must pass through several biological membranes in order to be absorbed, distributed into organs and tissues, and finally eliminated. Biological membranes are complex structures with a wide variety of compositions to serve many purposes. Membranes associated with the skin are concerned primarily with protection, and they also have a secretion component. The gastrointestinal (GI) epithelial lining membranes are concerned with selective absorption and secretion. Capillary membranes have a particular structure to facilitate diffusion of substances into and out of the systemic circulation. Membranes associated with the blood–brain barrier are designed to selectively protect the central nervous system from insult by foreign substances. Membranes of the proximal kidney tubules are concerned with selective secretion. Some membranes are sensitive to hormonal influences. For example, membranes associated with the distal renal tubules are acted upon by vasopressin from the posterior pituitary gland, resulting in water reabsorption, whereas membranes of the proximal renal tubule cells are unaffected by vasopressin.

The proportion of protein and lipid varies with different membranes, depending on membrane function. Some examples are given in Table 2.1. Typically, myelin has insulation properties and a high fat content, whereas mitochondria are associated with enzyme reactions and have a high protein content.

Despite the considerable diversity in membrane structure and function, there is general consensus regarding the basic structure of the cell membrane. Recent theories regarding the structure and function of the cell membrane, and also the relationship among membrane structure, physiological function, and mechanisms of transmembrane drug transport, have been reviewed (1).

The primary structure of the cell membrane, shown in Figure 2.1 is a 5 nm-thick bimolecular lipid film that separates intracellular and extracellular fluids.

The Cell Membrane

The lipid is composed mainly of the phospholipids phosphatidylserine and phosphatidylinositol and contains saturated and unsaturated fatty acids and sterols. The bimolecular lipid bilayer exhibits high permeability to hydrophobic molecules and low permeability to hydrophilic molecules (2).

The cell membrane is associated with intrinsic and extrinsic proteins. The intrinsic proteins are globular proteins that generally span the bilayer and are held within the membrane by hydrophobic interactions within the biolayer and by electrostatic interactions at the membrane surface. The proteins can form channels, carriers, or pumps that enable polar molecules to cross the membrane. The extrinsic proteins are attached to the cytoplasmic face of the cell membrane by electrostatic forces. These proteins do not appear to play a role in drug transport (2).

Membrane Transport

Several mechanisms have been identified for drug transport across membranes (1). One of them is passive, and the remainder entail some type of carrier mechanism.

Simple or Passive Membrane Transfer

This mechanism, based primarily on lipid solubility and concentration gradient, is responsible for membrane transport of the vast majority of drugs and other foreign substances.

The mechanism has also been termed solubility–diffusion because these characteristics control membrane permeability of passively transferred molecules. The direction and rate of transport, J (mass/time), are determined by the concentration gradient of a substance across the membrane, $C_o - C_i$ (mass/volume), where the subscripts o and i refer to the outside and inside of the membrane, the permeability of the substance in the biolayer, P (length/time), and the area of the membrane, A, as in equation 2.1.

$$J = P \cdot A \left(C_o - C_i\right) \tag{1}$$

The range of membrane permeabilities can be great. For example, for human erythrocytes, P values range from as low as 1×10^{-10} cm s^{-1} for erythritol to as high as 1×10^{-4} cm s^{-1} for propranolol. Typically, hydrophobic molecules have high partition coefficients whereas hydrophilic molecules and ions have low parti-

TABLE 2.1	**Membrane Protein and Lipid Content**	
Membrane	*Protein Content*	*Lipid Content*
Myelin	18	79
Human erythrocyte	49	43
Bovine retinal rod	51	49
Mitochondria (outer membrane)	52	48
Mycoplasma laidlawii	58	37
Sarcoplasmic reticulum	67	33
Gram-positive bacteria	75	25
Mitochondria (inner membrane)	76	24

Note: Units are percent dry weight.

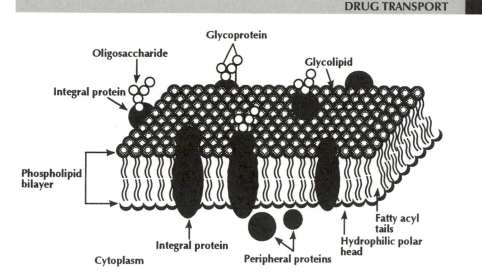

Glycoprotein

Oligosaccharide

Glycolipid

Integral protein

Phospholipid bilayer

Fatty acyl tails

Integral protein

Hydrophilic polar head

Cytoplasm

Peripheral proteins

A structural model for plasma membranes. The membrane is composed of a bimolecular leaflet of phospholipid with the polar head groups facing the extracellular and cytosolic compartments and the acyl tails in the middle of the bilayer. Sterols (e.g., cholesterol) are inserted between the acyl chains, but for simplicity are not shown. Integral membrane proteins are embedded in the lipid bilayer, and stability is ensured by hydrophobic bonding with membrane lipid and electrostatic bonding with polar head groups and the aqueous solution on each side of the membrane. The integral proteins are glycosylated on the exterior surface, and they may be phosphorylated on the cytoplasmic surface. Extrinsic membrane proteins, peripheral proteins (e.g., spectrin), are linked to the cytosolic surface of the intrinsic proteins by electrostatic interactions. (Reproduced with permission from reference 1.)

FIGURE 2.1

tion coefficients, the latter being particularly due to high dipole interactions and hydrogen bonding with water.

Hydrophobic molecules cross membranes more readily than hydrophilic molecules because membranes have a basic lipoidal structure. Molecules that are weak acids and weak bases cross membranes more readily when they are in the un-ionized form, that is, when the pH of fluids directly bathing the membrane surface is less than the pK_a of a weak acid or greater than the pK_a of a weak base. The pK_a values for a number of common acidic and basic drugs are shown in Table 2.2.

While passive membrane permeability is favored for the un-ionized form of acidic and basic drugs, aqueous solubility is favored for the ionized form. In order to be available to cross any membrane, a drug has to be in solution. A molecule cannot pass through a biological membrane in particulate form. This paradoxical requirement of both aqueous and lipid solubility is of particular concern in the area of drug absorption and presents a continuous challenge to those concerned with pharmaceutical formulation. Many potential drugs have failed because of poor aqueous solubility, and the pharmaceutical industry is currently replete with entrepreneurial organizations attempting to devise and utilize solubilizers, carriers, and stabilizers that will increase the aqueous solubility and absorbability of lipophilic molecules. This has been particularly important for many novel agents that, apart from their poor aqueous solubility or stability characteristics, show considerable therapeutic potential.

Membrane Transport Mechanisms

Many substances, particularly ions and polar molecules, cross membranes at rates greater than those predicted from solubility and permeability data alone. Some can cross membranes against a concentration gradient. For many years, scientists have attempted to understand the mechanisms for these "unexpected" characteristics. As a result of extensive studies during the 1950s and 1960s, and particularly advances in molecular biology leading to cloning and sequencing of transport proteins, many transport proteins have been cloned, sequenced, and studied. The technique of expression cloning (3) has been particularly important in this area. Current knowledge has permitted an operational definition of carrier proteins as channels, carriers, and pumps. The current state of the art in identification and characterization of these systems has been described by Wright (1).

Unstirred Water Layer

The total resistance of a solute across biological membranes is the sum of aqueous and membrane components, in series. A drug has to be in solution in order to reach the membrane surface to be transported and to cross the unstirred water layer adjacent to the membrane surface by diffusion. The nature of the unstirred water layer is described in Chapter 4; suffice it to say here that for hydrophobic compounds the

---

TABLE 2.2   **Ionization Constants (pKₐ) of Some Common Acidic and Basic Drugs**

Acids		Bases	
Drug	pK_a	Drug	pK_a
Cephalothin	2.5	Amphetamine	9.8
Ampicillin	2.5, 7.2	Ephedrine	9.6
Carbenicillin	2.6, 2.7	Imipramine	9.5
Flucloxacillin	2.7	Chlorpromazine	9.3
Methicillin	2.8	Erythromycin	8.8
Probenecid	3.4	Isoprenaline	8.6
Aspirin	3.5	Orphenadrine	8.4
Ibuprofen	4.4	Diphenhydramine	8.3
Chlorpropamide	4.8	Bupivacaine	8.1
Sulfafurazole	4.9	5-Fluorouracil	8.1
Cephalexin	5.2, 7.3	Mepacrine	7.7, 10.3
Dicoumarol	5.7	Codeine	6.0
Nitrofurantoin	7.2	Aminopyrine	5.0
Acetazolamide	7.2	Chlordiazepoxide	4.6
Pentobarbital	8.1	Diazepam	3.3
Phenytoin	8.3	Nitrazepam	3.2, 10.8
Ethosuximide	9.3	Caffeine	0.8

Note: Acidic and basic drugs are listed in order of decreasing strength. That is, cephalothin is the strongest acid, and ethosuximide is the weakest. Amphetamine is the strongest base, and caffeine is the weakest.

role of this layer can be significant in limiting intestinal absorption, regardless of the subsequent mechanism and efficiency of membrane transport.

Initial speculation on the existence of small aqueous pores in membranes was based on the high membrane permeability of some small polar molecules. For example, the permeability of water is 1000 times more, and that of urea is 10–100 times more, than predicted. These observations led to prediction of membrane aqueous channels with radii of approximately 4 Å (1).

Channels

Water Channels. Definitive evidence of water channels has now been described on the basis of successful cloning of proteins that increase membrane water permeability. These have been expressed in erythrocytes and cells of the renal tubule. Recent cloning of a separate urea transporter from kidney suggests different permeability enhancers for water and urea.

Ion Channels. Evidence for channels for ions in biological membranes started to be introduced in the 1970s. Possibly the most significant experiment was by Neher and Sakmann (4), who recorded single ion channel currents in muscle fibers using a patch-clamp technique. Many types of channels for sodium, potassium, calcium, and chloride ions have been reported, and each has a specific conductance, ion selectivity, and probability of opening. Ion channel opening may be controlled by voltage or by ligand, and channels are thus designated as voltage-gated or ligand-gated. Despite a certain structural and functional homology, each of the ion channels has a specific pharmacology. For example, sodium channels are 12 times more selective for sodium than for other cations, whereas calcium channels are 1000 times more selective for calcium than for other cations.

Facilitated diffusion is a simple mechanism that is proposed to explain membrane transport of water-soluble compounds. The principal characteristics of this transport system are that membrane permeability exceeds that predicted from solute partition coefficients, transport occurs down a concentration gradient, transport is saturable and stereospecific, and competition occurs between isomers. Isomers may be transported with different carrier affinities. A simple model for facilitated diffusion has been described by Wright (5) as in Figure 2.2.

Facilitated Diffusion

Facilitated diffusion has been used to explain cellular uptake of sugars and amino acids. As shown in Figure 2.2, the membrane protein can exist in two conformations, which can bind to a substrate on either side of the cell, and in a third conformation in which substrate is occluded within the cell.

The best documented carrier is the facilitated glucose transporter in human red cells (6). Since this transporter was first cloned in 1985, six human sugar transporters have been identified with different tissue distributions, substrate kinetics, and specificities. A number of facilitated amino acid transporters have also been identified in mammalian cells. System L, which transports neutral amino acids such as leucine and phenylalanine, is probably the best known of these.

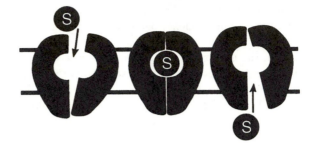

FIGURE 2.2   A simple model for facilitated diffusion of polar solutes across plasma membranes. Diffusion of a substrate S is mediated by a membrane protein that can exist in three different conformations: one in which S can bind from the outside of the cell; a second in which S is occluded; and a third in which the bound substrate is released to the cytoplasm. Transport can occur in either direction, and the direction of net transport depends simply on the substrate concentration gradient across the membrane. (Reproduced with permission from reference 1.)

Pumps

Pumps are proteins that can transport ions against their electrochemical potential gradients using adenosine 5′-triphosphate (ATP) as an energy source. Sodium–potassium pumps maintain intracellular sodium and potassium concentrations in animal cells and also control salt and water absorption by the epithelial cells in the intestine and kidney. The sodium–potassium pump transports three sodium ions out of the cell and two potassium ions into the cell at the cost of one molecule of ATP. This coupling of sodium and potassium transport indicates that the pump rate depends on intracellular sodium and extracellular potassium concentrations. The 3:2 coupling ratio results in a net loss of cations from the cell. This loss compensates for the passive leakage of sodium ions into the cell down an electrochemical gradient of approximately −150 mV, and it maintains cell volume. There is currently considerable activity attempting to elucidate the structures of the various isoforms and subunits of sodium–potassium pumps.

Cotransporters and Exchangers

There are many other examples of ionized ions and molecules being accumulated in cells against their electrochemical potential gradients, such as uptake of iodine by the thyroid gland, accumulation of acids in liver cells, and absorption of sugars and phosphate by the small intestine. Although these phenomena were originally attributed to some form of active transport, more recent studies have shown that they are governed by cotransport mechanisms. In all cases tested to date, sodium or hydrogen ion gradients are used to drive cotransporters, and these gradients are maintained by appropriate ion pumps. For example, glucose transport across the brush border of the small intestine is coupled with sodium transport, and the "uphill" sugar transport is driven by the sodium gradient.

Cotransporters.   Cotransporters use the sodium or hydrogen ion gradient to drive the transport of a substrate in the same direction as the ion flux. Many cotransporters have been identified, cloned, sequenced, and expressed in heterologous systems. The sodium–glucose cotransporter just described is one of these.

Other cotransporters facilitate the transport of other sugars, osmolytes, and amino acids. In humans, a disorder of intestinal glucose and galactose absorption is due to a defective sodium–glucose cotransporter.

Antiporters.   The best characterized antiporters, or exchangers, are the chloride–bicarbonate, sodium–hydrogen ion, and sodium–calcium exchangers. The cellular sodium–hydrogen ion exchanger controls cell volume, pH, growth, and sodium transport. Mammalian isoforms of these exchangers have been cloned and sequenced (7).

Sodium–calcium exchangers play a dominant role in the regulation of intracellular calcium, and thus the force of contraction in the heart. The therapeutic effect of cardiac glycosides is probably related to a decrease in sodium–calcium exchange in the heart caused by a decrease in the sodium gradient across the cell membrane (8).

P-Glycoprotein

P-glycoprotein (permeability glycoprotein, P-gp) is an ATP-dependent efflux pump responsible for pumping substances out of cells. It is associated with the development of drug resistance in tumor cells. In addition, the localization of P-gp in the apical membranes of intestinal, liver, and kidney cells, as well as at the blood–brain barrier, results in the potential for this pump to have a profound effect on drug absorption, disposition, elimination, and drug–drug interactions (9, 10).

Pinocytosis and Endocytosis

Pinocytosis is a nonspecific process by which substrate is taken up by invagination of the cell membrane to form an intracellular vesicle. Receptor-mediated endocytosis occurs when substrate binds to a specific membrane receptor before being internalized. Substrates ingested into cells by this mechanism are either stored in vesicles or degraded by lysosomes. Receptor-mediated endocytosis is responsible for cellular uptake of immunoglobulin, transferrin, and low-density lipoprotein.

Summary

In order to reach the site of action, a drug may have to pass through one or more biological membranes.

Cell membranes have diverse structures and functions, but are based on a common primary structure consisting of a bimolecular lipid bilayer with associated intrinsic and extrinsic proteins.

Drug transport across biological membranes may be passive or carrier-mediated.

References

1. Wright, E. M. In *Pharmacokinetics: Regulatory, Industrial, and Academic Perspectives,* 2nd ed.; Welling, P. G.; Tse, F. L. S., Eds.; Marcel Dekker: New York, 1995; pp 89–118.
2. Darnell, J.; Lodish, H.; Baltimore, D. *Molecular Cell Biology,* 2nd ed.; Scientific American: New York, 1990; pp 532–582.
3. Hediger, M. A.; Coady, M. J.; Ikeda, T. S.; Wright, E. M. *Nature* **1987,** *330,* 379–381.
4. Neher, E.; Sakmann, B. *Nature* **1976,** *260,* 799–802.
5. Wright, E. M. In *Pharmacokinetics, Regulatory, Industrial, and Academic Perspectives,* 2nd ed.; Welling, P. G.; Tse, F. L. S., Eds.; Marcel Dekker: New York, 1995; p 101.
6. Baldwin, S. A. *Biochim. Biophys. Acta.* **1993,** *1154,* 17–49.

7. Tse, M.; Levine, S.; Yun, C.; Brant, S.; Pouyssegur, J.; Donowitz, M. *J. Am. Soc. Nephrol.* **1993,** *4,* 969–975.

8. Philipson, K. D. In *Molecular Physiology and Pharmacology of Cardiac Ion Channels and Transporters;* Morad, M.; Ebashi, S.; Trautwein, W.; Kurachi, Y., Eds.; Kluwer Academic: Norvell, MA, 1996; pp 435–445.

9. Hunter, J.; Hirst, B. H.; Simmons, N. L. *Pharm. Res.* **1993,** *10,* 743–749.

10. Su, S.-F.; Huang, J. D. *Drug Metab. Dispos.* **1996,** *24,* 142–147.

Parenteral Routes of Drug Administration

3

 variety of methods and routes are used to administer drugs for systemic activity. Dosage routes may be divided into two major categories, enteral and parenteral. *Enteral* administration generally refers to drugs administered by way of the gastrointestinal (GI) system, whereas *parenteral* administration refers to drugs given by all other routes. However, the definition becomes obscure in some cases. For example, although buccal and rectal drug administration might be considered enteral insofar as the mouth and rectum are component parts of the GI tract, differences in the absorption mechanisms between these sites and the remainder of the GI tract permit them to be differentiated.

Compounds that are absorbed from regions of the GI tract ranging from the esophagus to the upper rectum are absorbed into the splanchnic circulation, which feeds into the portal vein, which passes directly into the liver before it reaches the general circulation via the hepatic vein. Thus all compounds absorbed by this route, with the exception of the very small quantities of compounds absorbed via the lymphatic system, have to pass through the liver and are thus potentially subject to hepatic as well as GI "first-pass" metabolism before entering the general circulation. Compounds that are absorbed from the buccal cavity and also from the lower rectum, on the other hand, pass directly into the local regions of the systemic circulation rather than the splanchnic circulation, and are thus not subject to possible first-pass hepatic metabolism or biliary clearance.

Thus, buccal and rectal administration (at least in the latter group, for compounds absorbed from the lower rectum) should be considered parenteral in the same way as intramuscular and intranasal administration and inhalation. Paradoxically, compounds that are administered by the intraperitoneal route, a common procedure used to administer compounds to experimental animals during drug development, enter the body largely via the splanchnic and portal circulations, so that this dosage route is essentially enteral in nature. This fact is not always appreciated in drug discovery and development, where compounds administered by intraperitoneal injection are often assumed to be 100% systemically bioavailable.

In this chapter the following parenteral routes of administration are discussed:

- Intraarterial
- Intrathecal
- Intravenous
- Intramuscular
- Buccal
- Intranasal
- Inhalation
- Transdermal
- Vaginal
- Rectal

Intraarterial Administration

The arterial route is used for regional delivery of drugs to various organs, for example, in cancer chemotherapy and in the use of vasopressin for GI bleeding. *Intraarterial* infusion increases drug delivery to the area supplied by the infused artery, while hopefully reducing access of the drug to the systemic circulation (1). This is particularly important for infusion of a highly cytotoxic drug into an artery that receives a small fraction of the cardiac output. In order to reduce entry of a toxic drug into the systemic circulation after arterial infusion, however, there must be loss of drug by metabolism, excretion, or chemical degradation during its passage through the region being perfused by the arterial blood.

Clinical reports have shown that intraarterial carmustine (*N, N'*-bis(2-chloroethyl)-*N*-nitrosourea, or BCNU) is effective for the treatment of metastatic brain tumor from lung carcinoma (2), and that local (pelvic) intraarterial actinomycin D is useful for malignant trophoblastic disease (3). Cason and Whaley (4) used a head-hunter catheter for intraarterial hepatic infusion of chemotherapeutic agents, which permits direct perfusion of primary and secondary liver tumors. Carotid artery infusion of vinblastine is more satisfactory than venous infusion of the drug in treating cerebral tumor in rats (5). One study has demonstrated the effectiveness of selective intraarterial vasopressin infusion in treating massive upper GI tract hemorrhage (6).

Use of the intraarterial route of drug administration is limited by potential dangers. Complications such as embolization, arterial occlusion, and excessive local drug toxicity are not uncommon, so maximum care is required in dose administration.

Intrathecal Administration

Injection directly into the cerebrospinal fluid ensures complete bioavailability of drugs that otherwise may have difficulty crossing the blood–brain barrier. This dosage route is therefore useful in the treatment of serious central nervous system (CNS) infections such as meningitis and ventriculitis. *Intrathecal* injection is also used to produce spinal anesthesia with such agents as mepivacaine and prilocaine, and in relieving chronic pain. Intrathecal administration of p-aminomethylbenzoic acid is effective in treating subarachnoid hemorrhages, while oral or intravenous administration of the drug fails to produce effective concentrations in the cerebrospinal fluid (7). In recent years, intrathecal chemotherapy has gained importance in the treatment of meningeal neoplasms.

Drugs administered by the intrathecal route are usually injected directly into the lumbar spinal subarachnoid space, into the subdural space, or into the ventricles, frequently via an implanted reservoir. The site of injection has a great effect on the availability of drug to various regions of the CNS. This effect is demonstrated in Table 3.1 (8). Intraventricular administration of ^{3}H-methotrexate in monkeys gave rise to significantly higher drug levels at 2 h in whole brain and in cervical and thoracic spinal cord, but lower levels in the lumbar spinal cord compared with levels obtained from a spinal catheter. At 4 h postdosing, brain levels from intraventricular administration were still 12 times greater than those obtained from the spinal catheter, although drug levels in the spinal cord were then significantly higher from the spinal catheter administration.

Drugs injected intrathecally are distributed initially into a much smaller volume (140 mL of cerebrospinal fluid) than those administered intravenously (3 L of plasma). The former route therefore generally provides much higher drug concentrations in the CNS with less risk of systemic toxicity. Intrathecally administered methotrexate produces concentrations in the cerebrospinal fluid that are 100-fold higher than simultaneous plasma concentrations. However, toxic methotrexate levels in plasma may be more prolonged following intrathecal injection compared with intravenous or oral doses, possibly because of slow release of the drug from the CNS into the systemic circulation (9).

Intravenous administration is the method used to introduce a drug directly into the venous circulation. Apart from intraarterial administration, intravenous administration is the only parenteral dosage form that introduces a compound directly into the systemic circulation without it having to cross one or more biological membranes.

Intravenous Administration

Mean Levels of ^{3}H-Methotrexate in Whole Brain and Spinal Cord Regions after Intraventricular and Intrathecal Spinal Catheter Administration to Monkeys

TABLE 3.1

Tissue	2 h Postinjection			4 h Postinjection		
	IVA[a]	SC[b]	IVA/SC Ratio	IVA	SC	IVA/SC Ratio
Whole brain	4.32	0.15	28.8	8.67	0.74	11.7
Cervical spinal cord	8.46	0.86	9.8	7.13	26.1	0.3
Thoracic spinal cord	7.92	1.36	5.8	4.29	60.2	0.1
Lumbar spinal cord	5.64	9.57	0.6	1.96	121.2	0.02

Note: Units are micrograms per gram of tissue.
[a]Intraventricular administration.
[b]Spinal catheter.
Source: Reproduced with permission from reference 8.

This method of administration, in common with intraarterial administration, results in complete absorption of drug into the systemic circulation. However the shape of the resulting drug profile is determined by the rate and duration of injection, as indicated in Figure 3.1.

The intravenous bolus route is particularly useful when prompt therapeutic effect is required, for example, in the use of intravenous lidocaine for emergency treatment of cardiac arrhythmia (*10*). A major disadvantage of intravenous administration is that, once injected, the dose cannot be recovered. It is therefore common practice to inject a drug slowly over a prolonged period to avoid excessively high drug levels in the circulation and also to permit administration to be stopped if intolerance occurs.

Because intravenous administration always results in 100% bioavailability of unchanged drug, this route of administration should be used in preliminary in vivo pharmacology screens for systemically acting compounds. If a compound is inactive after intravenous administration, it is unlikely to be active after administration by any other route unless presystemic metabolism of an enterally administered parent compound gives rise to a more pharmacologically active metabolite. In cases where low aqueous solubility precludes intravenous dosing, pharmacologic activity may be determined by oral or intraperitoneal doses. However, in these cases any pharmacologic activity (or lack thereof) is necessarily the combined result of intrinsic activity and systemic availability of a compound.

A major advantage of the intravenous dosage route in drug discovery, development, and clinical practice is the high degree of control that can be exercised regarding the dose size and rate of administration. By application of simple pharmacokinetic principles, an infinite variety of plasma drug profiles can be achieved. This is particularly useful when examining drug concentration–response relationships.

Apart from bolus injections, a number of methods are available for prolonged intravenous infusion. These may be based on mechanical devices using positive pressure or gravity flow. While quantitative delivery of a drug given by bolus injection is not in question, delivery efficiency from intravenous infusion devices may vary, both in terms of rate and extent (*11*). Variable delivery of drug during an infusion period has been reported with peristaltic systems and also with syringe pumps. With gravity flow systems using a roller clamp, drug delivery may decline with time, resulting in reduced drug delivery (*12*). For example, in one study the

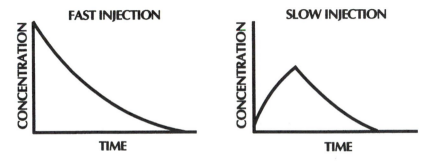

FIGURE 3.1   Profiles of drug concentration vs. time in blood after fast and slow intravenous injection.

tobramycin infusion rate was reduced by 26% after 5 min of infusion and by 45% after 95 min (*12*). This may be an exceptional case, and most reported variances in intravenous infusion rates are minor, but they may nonetheless be important, particularly for more potent drugs and in pediatric patients.

Intramuscular injection may be used for drugs that are not absorbed orally but for which rapid drug levels in blood, as obtained with intravenous bolus injection, are not required. Intramuscular injection generally, but not always, results in quantitative absorption of the drug into the systemic circulation. Slow or incomplete absorption after intramuscular injection has been reported for a number of compounds, including diazepam, digoxin, phenytoin, and phenobarbital. Although phenobarbital appears to be quantitatively absorbed after intramuscular injection in children, its absorption is only 80% of that seen in oral doses in adults (*13*).

Absorption of drug from intramuscular sites is affected by drug molecular weight, type of solvent, injection volume, and vascular perfusion. Sparingly soluble compounds may precipitate at the injection site. For example, quinidine base precipitates after intramuscular injection of quinidine hydrochloride solution. Drugs with low aqueous solubility may be administered in oily vehicles. In this case drugs may precipitate from these vehicles when introduced into the aqueous intramuscular environment, causing slow and possibly incomplete absorption. Drugs are generally absorbed more readily from small injection volumes than from large volumes perhaps because of mechanical compression of adjacent capillary beds, increasing volume–surface area ratio, a larger diffusion path from the center of the injection site, and a smaller concentration gradient with large injection volumes. Schriftman and Kondritzer (*14*) showed that atropine was absorbed faster from small injection volumes than from large injection volumes in guinea pigs.

Consistent with membrane theory, lipid-soluble molecules are more rapidly absorbed from intramuscular sites than water-soluble molecules, and absorption of the latter is dependent on molecular size. Small water-soluble compounds may enter the circulation directly via the capillaries, whereas larger molecules enter the circulation indirectly via the lymphatics. Thus absorption of the latter compounds is limited by lymph flow, which is approximately 0.1% of plasma flow.

Vascular perfusion is a rate-limiting factor for intramuscular drug absorption. Blood flows through muscular tissue at rates ranging from 0.02 to 0.07 mL/min per gram of muscle tissue. The faster the rate, the faster the absorption. Lidocaine is absorbed more rapidly from arm muscle than from the gluteus maximus, insulin is absorbed faster from arm muscle than from thigh muscle, and cefuroxime and cephradine are absorbed faster from intramuscular doses to men than to women. The last finding is presumably due to the greater proportion of adipose tissue in females.

Thus, although intramuscular drug administration may generally be more reliable than oral administration, it is not without its problems. Complete and prompt absorption cannot be assumed in every case. The intramuscular dosage route is often used for sustained-release medication, and considerable attention has focused on the use of aqueous suspensions, oily vehicles, complexes, microencapsulation, and liposomes to achieve delayed delivery. Although oral administra-

Intramuscular Administration

tion of liposomes has been unsuccessful, intramuscular administration has greater potential. The use of liposomes to target antitumor agents to cardiac tissue, the central nervous system, and other tissues is being actively investigated. Microcapsules consisting of small drug particles coated with a biodegradable polymer give rise to prolonged release after parenteral doses (*15*). Twofold increases in duration of activity of fluphenazine embonate and gold sodium thiosulfate have been achieved by intramuscular injection of microencapsulated drug (*16, 17*).

Buccal Administration

The first description of systemic effects following absorption of a compound through the oral mucosa was given by Sobrero in 1847 (*18*), who showed that sublingual administration of a small quantity of nitroglycerin caused severe and protracted headache. Some 30 years later, sublingual nitroglycerin was introduced into clinical practice for relief of angina pectoris (*19*).

Drugs can be absorbed from the oral cavity either from the buccal cavity itself (between cheek and gingiva) or sublingually (under the tongue) (*20*). Absorption from either route is rapid, sublingual being faster than buccal apparently because of greater permeability of membranes in the sublingual region, the thin mucous membrane, and the rich blood supply.

Absorption via the oral mucosa occurs mainly by passive diffusion into the lipoidal membrane (*21*). There is some evidence from animal experiments that water-soluble molecules with molecular volumes less than 80 cm^3/mole are absorbed from the oral cavity via membrane pores, while some large water-soluble molecules are absorbed paracellularly. After absorption, a drug enters the facial or sublingual veins, which then drain into the general circulation via the jugular vein, and thus bypasses hepatic first-pass metabolism. Absorption via the lymphatic system is relatively unimportant. The mean pH of saliva is 6 so that, as absorption is mainly passive in nature, absorption is favored for uncharged molecules, acids with pK_a values greater than 3, and bases with pK_a values less than 9.

Compounds that are administered or are being considered for administration by buccal or sublingual routes include organic nitrates, barbiturates, papaverine, trypsin, prochlorperazine, benzodiazepines, phenazocine, buprenorphine, captopril, isoprenalin, oxytocin, and nifedipine. Oxytocin is the only peptide currently marketed in a sublingual form. It is used for induction or stimulation of labor, tablets being placed in alternate check pouches every 30 min until a dosage is found that produces the desired response. Absorption of oxytocin from the buccal cavity is probably inefficient, as the buccal dose is about 100 times greater than intravenous or intramuscular doses. Swallowing the buccal tablets does not harm the patient because oxytocin is digested in the GI tract.

Sublingual administration of steroids has not been examined extensively. 17 β-Estradiol (E_2) for treatment of postmenopausal symptoms is readily absorbed sublingually. A 0.5-mg sublingual tablet gave rise to serum levels equivalent to those following a 2-mg oral dose of micronized E_2 (*22*). However, serum levels of E_2 are transient and are followed by more prolonged levels of estrone (E_1). Serum levels of E_1 and E_2 and changes in serum follicle stimulating hormone (FSH) and luteinizing hormone (LH) following a single 0.5-mg sublingual dose of E_2 are shown in Figure 3.2 (*22*). Sublingual administration of E_2 thus differs from vaginal

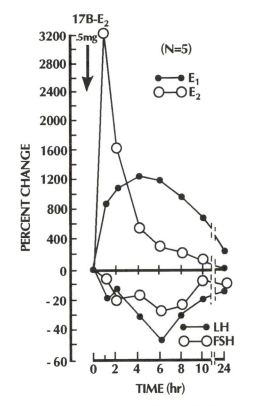

Relative changes in serum concentrations of 17β-estradiol (E₂), estrone (E₁), luteinizing hormone (LH), and follicle stimulating hormone (FSH) after sublingual administration of a single half-tablet containing 0.5 mg of micronized 17β-estradiol. (Reproduced with permission from reference 22).

FIGURE 3.2

administration, which, for reasons yet to be explained, does not appear to give rise to such extensive E_2–E_1 conversion.

Conventional buccal and sublingual dosage forms are short-acting because of the limited drug-exposure time to the oral mucosa. Although sublingual administration of prolonged-release dosage forms is impractical because of interference with eating and talking, prolonged buccal delivery has greater potential. Although a number of attempts have been made to exploit this novel drug delivery route by means of adhesive delivery devices, none has yet been commercially successful. Buccal and sublingual drug delivery has recently been reviewed (23–25).

Intranasal administration may be used for local or systemic effects (26). Local effects are achieved with corticosteroids, antihistamines, and other agents for such conditions as nasal allergy, rhinitis, and nasal decongestion. The mode of action of corticosteroids in nasal allergy appears to be associated with inhibition of basophilic cell accumulation at the mucosal surface. The use of intranasal delivery for systemic effects is less well established but is currently the focus of consid-

Intranasal Administration

erable activity. This interest has been prompted by a number of reports of efficient drug absorption by this route, predominantly in animal models. The results have been particularly promising for small peptide molecules that are inefficiently absorbed orally.

Despite the high interest in this area, no novel intranasal compounds have been successfully developed recently for systemic activity, and existing marketed forms are restricted to vasopressin analogs and oxytocin. Cocaine is taken illicitly by the nasal route for its euphoric effect. The systemic availability of intranasal cocaine may be no greater and may also be no faster than after oral administration (27). Perhaps the rapid "high" after nasal dosing is related to particular circulatory pathways connecting the nasopharynx, sinus, and brain (28).

In considering why the development of the nasal route for systemic drug effects has been relatively unsuccessful compared to the buccal and transdermal routes, it is useful to consider the anatomy and physiology of the absorbing surfaces. The lateral wall of the nasal cavity, shown in Figure 3.3 (29), consists of discrete anatomic functional regions. Region A in the figure is the skin of the nasal vestibule, B is the region of squamous epithelium without microvilli, C is the transitional epithelium with short variable microvilli, D is stratified epithelium with some ciliated cells, and E is pseudostratified epithelium with many ciliated cells.

The major function of the upper airways is to protect the delicate tissues of the lung from toxic agents in inspired air. The nose also acts as a humidifier and heat exchanger (30). The innervation of the nose is concerned primarily with regu-

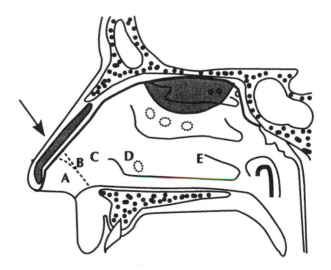

FIGURE 3.3   Lateral wall of the nasal cavity with the olfactory region (shaded area). The arrow indicates the internal ostium. The different types of epithelial cells are A, skin in nostril; B, squamous epithelium without microvilli; C, transitional epithelium with short microvilli of varying length; D, pseudostratified columnar epithelium with few ciliated cells; and E, pseudostratified columnar epithelium with many ciliated cells. (Adapted with permission from reference 29).

lation of nasal blood flow and secretions to enable it to perform its protective function. The mechanisms of the nose are exquisitely developed so that inspired or expired air can be handled, and the lungs protected from external environments.

The sophisticated structure and specialized function of the airways and membranes associated with the nasal cavity and the limited surface area of this region raise questions, at least in the writer's mind, as to the capacity of this region for drug delivery and also the ability of these structures to withstand the insult of chronic drug exposure. Although research in this area continues, evidence is gathering that local toxicity may be a major limiting factor in intranasal drug delivery, particularly in view of the apparent need for addition of surfactants to assist drug absorption by this route for systemic activity. Any advantage obtained by systemic absorption and avoidance of presystemic clearance may be offset by local toxicity.

Notwithstanding these qualifications, the physicochemical characteristics of compounds for optimal intranasal absorption appear to be similar to those for any other absorption route. The drug must be sufficiently water soluble to be administered as a solution or to dissolve rapidly in the fluids of the nasal mucosa, and it must also be sufficiently fat soluble to penetrate the lipid membranes of the nasal epithelial cells. Nasal absorption is facilitated by the high permeability of the small venules and capillaries associated with the nasal mucosa (31).

A drug may be delivered to the nasal mucosa in the form of a solution, formulated as drops or a nebulized spray, or as an aerosol. With nebulized sprays or particulate aerosols, the optimal particle size for nasal deposition is 5–10 µm. Smaller particles do not deposit as readily in the nasal mucosa but are deposited farther down the bronchial tree. A number of methods are available to determine particle-size distribution in aerosol systems. A variety of devices continue to be described for accurate delivery of metered drug doses from aerosols and nebulizers (32–34). Nasal absorption has been considered for a number of compounds, usually with limited success. The compounds examined include cardiovascular drugs, antimicrobials, anterior and posterior pituitary hormones, gonadotropin releasing hormones, adrenal and sex hormones, insulin, autonomic agents, CNS agents, prostaglandins, antihistamines, diagnostic agents, and inorganic salts.

Peptides

Peptides represent a major area of potential for intranasal drug delivery. Vasopressin, desmopressin, and oxytocin are commercially available as nasal sprays. Other compounds being actively investigated include thyrotropin releasing hormone (TRH), other vasopressin analogs, luteinizing hormone releasing hormone (LHRH) agonists and antagonists, adrenocorticotropic hormone (ACTH), and growth hormone releasing factor (GRF) (35). Intranasal administration of the immunomodulators normuramyl dipeptide and muramyl tripeptide phosphatidylethanolamine has been examined in mice (36). Intranasal doses resulted in accumulation of both compounds in the brain compared to intravenous administration.

Intranasal interferon has been examined principally for local treatment of rhinovirus. Although efficacy has been reported (37), side effects such as bleeding, nasal discharge, and superficial lesions were observed after long-term treatment (38, 39). Prolonged intranasal treatment does not appear to be feasible for prophylaxis of respiratory viral infection, at least not at the doses used in these studies.

The new LHRH agonist buserelin is effective for treatment of endometriosis after prolonged intranasal administration of 100–400 µg/day (40) and may be an effective contraceptive at intranasal doses of 400–600 µg/day. Intranasal doses were apparently well tolerated.

Insulin

The intriguing possibility of administering insulin intranasally, rather than by injection, has prompted considerable activity in this area, with variable results. A number of studies have demonstrated hypoglycemic activity with intranasal insulin, but with relatively low efficacy. By addition of surfactant, a hypoglycemic effect can be achieved that is approximately one-tenth of that from an intravenous dose (41–43). Among the surfactants, bile salts, such as sodium deoxycholate, Carbopol 934, and polyoxyethylene-9-lauryl ether (Laureth 9), appear to be promising. Even so, this route of administration is considered only as an adjunct to injectable insulin, and there is little information on long-term toxicity.

Steroids

Intranasal studies with sex hormones in experimental animals have produced some interesting results. The bioavailability of intranasal testosterone in the rat model was 99% and 90% from 25- and 50-µg doses, respectively, compared to intravenous doses. Intraduodenal bioavailability was only 1% (44). Mean blood testosterone concentrations from the three dosage routes examined in this study are shown in Figure 3.4. In the same rat model, progesterone was also shown to be quantitatively absorbed compared to the intravenous drug, with intraduodenal administration again yielding poor absorption (45); in a monkey model, however, the bioavailability of intranasal progesterone was significantly higher than that of intravenous or intramuscular doses (46, 47). It is unclear why intranasal progesterone would yield superior bioavailability to the parenterally administered compound.

A number of studies have attested to the potency of intranasal estrogens to inhibit spermatogenesis, and this may be related to higher brain concentrations attained intranasally compared with other routes. However, high cerebrospinal fluid concentrations may be a function of the animal model. High brain levels of estradiol relative to plasma levels were obtained after intranasal administration in one monkey model (48) but not in another (49).

Animal Models

The principal animal models used to examine intranasal drug delivery are the rat, sheep, dog, and monkey. These models appear to be useful to study nasal absorption from solutions, but not from aerosols. The latter have proven more difficult for two major reasons. The first is that it is not possible to train an experimental animal to sniff from a dispenser; the second is the natural reluctance of product development personnel to develop sophisticated aerosol delivery devices for animal testing at an early stage in drug development. A number of experiments in monkeys have utilized sprays (50). Gizurarson recently published an excellent review on animal models for intranasal drug delivery studies (51).

Inhalation

Inhalation, like intranasal administration, is used mainly for local effect. During the last 10 years, the understanding of the difficulties associated with aerosol drug

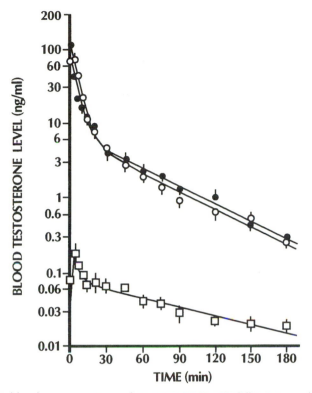

Mean (+SE) blood concentrations of testosterone in rats following nasal (O), intravenous (●), and intraduodenal (□) administration of 25 μg of testosterone per rat. (Reproduced with permission from reference 44).

FIGURE 3.4

delivery has increased, and devices have been described to improve drug delivery by this route. It is interesting that the lung offers a potential absorption service of 72 m^2, a much larger surface than the small intestine, and yet the lungs and associated airways are designed to deny access of administered compounds to this highly efficient absorption region. The respiratory system is designed to keep out particulate matter, and it does this very well. However, when compounds reach the central or peripheral regions of the lungs, absorption can be very efficient.

Respiratory Tract

The human respiratory tract is shown diagrammatically in Figure 3.5. The objective of inhalation drug delivery therapy is to reach the region of the respiratory tract where the drug may exert its maximum effect. The main problem lies in the uncertainty as to where that site is. Most proponents of inhalation drug delivery aim to achieve maximum drug penetration into peripheral lung air spaces, but the site for bronchodilatation is higher in the airways of the tracheobronchial tree.

The respiratory system consists of the upper respiratory tract (nose, nasal passages, sinuses, mouth, and larynx) and lower respiratory tract (tracheobronchial tree down to the pulmonary alveoli). Although the diameter of the airways decreases from about 2–3 cm in the pharynx to only 0.015 cm in the alveoli, air

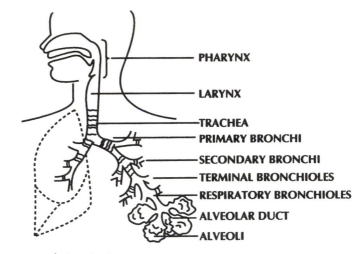

PHARYNX

LARYNX

TRACHEA
PRIMARY BRONCHI
SECONDARY BRONCHI
TERMINAL BRONCHIOLES
RESPIRATORY BRONCHIOLES
ALVEOLAR DUCT
ALVEOLI

FIGURE 3.5   The human respiratory tract.

flow also decreases from 50–100 cm/s in the pharynx to only 0.003 cm/s in the alveoli. This decrease is due to increased overall airway diameter in the alveolar region.

The degree to which inhaled substances penetrate into airway spaces is controlled by a number of factors, principally particle size and velocity. Optimum particle size for airway penetration is 3–5 μm. Particles of larger size tend to deposit in the upper respiratory tract by inertial impaction. Particles less than 2 μm tend not to deposit as effectively as larger particles, but there is some disagreement on this. Deposition of 3–5-μm particles in the lower respiratory tract is primarily by sedimentation.

Three major mechanisms are involved in removal of particulate material from the respiratory tract. Ciliated cells line the trachea and bronchial tree. The motion of the cilia, together with the viscous mucous blanket lining the surface of the larger airways, effectively removes particles from the airways. Other major mechanisms are coughing and also scavenging by alveolar macrophages. Mucociliary clearance may be markedly affected by inhaled drugs, being increased by bromhexine, aminophylline, and prednisolone and decreased by scopolamine. The commonly used β₂-adrenergic agents seem to have little effect.

Absorption from the lung appears to obey the same rules as those that govern absorption from the GI tract. Absorption is related to lipophilicity, and there is little evidence of a specialized transport system.

Systemic absorption of inhaled compounds may be inhibited by first-pass pulmonary metabolism. The lungs contain a variety of drug-metabolizing enzymes. These include microsomal mixed function and other oxidases, amine oxidase, monoamine oxidase, reductases, esterases, and a variety of conjugases. The primary defense mechanisms, the geometry of the bronchial tree, mucociliary and macrophage activity, and pulmonary metabolism combine to inhibit access of the drug to the peripheral pulmonary absorption sites and to the systemic absorption.

Most drug delivery devices for inhalation deliver approximately 10% of the dose to the lower respiratory tract, the balance being absorbed from the buccal cavity or swallowed. Of the 10% that penetrates the airways, approximately two-thirds deposit in the conducting airways and only one-third reaches the alveoli (52).

There have been several attempts to increase drug delivery to peripheral pulmonary vessels. Pressurized metered-dose inhalers (MDIs) have largely replaced such early devices as the Spinhaler. Even so, drug delivery to the lung from MDIs is inefficient, and a large number of spacer devices have been proposed to improve delivery (53). With one such device, 21% of the entered dose reached the lungs, and only 16% was deposited in the oropharynx (54); 56% of the dose was retained in the spacer. Lung deposition was thus increased while oropharyngeal deposition was reduced.

Despite these types of advances, drug delivery to the lung is still inefficient. Most compounds currently administered by inhalation are for local effects. They include ephedrine, isoproterenol, albuterol, terbutaline, metaproterenol, bitolteral, disodium chromoglycate, beclomethasone, and atropine.

Inhalation Delivery Systems

A variety of animal models have been used to study drug inhalation. These include the rat, rabbit, and dog. In dogs, rapid systemic absorption of the lipophilic molecule fluorescein anion was obtained after aerosol and intratracheal administration (55). Inhaled drug as liquid aerosols was absorbed more rapidly than when administered by intratracheal injection in rats (56).

Animal Models

Since the introduction of transdermal scopolamine (57), there has been considerable interest in transdermal drug delivery for systemic effects. The major advantages claimed for transdermal systems are constant release of drug to the systemic circulation, reduced presystemic clearance, facile drug withdrawal by removing the delivery device, and increased patient convenience and compliance. The major disadvantages are related to the barrier properties of the skin, sensitivity and other skin reactions, and dose size.

The skin has excellent barrier characteristics and effectively inhibits the penetration of all but a small range of lipophilic molecules. Even for molecules with appropriate physicochemical characteristics, absorption through the skin is inefficient. Transdermal delivery for systemic effects is a realistic option only for drugs that are given in small doses (≤10 mg) and that have optimum membrane penetration qualities. Only a small number of drugs are currently approved in the United States for transdermal delivery, namely clonidine, estradiol, nicotine, nitroglycerin, and scopolamine. Many other drugs are being examined. Most of these will probably fail.

Transdermal Administration

The anatomy, physiology, and biochemistry of the skin have been extensively characterized (58). The structure of the skin is shown schematically in Figure 3.6 (58). In order to reach the systemic circulation, a drug must pass through the stratum corneum and also the more hydrophilic epidermis. For compounds with molecular weights of less than 600 daltons, absorption is controlled predominantly by solubility. Absorption of larger molecules is limited primarily by their

The Skin

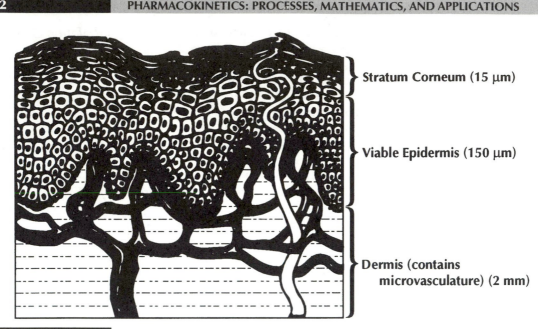

Stratum Corneum (15 μm)

Viable Epidermis (150 μm)

Dermis (contains
microvasculature) (2 mm)

FIGURE 3.6   Schematic of skin structure. (Reproduced with permission from reference 58).

bulk. Although transdermal delivery is often claimed to bypass presystemic metabolism, a large number of drug-metabolizing enzymes are present in skin. These enzymes may inhibit transdermal drug absorption, but they may also be useful in converting lipophilic prodrugs to active compounds during transdermal passage (*59*).

Absorption Models   In vitro systems are traditionally used as primary screens to test transdermal drug absorption (*60*). These systems generally use human cadaver skin, hairless mouse or rat skin, or synthetic membranes. These systems may provide useful preliminary data, but they may be poor predictors of actual in vivo absorption. In vivo animal models are more predictive of human absorption. Many species have been used, including the rabbit, monkey, dog, guinea pig, miniature pig, and rat. Of these, the miniature pig is probably the most predictive.

Factors Affecting   Attempts to improve the bioavailability of transdermally administered com-
Transdermal Drug   pounds have generated a great many delivery devices and promoters. The main
Delivery   objective of the devices is to maintain intimate contact between the drug and skin
                        in an occluded environment and to deliver the drug in a continuous fashion. The
                        sole objective of delivery enhancers is to influence the skin environment to facili-
                        tate drug passage, stopping short of solubilizing the skin. There is tremendous lit-
                        erature on this subject, and much of the material has been described in recent
                        reviews (*61*). Azone (1-dodecylazacycloheptan-2-one) is a typical promoter that
                        increases the absorption of both hydrophilic and hydrophobic compounds (*62*).
                        Many other enhancers are being developed and tested. Liposomes have been

examined to increase transdermal drug absorption (63). Results of in vitro studies show that absorption of some lipohilic drugs is increased from liposome vehicles. Liposomes themselves do not diffuse across skin.

Two major types of transdermal drug delivery systems are shown in Figures 3.7 and 3.8 (58). The membrane-modulated system (Figure 3.7) consists of a drug reservoir enclosed within an impermeable membrane, with a polymeric membrane on one side controlling drug transport from the reservoir to the skin via an adhesive layer. In the matrix device (Figure 3.8), the drug is dispersed throughout a polymeric matrix, and transport through the matrix controls release of drug to the skin. Despite the different mechanisms of drug release from these types of devices, and variations of these devices, the kinetics of drug release is similar. Penetration through the skin is often rate limiting. Much of the development of trans-

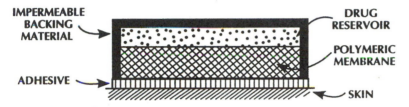

Diagram of a membrane-modulated transdermal drug delivery system. (Reproduced with permission from reference 58).   **FIGURE 3.7**

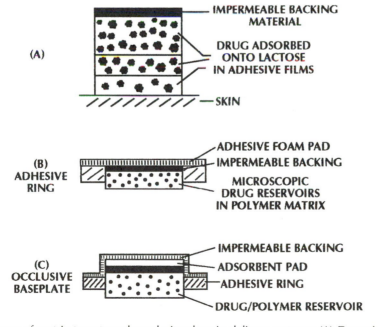

Diagrams of matrix-type transdermal nitroglycerin delivery systems: (A) Deponit-TTS; (B) Nitrodisc; and (C) Nitro-Dur. (Reproduced with permission from reference 58).   **FIGURE 3.8**

dermal delivery systems is still empirical, and there is need for a more systematic approach to their design and greater appreciation of the complexity of interplay between transdermal systems and the natural barrier to absorption afforded by the skin. Some compounds currently administered transdermally, and others under investigation, are summarized in the following sections.

Transdermally Delivered Drugs

Scopolamine.   Scopolamine was the first drug to be successfully administered transdermally for the treatment of motion sickness. A membrane-controlled system is used to provide initial fast release of drug, followed by controlled zero-order release at a rate of 3.8 µg/cm² h for 3 days (57). Scopolamine excretion patters following transdermal and intravenous infusion are compared in Figure 3.9, which demonstrates rapid release of the scopolamine loading dose during the 12–14 h after transdermal application. A number of studies have demonstrated the efficacy of transdermal scopolamine for motion sickness.

Nitroglycerin.   Nitroglycerin was early recognized as an ideal candidate for transdermal drug delivery owing to its physical properties, low dose, and poor oral systemic bioavailability. A number of transdermal systems are available. Commercial formulations, which include both matrix and membrane-modulated systems, include Deponit-TTS, Nitrodisc, and Nitro-Dur (matrix systems), and Transderm-Nitro (membrane). All of these devices give rise to similar nitroglycerin plasma profiles (58). Most preparations claim a 24-h duration of effect. Onset of action is generally within 30 min of application, and effects continue for about 30 min after removal. Several studies have examined the mechanisms of nitro-

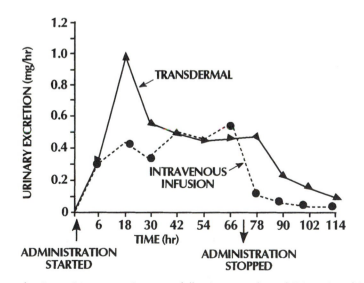

FIGURE 3.9   Mean scopolamine urinary excretion rates following transdermal ($N = 7$) and intravenous ($N = 6$, 3.7–6.0 µg/h) administration. Drug input was stopped at 72 h. (Adapted with permission from reference 58).

glycerin release from transdermal systems (64) and the kinetics of transdermal nitroglycerin delivery (65).

Despite the popularity of transdermal nitroglycerin, questions regarding the possible development of tolerance to continuous drug exposure with these dosage forms for prophylaxis against angina are unresolved. Similar concerns have been expressed for the closely related compound isosorbide dinitrate (66). Topical nitroglycerin has been associated with contact dermatitis, as have the transdermal dosage forms of a number of other compounds.

Clonidine.   The low oral therapeutic dose of this β-adrenergic antihypertensive agent makes it an excellent candidate for transdermal dosage. Devices with surface areas of 2.5–10.5 cm^2 have yielded plasma concentrations of 0.6–1.2 ng/mL (67). The efficacy of transdermal clonidine has been shown to be equivalent to that of conventional oral doses. Skin reactions, including irritation and sensitization, have been reported with transdermal clonidine.

Estradiol.   The efficacy of transdermal estradiol for treatment of menopausal symptoms, at delivery rates of 0.025–0.1 mg/day, is equivalent to 20–40-fold higher oral doses (68). A membrane-controlled device, Estraderm, was shown to yield dose-related serum estradiol levels, as shown in Figure 3.10 (69). Other studies have examined simultaneous skin permeation and metabolism of lipophilic estradiol esters to estradiol (70). This is an excellent example of the use of lipophilic prodrugs to enhance transdermal delivery.

Nicotine.   Nicotine replacement therapy has been shown to be effective in alleviating withdrawal symptoms in individuals attempting to stop smoking (71). Transdermal nicotine delivery provides predictable plasma nicotine concentrations, and efficacy trials have shown that it offers a useful aid when reducing or discontinuing smoking (72, 73). Circulating plasma levels of nicotine and its major metabolite cotinine have been shown to be dose proportional with linear pharmacokinetics (74) and to be similar in normal-size men and women, but lower in obese men (75).

Despite the encouraging results obtained with transdermal nicotine for smoking withdrawal, and its uncomplicated and predictable pharmacokinetics, the transdermal nicotine dosage form has not achieved the market acceptance that was originally anticipated. This lack of acceptance is most likely due to the multiplicity of behavioral and psychological factors that contribute to smoking satisfaction in addition to circulating levels and action of nicotine.

β-Adrenergic Antagonists.   The success achieved with transdermal delivery of the cardiovascular agents nitroglycerin and clonidine prompted interest in other cardiovascular agents, particularly β-adrenergic blocking agents. Preliminary human screenings with timolol supported earlier positive results obtained in animal models (76). Other studies have examined propranolol absorption from ethanol–propylene glycol gel ointments (77) and the use of iontophoresis to induce transdermal delivery of metoprolol in human volunteers (78). In the latter study,

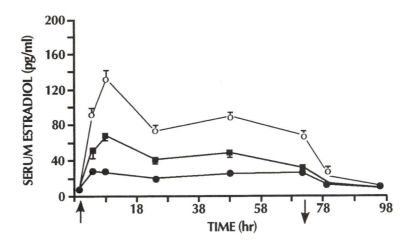

FIGURE 3.10 Mean (±1 standard error) serum concentrations of estradiol resulting from single 3-day topical applications of Estraderm 0.025 (●); Estraderm 0.05 (■); and Estraderm 0.1 (○). Corresponding surface areas are 5, 10, and 20 cm³, respectively. Arrows indicate times of patch application (↑) and removal (↓) (Reproduced with permission from reference 69).

delivery of the drug from a 50-cm² electrode pad on the forearm yielded therapeutic plasma concentrations that were maintained for a longer period than after conventional oral administration.

Other Drugs.   Many other drugs are being investigated for transdermal delivery. For some, preliminary data are appearing in the literature. Others may remain unannounced for a while. Drugs mentioned in published studies include ephedrine, theophylline, indomethacin, and mitomycin C. Because of its greater lipophilicity, ephedrine has a transdermal permeability 120 times greater than that of scopolamine. A transdermal device has been described to deliver ephedrine at a rate of 370–380 μg/h from a 10-cm² area, and urinary excretion profiles indicate that steady state is achieved by 9–10 h in humans (79).

Serum theophylline levels within the therapeutic range of 4–12 μg/mL were obtained from a transdermal device applied to the upper abdomen in 11 of 13 infants (80). Transdermal theophylline may thus be a useful alternative to oral theophylline in preterm infants with apnea in whom oral absorption is erratic.

In vitro release of indomethacin has been shown to vary from different ointment bases, and in vitro release rates correlate with in vivo release through rabbit skin (81). In this study, indomethacin was absorbed rapidly from ointment bases, yielding peak plasma concentrations of 1.1–2.7 μg/mL at about 2 h from 1–5% ointments.

Prodrug approaches have been used to improve transdermal delivery of a number of anticancer agents, including vidarabine, 5-fluorouracil, 6-thiopurines, and mitomycin C. Most of these approaches use lipophilic prodrugs that are converted to the parent drug in the skin by enzymatic or chemical lability. For example, lipophilic prodrugs of mitomycin C have enhanced transdermal delivery between 3- and 5-fold.

The tremendous investment in transdermal drug delivery systems has resulted in marketed products for only five drugs in the United States. This modest success rate reflects the problems inherent in this delivery approach. Transdermal delivery is limited to lipophilic molecules with excellent skin permeability that are not toxic to the skin. The prodrug approach is being used to increase the number of drugs that can be given by this route.

Vaginal Administration

The human vagina is a fibromuscular tube 4 to 6 in. long, directed upward and backward, extending from the vulva to the lower part of the uterine cervix. Blood supply to the vagina is via the uterine and pudendal arteries, which arise from the iliac artery. The vagina is drained by a rich plexus, which empties into the internal iliac veins. The epithelial lining of the vaginal wall consists of lamina propria and a surface epithelium, which is composed of noncornified, stratified squamous cells. The surface of the vaginal epithelium is kept moist by a cervical secretion, whose composition and volume vary with age, stage of the menstrual cycle, and degree of sexual activity. The pH of vaginal fluid varies between 4 and 5, depending on the stage of the cycle and location. The pH is lowest around the anterior fornix and highest around the cervix.

The vaginal drug delivery route is used mainly for local effects. However, a number of studies have shown that vaginal administration can give rise to rapid and complete systemic availability. Good systemic absorption via the vaginal route and also the ability of the vagina to retain delivery devices for reasonably prolonged periods have given rise to a number of vaginal dosage forms for steroid contraceptives. This dosage route clearly lends itself to controlled and sustained delivery of medication.

Two major types of intravaginal controlled-release systems are available: vaginal rings and microcapsules. There are two common types of vaginal rings: homogeneous and shell. A burst effect of drug release on insertion and a declining release rate after extended wear commonly occur with homogenous rings. Shell rings apparently minimize the burst effect and are able to maintain a steady drug release rate. For most vaginal rings, the rate of vaginal drug absorption shortly after insertion is controlled by either an aqueous hydrodynamic diffusion layer or by the vaginal wall. At later times, the rate of vaginal absorption is determined by the drug release rate from the ring (82). Because of problems associated with erosion of the vaginal wall, ring expulsion, interference with coitus, and storage and sanitation, vaginal rings have received only moderate market acceptance.

One example of the ease with which compounds may be absorbed from the vaginal dosage route is shown in Figure 3.11 (83). This figure shows plasma levels of estrone and estradiol in a healthy 53-year-old woman following oral and vaginal doses of conjugated equine estrogens. The levels of both estradiol and estrone are much higher following the vaginal dose rather than the oral dose. Four out of five women studied in this experiment had similar results.

An alternative intravaginal contraceptive system, free of most of these problems, is the biodegradable microsphere. The rationale for its development is that

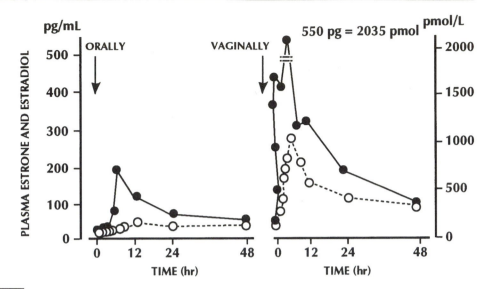

FIGURE 3.11   Plasma estrone (●) and estradiol (○) levels following 1.25-mg oral and vaginal doses of conjugated equine estrogens in a 53-year-old female patient. (Reproduced with permission from reference 83).

particles can migrate from the vagina across the cervix into the fallopian tube or the perimetrial lining of the uterus without causing erosion of the vaginal wall by virtue of its small size.

Rectal Administration

Rectal drug administration is a common alternative when oral administration of drugs is not practical. This route is now well accepted for delivery of a number of drugs, including anticonvulsants, narcotic and non-narcotic analgesics, theophylline, antiemetics, and antibacterial agents, as well as for inducing anesthesia in children (84).

The human rectum is approximately 15–20 cm long. In the resting state it has no active motility. The rectum is normally empty and contains only 2–3 mL of mucous fluid (pH 7–8). There are no villi or microvilli on the rectal mucosa, and only a limited surface area (200–400 cm²) is available for absorption. Both blood and lymphatic vessels are abundant in the submucosal region of the rectal wall. The upper veins drain into the portal circulation, while the lower and middle veins drain directly into the inferior vena cava. However, there are extensive anastomoses among these veins, so that clear-cut anatomical differentiation is difficult. Nonetheless it is well established that drugs absorbed from the lower colon, in contrast to those absorbed from the upper colon, tend to avoid hepatic first-pass metabolism (85).

In general, rectal absorption occurs at a slower rate and to a lesser extent than oral drug absorption. However, for a number of drugs, rectal absorption has been reported to exceed oral absorption, which may reflect partial avoidance of hepatic first-pass metabolism after rectal delivery. This result has been reported for morphine, metaclopramide, ergotamine, lidocaine, and propranolol. Rectal bioavailability of the highly metabolized drug lidocaine in humans is 65% compared with oral bioavailability of 30% (86). The extent of first-pass metabolism after rectal administration may be influenced by the site of drug administration in the rectum.

For poorly water-soluble drugs, the rectal absorption rate is determined by release surface area rather than by drug concentration in the dosage form. Absorption from aqueous and alcoholic solutions is in general much faster than from suppositories, which are dependent on the particle size of active ingredient and the nature of the suppository base, surfactants, and other additives. Nonsurfactant adjuvants, such as salicylates, have been shown to increase rectal absorption of water-soluble drugs and also of high-molecular-weight drugs such as insulin, heparin, and gastrin.

The development of rectal controlled-release drug delivery systems is limited by a number of factors, including interruption of absorption by defecation and, in certain geographic locations, lack of patient acceptance of this route. Local irritation is increasingly being acknowledged as a possible complication of rectal drug therapy. Long-term medication with rectal ergotamine and acetylsalicylic acid, for example, may result in rectal ulceration. Irritation after a single administration of a number of drugs and formulations has also been described.

Summary

Parenteral routes of drug administration may be divided into those involving penetration of one or more layers of skin, for example, intradermal and transdermal administration; those involving penetration of serous membranes, for example, inhalation, intraperitoneal administration, and vaginal doses; direct administration into the systemic circulation, for example, intravenous injection; and those involving direct targeting of the drug to particular organs or tissues, for example, intraarterial injection.

All parenteral dosage routes except intraperitoneal, intrathecal, and possibly intraarterial have the property that the drug passes from the route of administration directly into the general circulation. The fact that these drugs enter the general circulation directly is one of the major properties that differentiate parenteral from enteral dosage routes.

Some parenteral dosage routes, for example, intravenous, intraarterial, and intrathecal injection, lead to rapid drug availability in the body. Others, for example, transdermal and intradermal administration, lead to slower absorption. Intramuscular and intraperitoneal injection have intermediate properties.

Parenteral routes of administration are varied in nature and in their effect on drug absorption. Each method of parenteral administration has advantages and disadvantages. Advantages are related mainly to efficient absorption, accurate dosing, efficient organ targeting, absorption directly into the systemic circulation, or combinations of these. Disadvantages are related mainly to the need for sterile conditions, difficulty in retrieving a dose, and inconvenience. Many new approaches for drug delivery by these routes are being examined.

References

1. Eckman, W. W.; Patlack, C. S.; Fenstermacher, J. D. *J. Pharmacokinet. Biopharm.* **1974,** *2,* 257–285.
2. Yamada, K.; Bremer, A. M.; West, C. R.; Ghoorah, J.; Park, H. C.; Takita, H. *Cancer* **1979,** *44,* 2000–2007.
3. Goldman, J. A.; Peleg, D.; Agmon, M.; Shapiro, G. *Acta Obstet. Gynecol. Scand.* **1979,** *58,* 415–416.
4. Cason, W. P.; Whaley, R. A. *Radiology,* **1980,** *134,* 247.
5. Norell, H. A.; Wilson, C. B. *Surg. Forum.* **1965,** *16,* 429–431.
6. Mallory, A.; Schaefer, J. W.; Cohen, J. R.; Holt, S. A.; Norton, L. W. *Arch. Surg.* **1980,** *115,* 3032.
7. Maher, R.; Mehta, M. In *Persistent Pain;* Lipton, S., Ed.; Academic Press: London, 1977; Vol 1, pp 61–99.
8. Yen, J.; Reiss, F. L.; Kimbelberg, H. K.; Bourke, R. S. *J. Neurosurg.* **1978,** *48,* 894–902.
9. Jacobs, S. A.; Bleyer, W. A.; Chabner, B. A.; Johns, D. G. *Lancet* **1975,** *1,* 465–466.
10. Tse, F. L. S.; Welling, P. G. *J. Parenter. Drug Assoc.* **1980,** *34,* 409–421.
11. Nahata, M. C. *Clin. Pharmacokinet.* **1993,** *24,* 221–229.
12. Nahata, M. C. *Am. J. Hosp. Pharm.* **1986,** *43,* 2237–2239.
13. Viswanathan, C. T.; Booker, H. E.; Welling, P. G. *J. Clin. Pharmacol.* **1978,** *18,* 100–105.
14. Schriftman, H.; Kondritzer, A. *Am. J. Physiol.* **1957,** *191,* 591–594.
15. Lee, V. H. L.; Robinson, J. R. In *Sustained and Controlled Release Drug Delivery Systems;* Robinson, J. R.; Lee, V. H. L., Eds.; Marcel Dekker: New York, 1978, pp 123–209.
16. Florence, A. T.; Jenkins, A. W.; Loveless, A. H. *J. Pharm. Pharmacol.* (Suppl), **1973,** *25,* 120p–121p.
17. Scheu, J. D.; Sperandio, G. J.; Shaw, S. M.; Landolt, R. R.; Peck, G. E. *J. Pharm. Sci.* **1977,** *66,* 172–177.
18. Sobrero, A. *Compt. Rend.* **1847,** *24,* 247–248.
19. Murrell, W. *Lancet* **1879,** *1,* 80–81, 113–115, 151–152, 225–227.
20. Li, V. H. K.; Robinson, J. R.; Lee, V. H. L. In *Controlled Drug Delivery: Fundamentals and Applications;* Robinson, J. R.; Lee, V. H. L., Eds.; Marcel Dekker: New York, 1987; pp 3–94.
21. Siegel, I. A. In *The Structure and Function of the Oral Mucosa;* Meyer, J.; Squier, C. A.; Gerson, S. J., Eds.; Pergamon: Oxford, England, 1984; pp 95–108.
22. Burnier, A. M.; Martin, P. L.; Yen, S. S. C.; Brooks, P. *Am. J. Obstet. Gynecol.* **1981,** *140,* 146–149.
23. deVries, M. E.; Bodde, H. E.; Verhoef, J. C.; Junginger, H. E. *Clin. Rev. Ther. Drug Carrier Syst.* **1991,** *8,* 271–303.
24. Motwani, J. G.; Lipworth, B. J. *Clin. Pharmacokinet.* **1991,** *21,* 83–94.
25. Harris, D.; Robinson, J. R. *J. Pharm. Sci.* **1992,** *81,* 1–10.
26. Welling, P. G. In *Pharmacokinetics: Regulatory, Industrial, Academic Perspectives;* Welling, P. G; Tse, F. L. S., Eds.; Marcel Dekker: New York, 1988; pp 97–157.
27. Wilkinson, P.; Van Dyke, P.; Jatlow, P.; Barash, P.; Byck, R. *Clin. Pharmacol. Ther.* **1980,** *27,* 386–394.
28. Fogler, W. E.; Wade, R.; Brundish, D. E.; Fidler, I. J. *J. Immunol.* **1985,** *135,* 1372–1377.
29. Mygind, N.; Pederson, M.; Nielsen, M. H. In *The Nose, Upper Airway Physiology and the Atomspheric Environment;* Proctor, D. F.; Anderson, I. B., Eds.; Elsevier: New York, 1982; pp 71–97.

30. Eccles, R. In *The Nose, Upper Airway Physiology and the Atmospheric Environment*; Proctor, D. F.; Anderson, I. B., Eds.; Elsevier: New York, 1982; pp 191–214.

31. Watanabe, K.; Saito, Y.; Watanabe, I.; Mizuhira, V. *Ann. Otol.* **1980,** *89*, 377–382.

32. Su, K. S. E.; Campanali, K. M. In *Transnasal Systemic Medications*; Chien, Y. W., Ed.; Elsevier Scientific: Amsterdam, 1985; pp 139–159.

33. Sato, Y.; Hyo, N.; Sato, M.; Takano, H.; Okuda, S. Z. *Erkr. Atmungsorgane.* **1981,** *157*, 276–280.

34. Petri, W.; Schmiedel, R.; Sandow, J. In *Transnasal Systemic Medications*; Chien, Y. W., Ed.; Elsevier Scientific: Amsterdam, Netherlands, 1985; pp 161–181.

35. Sandow, J.; Petri, W. In *Transnasal Systemic Medications*; Chien, Y. W., Ed.; Elsevier Scientific: Amsterdam, Netherlands, 1985; pp 183–199.

36. Fogler, W. E.; Wade, R.; Brundish, D. E.; Fidler, L. T. *J. Immunol.* **1985,** *135*, 1372–1377.

37. Scott, G. M.; Phillpotts, R. J.; Wallace, J.; Secher, D. S.; Cantell, K.; Tyrrell, D. A. J. *Br. Med. J.* **1982,** *284*, 1822–1825.

38. Samo, T. C.; Greenberg, S. B.; Couch, R. B.; Quarles, J.; Johnson, P. E.; Hook, S.; Harmon, M. W. *J. Infect. Dis.* **1983,** *148*, 535–542.

39. Hayden, F. G.; Mills, S. E.; Johns, M. E. *J. Infect. Dis.* **1983,** *148*, 914–921.

40. Shaw, R. W.; Fraser, H. M.; Boyle, H. *Br. Med. J.* **1983,** *287*, 1067–1069.

41. Hirai, S.; Yashiki, T.; Mima, H. *Int. J. Pharm.* **1981,** *9*, 165–172.

42. Hirai, S.; Yashiki, T.; Mima, H. *Int. J. Pharm.* **1981,** *9*, 173–184.

43. Moses, A. C.; Gordon, G. S.; Carey, M. C.; Flier, J. S. *Diabetes* **1983,** *32*, 1040–1047.

44. Hussain, A. A.; Kimura, R.; Huang, C. H. *J. Pharm. Sci.* **1984,** 73, 1300–1301.

45. Hussain, A. A.; Hirai, S.; Bawarshi, R. *J. Pharm. Sci.* **1981,** 70, 466–467.

46. David, G. F. X.; Puri, C. P.; Anand Kumar, T. C. *Experientia* **1981,** *37*, 533–534.

47. Anand Kumar, T. C.; David, G. F. X. *Proc. Nat. Acad. Sci.* **1982,** *79*, 4185–4189. 55.

48. Anand Kumar, T. C.; David, G. F. X.; Umberkomen, B.; Saini, K. D. *Curr. Sci.* **1974,** *43*, 435–439.

49. Ohman, L.; Hahnenberger, R.; Johansson, E. D. B. *Contraception* **1980,** *22*, 349–358.

50. Anand Kumar, T. C.; David, G. R. X.; Puhi, V. *Nature* **1977,** *270*, 532–533.

51. Gizurarson, S. *Acta Pharm. Nord.* **1990,** *2*, 105–122.

52. Newman, S. P.; Pavia, D.; Moren, F.; Sheahan, N. F.; Clarke, S. W. *Thorax* **1981,** *36*, 52–55.

53. Toogood, J. H.; Baskerville, J.; Jennings, B.; Lefcoe, N. M.; Johansson, S. *Am. Rev. Respir. Dis.* **1984,** *129*, 723–729.

54. Newman, S. P.; Millar, A. B.; Lennard-Jones, T. R.; Moren, F.; Clarke, S. W. *Thorax* **1984,** *39*, 935–941.

55. Clark, A. R.; Byron, P. R. *J. Pharm. Sci.* **1985,** *74*, 939–942.

56. Brown, R. A., Jr.; Schanker, L. S. *Drug Metab. Dispos.* **1983,** *11*, 355–360.

57. Chandrasekaran, S. K. *Drug Dev. Ind. Pharm.* **1983,** *9*, 627–646.

58. Guy, R. H.; Hadgraft, J. *J. Controlled Release* **1987,** *4*, 237–259.

59. Bucks, D. A. W. *Pharm. Res.* **1984,** *1*, 148–153.

60. Guzek, D. B.; Kennedy, A. H.; McNeill, S. C.; Wakshull, E.; Potts, R. O. *Pharm. Res.* **1989,** *6*, 33–39.

61. Ranade, V. V. *J. Clin. Pharmacol.* **1991,** *31*, 401–418.

62. Stoughton, R. B.; McClure, W. O. *Drug Dev. Ind. Pharm.* **1983,** *9*, 725–744.

63. Touitou, E.; Junginger, H. E.; Weiner, N. D.; Nagai, T.; Mezei, M. *J. Pharm. Sci.* **1994,** *83*, 1189–1203.

64. Keshary, P. R.; Chien, Y. W. *Drug Dev. Ind. Pharm.* **1984,** *10*, 1663–1699.

65. Wolff, M.; Cordes, G.; Luckow, V. *Pharm. Res.* **1985,** *2*, 23–29.

66. Parker, J. O.; von Koughnett, K. A.; Fung, H-L. *Am. J. Cardiol.* **1984**, *54*, 8–13. 56.

67. MacGregor, T. R.; Matzek, K. M.; Keirns, J. J.; van Wayjen, R. G. A.; van den Ende, A.; van Tol, R. G. L. *Clin. Pharmacol. Ther.* **1985**, *38*, 278–284.

68. Padwick, M. L.; Endacott, J.; Whitehead, M. I. *Am. J. Obstet. Gynecol.* **1985**, *152*, 1085–1091.

69. Good, W. R.; Powers, M. S.; Campbell, P.; Schenkel, L. *J. Controlled Release* **1985**, *2*, 89–97.

70. Chien, Y. W.; Valia, K. H.; Doshi, U. B. *Drug Dev. Ind. Pharm.* **1985**, *11*, 1195–1212.

71. Daughton, D. M.; Heatley, S. A.; Prendergast, J. J.; Causey, D.; Knowles, M.; Rolf, C. N.; Cheney, R. A.; Hatlelid, K.; Thompson, A. B.; Rennard, S. I. *Arch. Int. Med.* **1991**, *151*, 749–752.

72. Dubois, J. P.; Sioufi, A.; Muller, P. H.; Mauli, D.; Imhof. P. R. *Methods Find. Exp. Clin. Pharmacol.* **1989**, *11*, 187–195.

73. Tonnesen, P.; Norregaard, J.; Simonsen, K.; Sawe, U. *N. Engl. J. Med.* **1991**, *325*, 311–315.

74. Gorsline, J.; Gupta, S. K.; Dye, D.; Rolf, C. N. *J. Clin. Pharmacol.* **1993**, *33*, 161–168.

75. Prather, R. D.; Tu, T. G.; Rolf, C. N.; Gorsline, J. *J. Clin. Pharmacol.* **1993**, *33*, 644–649.

76. Vlasses, P. H.; Ribeiro, L. G. T.; Rotmensch, H. H.; Bondi, J. V.; Loper, A. E.; Hichens, M.; Dunlay, M. C.; Ferguson, R. K. *J. Cardiovasc. Pharmacol.* **1985**, *7*, 245–250.

77. Nagai, T.; Santoh, Y.; Nambu, N.; Machida, Y. *J. Controlled Release* **1985**, *1*, 239–246.

78. Okabe, K.; Yamaguchi, H.; Kawai, Y. *J. Controlled Release* **1986**, *4*, 79–85.

79. Shaw, J. E.; Chandrasekaran, S. K.; Michaels, A. S.; Taskovich, L. In *Animal Models in Dermatology*; Maibach, H., Ed.; Churchill Livingstone: Edinburgh, 1975; pp 138–146.

80. Evans, N. J.; Rutter, N.; Hadgraft, J.; Parr, G. *J. Pediatr.* **1985**, *107*, 307–311.

81. Kazmi, S.; Kennon, L.; Sideman, M.; Plakogiannis, F. M. *Drug Dev. Ind. Pharm.* **1984**, *10*, 1071–1083.

82. Chien, Y. W.; Mares, S. Y.; Berg, J.; Huber, S.; Lambert, H. J.; King, K. F. *J. Pharm. Sci.* **1975**, *64*, 1776–1781.

83. England, D. E.; Johansson, E. D. B. *Br. J. Obstet. Gynecol.* **1978**, *85*, 957–964. 57.

84. Van Hoogda em, E. J.; deBoer, A. G.; Breimer, D. D. *Clin. Pharmacokinet.* **1991**, *21*, 11–26.

85. Müller, B. W.; Bremer, U. *Suppositorien: Pharmakologie, Biopharmazie und Galenik rektal und vaginal anzuwendender Arzneiformen*; Wissenschaftliche Verlagsgesellschaft: Stuttgart, Germany, 1986; p 298.

86. Ritschel, W. A.; Elconin, H.; Alcorn, G. J.; Denson, D. D. *Biopharm. Drug Dispos.* **1985**, *6*, 281–290.

Enteral Routes of Drug Administration

4

nteral routes of drug administration are from the stomach and the small and large intestine. As described in the previous chapter, intraperitoneal absorption may also be considered to be enteral because compounds absorbed from the peritoneal cavity enter the body by the same route as those from other enteral sites. Substances absorbed from enteral sites enter the splanchnic circulation and pass through the liver before entering the systemic circulation.

The gastrointestinal (GI) tract is the site of absorption for most nutrients. Unlike many parenteral absorption sites, such as the skin and the buccal and nasal cavities, the GI site is designed or has evolved to facilitate absorption of substances. The peristaltic action of the stomach, secretion of various enzymes and hydrochloric acid, the villi and microvilli of the intestine, and also the rich blood supply and lymphatics in this region are all designed to facilitate absorption of substances into the body. Enteral absorption is generally by far the most efficient drug delivery route, and whenever possible drugs are administered this way.

Despite the many characteristics of the GI tract that favor absorption, any orally ingested compound is exposed to an absorption environment that is both friendly and hostile, depending on the nature of the compound and the patient, and one that changes dramatically from one region of the GI tract to another. It is important at this point to review the structure and physiology of the GI tract relative to drug absorption.

The pH of various regions of the GI tract is shown in Figure 4.1. After leaving the slightly acidic region of the mouth, the compound enters the very acidic region of the stomach. High gastric acidity is a consequence of hydrochloric acid secretion by the parietal cells of the stomach. Acid secretion plays an important role in food digestion by facilitating conversion of pepsinogens and zymogens to active proteolytic enzymes. Drug absorption is also profoundly affected.

Physiology of the GI Tract

Gastrointestinal pH

43

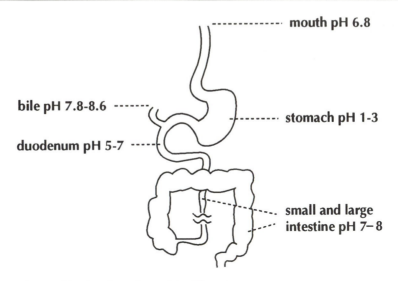

FIGURE 4.1   Approximate pH values in regions of the GI tract.

The acidic environment of the stomach tends to favor absorption of acidic drugs through the gastric epithelium, provided the acidic drugs can dissolve or remain in solution in the acidic fluids of the stomach. On the other hand, basic drugs tend to dissolve quite readily in the stomach, but absorption may be prevented because most of the drug will be in the ionized state and therefore not sufficiently fat-soluble for efficient membrane transport to occur.

The acidic environment of the stomach may also cause reduced drug absorption because of acid-catalyzed degradation. Some organic bases are unstable in acid and undergo rapid chemical degradation unless they are protected in some way. An example is the antibiotic erythromycin. Erythromycin is used for a variety of infections commonly associated with Gram-positive organisms. Erythromycin is efficiently absorbed from oral doses and is generally administered by that route, although an intravenous form does exist. Erythromycin is a base and is susceptible to acid-catalyzed degradation in the stomach, which may lead to poor and erratic absorption.

Two approaches have been taken to prevent erythromycin degradation. The first is to formulate erythromycin base or its stearate salt in tablets or capsules with acid-resistant coatings. The coating remains intact while the tablet or capsule is in the acidic environment of the stomach, but dissolves to release active ingredient after the tablet passes from the stomach into the relatively alkaline region of the small intestine.

The second approach to prevent degradation is to convert erythromycin to an acid-stable esterified derivative. The esterified forms of erythromycin are also more lipophilic than the parent drug and cross the membranes of the GI epithelium more efficiently.

If a drug dissolves in the stomach or is a liquid, and if it is fat-soluble and acid-stable, then the drug is likely to be absorbed efficiently from the stomach. A good example is ethyl alcohol. Alcohol is a liquid that is completely miscible with water and sufficiently lipophilic to cross membranes, and it is therefore efficiently absorbed from the stomach.

After passing through the pyloric sphincter, a compound reaches the duodenum and then the jejunum and the ileum. These regions of the small intestine differ from the stomach with respect to pH, the presence of digestive enzymes, and the absorptive surface area. Excretion of alkaline bile, which has a pH of 7.8–8.6, into the duodenum raises the pH of the duodenal and subsequent intestinal contents to approximately 5–7.

The abrupt change in pH from acidic to neutral on passing from the stomach into the proximal small intestine causes many changes. Enteric coatings that were impermeable in the stomach will dissolve. Acidic drugs will also dissolve in the relatively basic environment, and the pH will not be raised sufficiently to prevent dissolution or cause precipitation of most weakly basic drugs.

The neutral to slightly alkaline pH of the small and large intestine will clearly affect absorption of ionizable substances, although other factors such as absorption surface also play an important role. Strong acids and bases are ionized at neutral pH so that, although they might remain in solution, they are unlikely to be absorbed efficiently across intestinal membranes. One might safely assume that, provided they have adequate solubility, compounds that are least ionized at pH 7–8 will likely be most efficiently absorbed.

GI Structure and Motility Factors

The stomach is a pouch-like structure lined with a relatively smooth epithelial surface. Although compounds can be absorbed quite efficiently from the stomach, the overall contribution of this region to drug absorption is modest. Absorption of aspirin and ethanol from the human stomach after oral administration of aqueous solutions has been estimated to be 10% and 30%, respectively.

The absorptive properties of the proximal small intestine are superior to those of the stomach or any other part of the GI tract, and most compounds are more efficiently absorbed from this region. In order to reach the small intestine, a drug must pass through the pyloric sphincter, which separates the stomach from the duodenum. The rate at which substances pass from the stomach into the small intestine is one of the major rate-limiting steps controlling drug absorption.

Stomach motility is a complex phenomenon that is influenced by a variety of nervous and hormonal stimuli. The stomach-emptying rate is a function of rhythmic contractions, which are quite vigorous, have a frequency of approximately three per minute in a hungry person, and become diminished when food enters the stomach. Food passes from the stomach into the duodenum as a result of these rhythmic contractions. The heavier the meal and the higher the fat content, the longer it will take for the meal and any drug ingested with the meal to pass into the small intestine. The reduced number of contractions and force of gastric rhythms, and the consequent decrease in gastric emptying rate after meals, appear to result from nervous and hormonal feedback mechanisms based on activation of receptors situated in the duodenum: the fat receptors, osmoreceptors, and acid receptors. This finely tuned process can be regarded as a defense mechanism by which substances are prevented from entering the proximal small intestine and injuring the delicate absorptive surface of this region until the substances have been reduced to a suitable consistency in the stomach.

Although solid food tends to delay stomach emptying, liquids tend to accelerate the process. In fact, increased fluid volume is the only known natural stimulus to stomach emptying. This stimulation results from activation of stretch receptors in the stomach wall. When the fluid is water, subsequent activation of the inhibitory receptors in the duodenum is stopped, and the net result is rapid emptying of the stomach contents.

From a physical viewpoint, drugs should be absorbed better from concentrated solutions than from diluted solutions because of a greater mucosal to serosal concentration gradient. However, the reverse is often the case. Studies in animals and humans have shown that drug absorption from the GI tract increases when a drug is administered as a solid dosage form or as a solution in a large fluid volume. Increased drug absorption results from the combined effects of accelerated stomach emptying, rapid exposure of dissolved drug molecules to a larger GI surface area, and faster dissolution of solid dosage forms.

By far, the most important difference between the proximal small intestine and the stomach is the nature of the mucosal surface of the epithelium. As shown in Figure 4.2 (1) the mucosal surface of the small intestine has a unique morphology because its surface area is increased by finger-like projections, or villi, arising from the folds of Kerckring, which are folds in the intestinal mucosa, and by microvilli in turn arising from the villi. These various invaginations and projections increase the surface area of the mucosa approximately 600-fold. The total surface area of the small intestine has been calculated to be approximately 200 m^2, and an estimated 1.0–1.5 L of blood passes through the intestinal capillaries each minute. The corresponding values for the stomach are only 100 m^2 of surface area, and a blood flow rate of 150 mL/min. Thus, the intestine has a surface area approximately double that of the stomach but a blood perfusion rate that is 6–10 times faster. Both factors—surface area and blood perfusion rate—strongly favor more efficient absorption from the intestine than from the stomach.

The villi and microvilli of the intestine, like many other tissues in the body, are lined by a sulfated mucoprotein, glycocalyx, also known as the fuzzy-coat. Fluid that is trapped within the glycocalyx is stationary, and a series of thin layers, each progressively more stirred, extends from the epithelial cell surface to the bulk phase of the intestinal lumen. This series of thin layers is known as the "unstirred layer" and has an effective thickness of 0.01–1.0 mm.

Molecules move within the unstirred layer only by diffusion, the rate of which is inversely proportional to the square root of molecular weight below 450, and inversely proportional to the cube root of molecular weight above 450.

The glycocalyx on the microvilli is negatively charged, and the counterions are in the unstirred layer. If a substantial portion of these cations is composed of hydrogen ions, as is often the case, then the microclimate within the brush border of the epithelium is likely to be acidic relative to the bulk phase of the intestinal lumen. This may influence drug ionization at the membrane surface and provide a basis for the "acid microclimate" frequently associated with the GI mucosa.

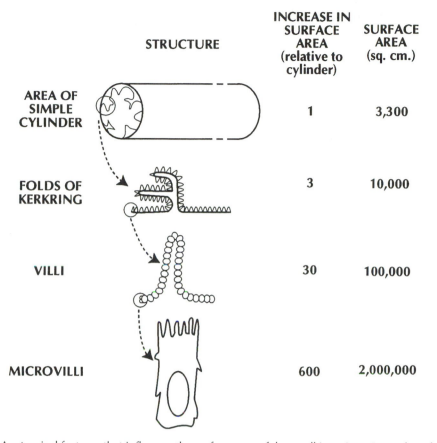

STRUCTURE		INCREASE IN SURFACE AREA (relative to cylinder)	SURFACE AREA (sq. cm.)
AREA OF SIMPLE CYLINDER		1	3,300
FOLDS OF KERKRING		3	10,000
VILLI		30	100,000
MICROVILLI		600	2,000,000

Anatomical features that influence the surface area of the small intestine. (Reproduced with permission from reference 1.)

FIGURE 4.2

The length of time during which material stays in the small intestine varies, but a reasonable estimate is 5 min in the duodenum, 2 h in the jejunum, and 3–6 h in the ileum. After passing through these regions, material enters the large intestine.

The large intestine is similar to the stomach because it does not have villi or microvilli at its mucosal surface, but it is different from the stomach in terms of pH. The stomach contents are strongly acidic, whereas the large intestine tends to be neutral or alkaline. Absorption of drugs from the large intestine tends to be less efficient than from the small intestine. The large intestine and colon contain an active bacterial microflora, which can degrade foreign molecules and which tends to reduce the absorption of drugs present in this region of the GI tract.

At the extreme distal end of the GI tract is the rectum. The rectum contains no microvilli and apparently no sites for active absorption, it has an active bacterial microflora, and it usually contains fecal material. A combination of these factors would suggest that the rectum is a very poor absorption site. However, this is not necessarily the case. Drugs are frequently administered in the form of suppositories or enemas, and rectal absorption efficiency is often good, if erratic. Some drugs given rectally for systemic action include acetaminophen, aspirin, indomethacin, perchlorperazine, phenobarbital, promethazine, and theophylline.

GI Secretions

Acid Secretion. Secretion of hydrochloric acid by the parietal cells of the stomach was mentioned briefly in the previous section. The rate of acid secretion is controlled by receptors on the plasma membrane of parietal cells that bind acetylcholine, histamine, and gastrin and thereby directly stimulate the parietal cells to secrete HC1. Acetylcholine is released near parietal cells by cholinergic nerve terminals, whereas gastrin, a hormone produced in the mucosa of the gastric antrum and duodenum, reaches parietal cells through the bloodstream. Histamine, released from cells in the gastric mucosa, diffuses into the parietal cells. Acetylcholine, histamine, and gastrin are important for regulation of HC1 secretion and act directly on parietal cells to enhance the acid secretion rate.

Gastrin is not as potent a direct stimulant of parietal cells as acetylcholine and histamine. Muscarinic antagonists such as atropine, and H_2-receptor blockers such as cimetidine, do not block the direct actions of gastrin, and the second messenger for gastrin action is unknown. However, cimetidine attenuates the physiological response to elevated levels of gastrin in the blood, suggesting that the response to gastrin may be due to gastrin-stimulated histamine release (2).

Phases of Gastric Acid Secretion. The basal rate of HC1 secretion varies diurnally, being highest in the evening and lowest in the morning (2). When the stomach has been empty for several hours, HC1 secretion is approximately 10% of its maximal rate.

After ingestion of a meal, the rate of acid secretion by the stomach increases rapidly. The three phases of increased acid secretion in response to food are the cephalic phase (elicited before food reaches the stomach), the gastric phase (elicited by the presence of food in the stomach), and the intestinal phase (elicited by input originating in the duodenum and upper jejunum).

Cephalic Phase. This phase of gastric secretion is elicited by the sight, smell, and taste of food. The acid secretion rate during this phase can be as much as 40% of the maximum rate. Other stimuli sensed in the brain, in addition to those related to the presence of food, may revoke acid secretion through vagal impulses. Decreased concentration of glucose in the cerebral arterial blood elicits acid secretion. A low pH in the antrum will diminish the amount of HC1 secreted during the cephalic phase, and the amount of acid secreted is limited by a direct effect on parietal cells.

Gastric Phase. The presence of food in the stomach evokes gastric acid secretion. The principal stimuli include distension of the stomach and the presence of amino acids and peptides. Distension of the body or the antrum of the stomach stimulates mechanoreceptors that bring about secretion of acetylcholine, HCl, and gastrin.

Intestinal Phase. The presence of chyme in the duodenum results in neural and endocrine responses that first stimulate and later inhibit secretion of acid by the stomach. The stimulatory influences dominate when the pH of gastric chyme is above 3. However, when the buffer capacity is exhausted and pH falls below 2, inhibitory influences dominate.

Gastric acid secretion may be regulated by several brain peptides, some of which may affect pathways that enhance secretion, whereas others may act centrally to inhibit acid secretion. Norepinephrine appears to be the neurotransmitter in some central pathways that inhibit gastric acid secretion, whereas γ-aminobutyric acid is the transmitter for certain pathways that stimulate acid secretion.

Gastric Juice

In addition to hydrochloric acid, gastric juice contains a mixture of secretions from surface epithelial cells and gastric glands. Salts, water, pepsins, intrinsic factor, and mucus are main components of gastric juice. Gastric secretions increase after a meal. The ionic composition of gastric juice is related to the rate of secretion. The higher the secretory rate, the higher the hydrogen ion concentration $[H^+]$. At lower secretory rates (<1 mL/min), $[Na^+]$ increases and $[H^+]$ diminishes. The major anion in gastric juice at all rates of secretion is chloride. The rate of gastric acid secretion varies considerably among individuals, possibly because of variations in the number of parietal cells. In humans, the basal (unstimulated) rate of gastric acid production typically ranges from approximately 1 to 5 mEq/h. With histamine or pentagastrin stimulation, maximum acid output rises to 6–40 mEq/h. In general, patients with gastric ulcers secrete less HCl, and patients with duodenal ulcers secrete more HCl than normal individuals. Gastric HCl secretion is reported to be less in patients with gastric cancer than in healthy persons (3).

Other Secretions

Bile, pH 7.8–8.6, is produced continuously in humans. Hepatic bile is concentrated and stored in the gall bladder between meals, being ejected from the gall bladder and flowing into the duodenum when food enters the intestine. Bile formation results from activity of (1) the reticuloendothelial cells that line the liver sinusoids and (2) the liver parenchymal cells. The major constituents of bile are bile salts; bilirubin, the end product of hemoglobin breakdown; the electrolytes sodium, chloride, and bicarbonate; and smaller amounts of cholesterol, phospholipids, and lecithin (4). Bile is secreted at rates varying from 250 to 1000 mL/day in humans. The gall bladder contracts within 30 min after eating because of the liberation of cholecystokinin; the most effective stimuli are foods high in fat.

Bile salts, which are surface-active, promote dissolution of lipophilic drugs and also lipophilic drug formulations, enteric coating, and waxy drug matrices. Bile salts may also promote membrane permeability of hydrophobic molecules through micelle formation and solubilization. On the other hand, bile salts have been known to form insoluble complexes with some drugs including neomycin, kanamycin, and vancomycin. Insoluble complex formation could reduce the systemic availability of these drugs.

Pancreatic juice has two major components, an alkaline fluid and enzymes, both of which empty into the duodenum. The alkaline pH contributes, together with bile, to neutralize the acid emptying from the stomach, while the enzymes amylase, lipase, trypsin, and chymotrypsin play major roles in digestion of carbohydrate, fat, and protein. Trypsin and the chymotrypsins are secreted as the inactive precursors trypsinogen and chymotrypsinogens and are then converted to their active forms enzymatically. Pancreatic secretion is probably continuous in humans. After a meal, secretion may amount to 1500 mL and may reach a maximum of 2700 mL per day (4).

As a result of proteolytic enzyme secretion into the duodenum, protein or polypeptide drugs such as corticotropin, vasopressin, and insulin are rapidly degraded and generally cannot be given orally. The hormonal agents progesterone, testosterone, and aldosterone are similarly unstable in the intestine and are generally given by the buccal or other parenteral routes.

Secretory activity of the pancreas is under both hormonal and nervous control. The hormone secretin stimulates secretion mainly of water and bicarbonate, whereas the hormone pancreozymin stimulates secretion of enzymes and their precursors without affecting the volume or the bicarbonate component of pancreatic juice.

Intestinal secretions do not exist in the same sense as gastric, pancreatic, or biliary secretions. Nonetheless, large fluid fluxes take place throughout the small and large intestine. Net water flux in the duodenum is from serosal to mucosal to render the luminal contents isoosmotic with blood. For the remainder of the small intestine, net flux is mucosal to serosal. Other secretions occur from duodenal glands, the small intestine, and the colon. Duodenal secretions do not appear to participate directly in digestion but may play a role in protecting the duodenal lumen from damage by gastric acid. Secretions from the intestinal mucosa appear to have predominantly lubricant and protective effects.

Gastrointestinal Blood Flow in Relation to Drug Absorption

Once a drug crosses from the mucosal to the serosal side of the GI epithelium, the drug may be transported away from the serosal side by one or both of two mechanisms. The GI tract is richly supplied by a blood-capillary network that is associated with the splanchnic circulation. This network is close to the GI epithelium and provides a perfect route for drug absorption. In addition to the blood supply, the GI tract is extensively supplied by the lymphatic system, and absorbed material can be taken up by the lymphatic vessels in the GI epithelium and carried by the lymphatic system that drains the abdominal area to the thoracic duct. Material absorbed by this route joins the venous system of the general circulation at the junction of the left internal jugular vein and the subclavian vein. Any drug that is absorbed via the lymphatic system thus enters the systemic circulation directly, does not have to pass through the liver, and is therefore not susceptible to any first-pass presystemic hepatic metabolism. Despite the presence of both the blood-capillary network and the lymphatic system, absorption of the great majority of drugs appears to occur almost exclusively via the capillary system associated with the splanchnic circulation.

Why should GI drug absorption be so selective for the splanchnic circulation? The answer appears to lie, for the most part, in the relative flow rates of blood and lymph. The rate of blood flow in the splanchnic circulation bathing the abdominal area is 1.0–1.5 L min^{-1}, representing approximately 30% of cardiac output. This rate may increase to 2 L/min after a meal. Lymph flow through the same region is only 1–2 mL/min, although this may increase to 5–20 mL/min after a meal. Lymph flow in this region is thus approximately 500–700 times slower than blood flow. Relatively fast splanchnic blood flow establishes virtually sink conditions on the serosal side of the GI epithelium and ensures a steep concentration gradient across the membrane. These conditions promote efficient absorption into the bloodstream. This situation does not exist with the very sluggish lymph flow. Therefore, from a purely physical viewpoint, the efficient absorption that occurs into the splanchnic circulation would not be expected to occur via the lymphatic system.

Thus, for most drugs absorption from the GI tract occurs via the bloodstream. Only a very small fraction of drug molecules may be absorbed via the lymphatic system. These include drugs with very high molecular weights that have difficulty entering the capillaries, and also some specific molecules such as steroids.

Special Transport Mechanisms

Despite the existence of several special transport mechanisms (see Chapter 2), the vast majority of compounds are passively absorbed from the GI tract, so that absorption efficiency is controlled both by solubility and stability within the fluids of the GI tract and the ability of a compound to passively pass through the intestinal and associated capillary membranes.

However, with increased interest in peptide and protein drugs, and also in prodrugs of poorly absorbed compounds, there is increasing awareness and exploitation of specialized transport processes. The flurry of activity in this area during recent years has unearthed several specialized transport processes and has led to further developments to improve oral systemic availability of these types of compounds. This topic has recently been reviewed (5, 6). Absorption of peptide and protein compounds, important for several therapeutic classes, is generally poor. Significant pharmacologic effects of peptide and protein drugs have been observed even though oral bioavailability is in the single-digit range (7). Trace amounts of intact protein molecules may be absorbed into the systemic circulation (8–10). Cyclosporin is an orally active peptide, and significant membrane transport of enkephalin and renin inhibitors was observed when peptidase inhibitors were used. Oral efficacy of insulin and vasopressin was improved, though still at levels too low to be useful, using peptidase inhibitors or stable analogs. These results indicate that oral delivery of these types of molecules is feasible if proteolysis is avoided (11–15).

Peptide and protein drugs are transported across the intestinal epithelium by various mechanisms, depending on their physicochemical properties. Di- and tripeptides are absorbed in the mammalian intestine by a common carrier-mediated process (16). Di- and tripeptide drugs such as β-lactam antibiotics and angiotensin-converting enzyme (ACE) inhibitors are efficiently absorbed through the peptide transporter (17–19). Targeting the peptide carrier can achieve significant oral availability of di- and tripeptide drugs or drugs with low membrane permeability in the form of di- and tripeptide prodrugs (*19*). The actual nature of transport processes of polypeptide and protein drugs is not yet defined. Both energy-dependent and passive transport processes for peptide and protein drugs have been reported (*20–23*). It is not clear whether these are due to distinct transport mechanisms or to laboratory variations. A brief discussion of observations made on two therapeutic drug classes will indicate the type of progress being made in this area.

β-*Lactam Antibiotics*

Amino-β-lactam antibiotics with free N-terminal amino and C-terminal carboxyl groups, including cyclocillin, amoxacillin, ampicillin, cefaclor, cefadroxil, and cephalexin, and some β-lactam antibiotics without a free N-terminal amino group, including cefixime, ceftibuten, and FK 089, are absorbed by means of the peptide transporter (*24–27*). Absorption is inhibited by other di- and tripeptides but not by amino acids. As shown in Figure 4.3 (*24*), antibiotic absorption exhibits mutual inhibition. Absorption of D-cephalexin is saturable and is inhibited by its L-isomer, which is absorbed by the peptide transporter (*28*).

Angiotensin-Converting Enzyme Inhibitors

Orally active ACE inhibitors represent one of the newest and most therapeutically significant classes of cardiovascular and antihypertensive agents (*29*). Recent findings suggest that the ACE inhibitors captopril, SQ 29852, enalapril, lisinopril, benazepril, and quinapril are transported by the peptide transporter (*30–32*). Their uptake is significantly inhibited by small peptides and cephradine. The diacid ACE inhibitors enalaprilate, quinaprilate, and benazeprilate are poorly absorbed, whereas their ester prodrugs, including enalapril, quinapril, and benazepril, are absorbed by means of the transporter. Lisinopril, which has two free carboxyl groups at similar positions as enalaprilat, is poorly absorbed by the transporter as is FK 089, a β-lactam antibiotic. Further work is needed to explain the observed specificity of the peptide transporter for these compounds.

Animal Models for Oral Drug Absorption

During drug discovery and development, extensive use is made of experimental animal models to screen for pharmacologic and toxicologic activity after oral doses and for oral absorption, elucidate sites and mechanisms of absorption, and to explore interactions, formulations, and other factors that might affect the rate and extent of oral absorption. Decisions on whether to take a new drug or formulation candidate into the clinic may be profoundly influenced by its systemic availability in animal species. There has also been debate on the feasibility of using (inexpensive) nonclinical models to assess oral drug availability as surrogates for (expensive) clinical trials. Despite extensive use of nonhuman models to study these many aspects of absorption, it is important to recognize the differences that exist and may therefore give rise to disparate results among various animal species, and between these and humans.

The main species used in nonclinical absorption studies are the mouse, rat, dog, and monkey. Other species may be used for particular drugs or drug classes. During early development of new chemical entities as drug candidates, the rat and dog, and less frequently the monkey, are used to provide initial estimates of absorption and systemic availability. Despite quite large differences in GI anatomy

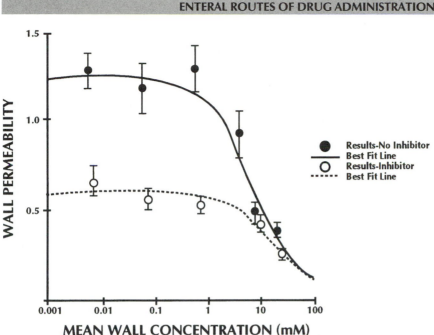

Plot of wall permeability of cephalexin perfused alone and in the presence of a competitive inhibitor, cefadroxil (7 mM). Results are reported as mean wall permeability ± SEM. (Adapted from reference 24).

FIGURE 4.3

and physiology among these species, and between them and humans, animals often turn out to be quite reasonable predictors of the human situation. A compound that has poor bioavailability in the rat is likely to have poor bioavailability in humans. Conversely, a compound that is well absorbed in the rat is likely to be similarly well absorbed in humans. Of course, there are many exceptions to this generalization, and drug company files are replete with examples of animals being poor predictors of drug bioavailability in humans. This writer has personal experience with a new compound that exhibited good bioavailability in two animal species, yielding circulating levels that were proportional through a wide range of doses. However, in initial Phase 1 studies in humans the compound was poorly absorbed, and circulating levels of the compound failed to increase with increasing doses. The compound was dropped from development. This example illustrates that, although nonclinical data may be useful for predicting drug absorption in humans, such data need to be interpreted cautiously.

Although small rodents can provide bioavailability data during early drug discovery and development, and are extensively used for this purpose, they are not useful for examining drug dosage forms. They are simply not big enough to accommodate the human dose size. Species that have been used extensively for such studies are the dog, monkey, pig, and, to a lesser extent, rabbit. The reduced cost of conducting animal studies has led many to advocate the use of animals to determine drug availability as a surrogate to humans. Although the best test of this hypothesis is to compare animal and human data directly, and many such studies have been conducted with equivocal results, it is informative to briefly compare the different species and humans regarding GI anatomy and physiology.

The human, monkey, pig, and dog have grossly similar GI anatomy and physiology in that they all have a single stomach with acid pH, a small intestine with villi, a cecum, and a colon, and they have similar biliary and pancreatic secretions. However, there are great differences in anatomical dimensions and physiology among the species (33). For example, the domestic pig has a far longer small intestine than humans, whereas the small intestine in the dog is much shorter. Intestinal villi in humans are far more filamentous than in other species. Each of the species has a gall bladder from which bile is secreted into the proximal small intestine. Bile salt composition is somewhat different in pigs and dogs than in humans, and the ability of the gall bladder to concentrate bile is lower in pigs than in other species. In humans, the colon consists of three parts, ascending, transverse, and descending, with a total length of about 1.0-1.5 m. In dogs and monkeys, the colon is relatively short, whereas in pigs the colon is much longer, 4.0–4.5 m. The upper GI interdigestive migrating motility complex (IMMC), a strong pattern of contractions that occurs periodically in the fasting state and is thought to be responsible for clearance of nondisintegrating dosage forms from the stomach, appears to be common to all species. The differences in anatomy and physiology between the species and humans have no doubt contributed to inconsistencies in drug absorption. Some examples may be illustrative.

Rabbit

Several physiological and anatomical differences between the rabbit and human limit the use of the rabbit as a model for oral absorption (33). The major problem is size. In a study by Aoyagi (34), enteric-coated aspirin tablets (diameter 11 mm) did not produce significant salicylate levels in the rabbit, although 1-mm enteric-coated granules produced salicylate levels within 1 h of administration. Serum salicylate levels obtained in the human, dog, and rabbit in this study are shown in Figure 4.4. These results highlight the limitations of the rabbit gastric dimensions, as salicylate levels were quickly achieved in humans and dogs after administration of both the tablet and pellet formulations. Subsequent studies showed that there was no gastric emptying of 11-mm enteric-coated tablets from the rabbit stomach within 24 h of administration and that emptying of 1-mm particles ranged between 20 and 100% in the 24-h period. The results of this and several other studies suggest that the rabbit is an unreliable model for bioavailability testing of human dosage forms.

Monkey

Many species of monkey are used in drug research. Probably the most useful species for bioavailability is the rhesus (*Macaca mulatta*). Although the rhesus monkey is much smaller than the human, the rhesus GI tract is proportionately larger than that of humans, so that it can accommodate most human-scale dosage forms (33). Under natural conditions gastric acid production in monkeys is similar to that in humans. However, in Old World monkeys, including the macaque family, secretion of gastric acid is almost completely shut down when the animal is restrained, and the pH of the gastric contents may be neutral or even slightly alkaline. Gastric juice secretion is also reduced, and GI motility may be inhibited. New World mon-

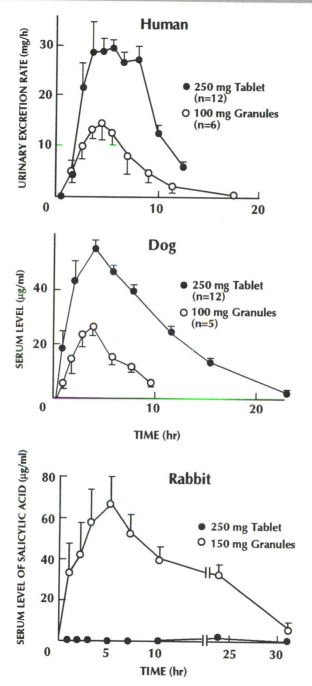

Salicylate levels, expressed as total salicylic acid urinary excretion rate or mean serum levels, after administration of enteric-coated tablets (diameter 11 mm) or granules (diameter 1 mm) to human, dog, and rabbit. In all experiments subjects were fasted overnight. Vertical bars show standard errors. (Reproduced with permission from reference 34.)

FIGURE 4.4

keys, by contrast, appear to secrete greater amounts of acid when restrained. Because it is usually recommended that monkeys over 10 kg be restrained for all procedures, the larger monkeys do not appear to be particularly suitable for testing oral dosage forms because of the likelihood of stress-induced changes.

The duodenum in the rhesus monkey has a length of about 5 cm, much shorter than in humans. The diameter of the duodenum and the small intestine is a little larger than might be expected from the overall size of the animal, and this may compensate, to some extent, for the shorter length in terms of absorptive capacity. Intestinal villi in the rhesus monkey tend to be broad and leaf shaped, as opposed to the filamentous appearance of human villi. Enzymes in the brush border are qualitatively similar in the rhesus monkey and in humans. There is a concentration in the upper small intestine of phosphatases and esterases, whereas leucine aminopeptidase is found mainly in the jejunum and proximal ileum in the rhesus monkey (35).

The pancreatic juice in monkeys contains both digestive enzymes and bicarbonate and appears to be secreted at fairly similar rates. The pH has been reported to be between 7 and 9, quite alkaline relative to humans. The overall length of the colon in the rhesus monkey is about 40–50 cm, about one-third that of the human colon. In section, the colon of the rhesus monkey is virtually indistinguishable from that of humans.

The literature contains only a few studies that have directly compared absorption of drugs in monkeys and humans. Nadolol, methyldopa, and acyclovir (36–38) are three drugs that are incompletely absorbed and yet are absorbed to similar extents in monkeys and humans. Comparable bioavailability of these drugs in humans and monkeys, which is significantly less than in the dog, is shown in Table 4.1 (36–38). Although these results are interesting, it cannot be assumed that monkeys are a good general model for human drug absorption, particularly for testing controlled- or sustained-release dosage forms. Even if the correlation in bioavailability with humans were proven, problems associated with the use of primates, ranging from risks to personnel to ethical issues and also cost, will probably continue to limit their use for oral drug absorption studies.

Pig

There are several differences in the anatomy and physiology of the pig as compared to the human. The pig stomach is larger than the human stomach, and also contains a diverticulum, or pouch, possibly a site of microbial metabolism (33). Acid output in the pig is similar to that of humans, but reports of gastric pH val-

TABLE 4.1   **Bioavailability of Nadolol, Acyclovir, and Methyldopa in the Dog, Human, and Monkey**

Drug	Bioavailability (%)		
	Dog	Human	Monkey
Nadolol	88–104	20–33	12–44
Acyclovir	75	18	4
Methyldopa	92, 83	25, 29	18

Source: Data from references 36–38.

ues vary. Pigs exhibit the IMMC when fasting, as do humans, but unlike humans emptying of food from the stomach is bimodal, with about 30–40% emptying in the first 15 min followed by sustained emptying after about 1 h (*39*). Stomach emptying also appears to be incomplete, so that it is probably not possible to conduct a drug-absorption study in a truly fasted-state pig.

The small intestine of the pig is approximately twice as long as in humans, so that the absorption surface area is larger. The pig has less capacity to concentrate bile in the gall bladder, and bile salts are derived from hydrocholic acid rather than cholic acid. Differences in bile concentration, and also the nature of bile salts, may produce differences in solubilizing and wetting characteristics in GI fluids, possibly affecting drug dissolution and absorption. Consistent with the altered bile concentration and type, a study in minipigs demonstrated longer t_{max} values than in humans after oral doses of four griseofulvin formulations. Prolonged t_{max} values obtained in this study might also have been related to delayed gastric emptying. The time of t_{max} values is shown in Figure 4.5 (*40*). The cecum of the pig is considerably larger than in humans, which is consistent with the greater herbivore nature of the normal pig diet.

An additional problem with pigs or minipigs is the deep-seated nature of peripheral veins. Although this is unrelated to GI anatomy and physiology, it is nonetheless an important factor requiring prolonged cannulation in pigs for the duration of any drug absorption or bioavailability study (*33*).

Dogs have been used extensively to investigate oral drug absorption. Gastric dimensions in dogs are very similar to those in humans. Basal gastric acid output

Dog

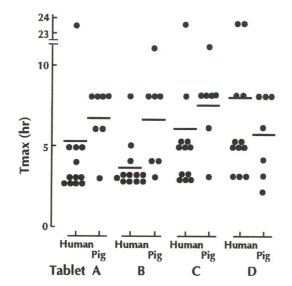

Times of peak blood levels of griseofulvin after administration of four different tablet formulations to humans ($N = 12$) and minipigs ($N = 7$). (Adapted with permission from reference 40.)

FIGURE 4.5

in dogs many be lower than in humans, so that gastric pH is generally higher than in humans and is sometimes indistinguishable from that in the small intestine (33). The fasted-state motility cycle follows a similar pattern and periodicity in dogs and humans. Motility response to feeding is also similar in the two species, with solids and nutrient liquids following zero-order emptying. However, food seems to empty from the stomach at a slower rate in dogs (41).

The small intestine of the dog is only about one-half as long as in humans. This difference in length correlates well with observed transit times in the two species. In studies with beagle dogs and humans, the mean residence time of a Heidelberg capsule in the small intestine was observed to be approximately 2 h in dogs, as compared with almost 4 h in humans (41).

This difference appears to be true for dosage forms as well. Studies by Ueda and co-workers (42) and Davis and co-workers (43) indicate that, especially for pellets or granules, intestinal transit times are considerably shorter in dogs than in humans. The average intestinal transit time for pellets was 1.3 h in dogs and 3.8 h in humans. For tablets, the values were 2.6 h in dogs and 3.1 h in humans. Thus the contact time between the drug and the absorbing mucosa will be substantially shorter in dogs, possibly causing reduction of absorption, particularly with controlled-release dosage forms. Dog pancreatic juice contains digestive enzymes and bicarbonate and appears to be secreted at a rate similar to that in humans. The duodenal pH in the fasted state is 0.5–1 pH unit higher in dogs than in humans.

It has been suggested in the literature that transport systems may have different efficiencies or specificities in the dog. For example, several compounds that are poorly absorbed in humans, such as acyclovir, methyldopa, and nadolol, are virtually completely absorbed in the dog. The mechanistic basis of these differences in absorption is unknown (33).

Ogata (44) established correlations between absorption of several drugs from conventional dosage forms in dogs and in humans. These results are summarized in Table 4.2. For all of the compounds studied, diazepam, griseofulvin, nalidixic

TABLE 4.2   **Correlation Coefficients Between Dog and Human Pharmacokinetic Data for Five Compounds**

| Drug | Dosage Form | N | R (correlation coefficient) | | |
			C_{max}	T_{max}	AUC
Diazepam	Uncoated tablet	4	0.627	0.310	0.990[b]
Griseofulvin	Uncoated tablet	4	0.388	0.711	0.306
Nalidixic acid	Uncoated tablet	5	0.895[a]	0.654	0.690
Flufenamic acid	Capsule	5	0.648	0.228	0.121
Metronidazole	Sugar-coated tablet	5	0.863	—	0.793

[a] $p < 0.05$

[b] $p < 0.01$

Note: N is the sample size, C_{max} is the maximum concentration, T_{max} is the time to reach the maximum concentration, and AUC is area under the curve.

Source: Data are from reference 44.

acid, flufenamic acid, and metronidazole, poor correlations were obtained for maximum concentrations, times of maximum plasma concentrations, and areas under plasma level curves between the two species.

Two other examples of poor predictability of dog absorption data to man, particularly for modified-release dosage forms, are provided by valproic acid and aminorex. The percentage of valproic acid absorbed from orally administered immediate-release and two different sustained-release dosage forms is shown in Figure 4.6 (45). Although the rank order of absorption is the same in the two species, the relative bioavailability of the sustained-release formulations of valproic acid were much lower in the dog.

In the case of aminorex, the bioavailability of a sustained-release dosage form was equivalent to that of an immediate-release formulation in humans, but the percentage absorbed from the sustained-release formulation was only two-thirds of that of the immediate-release formulation in dogs (45). Plots of the percentage of drug absorbed from the two dosage forms in the two species are shown in Figure 4.7 (45).

Although some other studies have demonstrated similar absorption from different dosage forms in dogs and humans, there are sufficient examples of differences between the species to cast doubt on the general predictability of the dog to the human situation.

Animal species will continue to play an important role in examining the nature and also the mechanisms of drug absorption during drug discovery and development. Larger species may also be useful for preliminary investigation of new dosage forms. However, differences in GI physiology and anatomy between animal species and humans generally prevent the use of animal species to accurately predict drug formulation performance in humans.

Summary

The stomach-emptying rate may influence drug absorption and GI residence time.

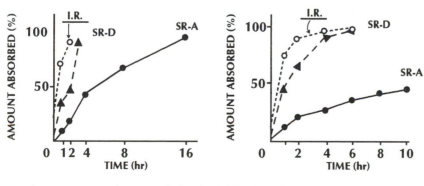

Cumulative percent valproic acid absorbed following administration of immediate release (IR) and sustained release (SR) dosage forms to humans and dogs. (Redrawn from data in reference 45.)

FIGURE 4.6

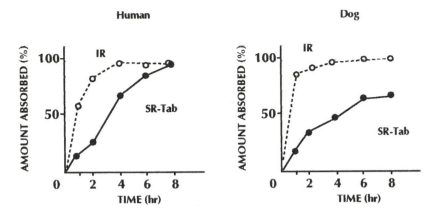

FIGURE 4.7   Cumulative percent aminorex absorbed after administration of immediate release (IR) and sustained release (SR) dosage forms to humans and dogs. (Reproduced with permission from reference 45.)

■ Factors that may influence drug absorption include secretion of acid, bile, and digestive enzymes, the unstirred layer, the glycocalyx, and drug degradation by intestinal microflora.

■ Drugs may be absorbed from the GI tract via the splanchnic circulation or the lymphatic system. The splanchnic route is the most efficient but is accompanied by the potential problem of first-pass hepatic metabolism.

■ Most drugs are passively absorbed from the GI tract into the systemic circulation. However, absorption of some compounds, including peptides, ACE inhibitors, and β-lactam antibiotics may involve active carrier-mediated mechanisms.

■ Animal models are useful for examining the nature and mechanisms of drug absorption. However, differences in GI anatomy and physiology among species may limit the use of animals to predict drug product performance in humans.

References

1. Wilson, T. H. *Intestinal Absorption*; Saunders: Philadelphia, PA, 1962; p 2.
2. Kutchai, H. C. In *Physiology*, 2nd ed.; Berne, R. M.; Levy, M. N., Eds.; C. V. Mosby: St. Louis, MO, 1988; pp 682–717.
3. Davenport, H. W. *Physiology of the Digestive Tract*, 4th ed.; Year Book Medical: Chicago, IL, 1978; p 112.
4. Texter, E. C.; Chou, C. C.; Laureta, H. C.; Vantrappen, G. R. *Physiology of the Gastrointestinal Tract*; C. V. Mosby: St. Louis, MO, 1968; pp 188–198.
5. Bai, J. P. F.; Stewart, B. H.; Amidon, G. In *Pharmacokinetics of Drugs*; Welling, P. G.; Balant, L. P., Eds; Springer-Verlag: Heidelberg, Germany, 1994; pp 189–206.
6. *Peptide-Based Drug Design*; Taylor, M. D.; Amidon, G. L., Eds.; American Chemical Society: Washington, DC, 1995; p 567.

7. Lee, V. H. L.; Dodda-Kashi, S.; Grass, G. M.; Rubas, W. In *Peptide and Protein Drug Delivery*; Lee, V. H. L., Ed.; Marcel Dekker: New York, 1991; pp 691–738.

8. O'Hagan, D. T.; Palin, K. J.; Davis, S. S. *CRC Crit. Rev. Ther. Drug Carrier Syst.* **1987,** *4*, 197–221.

9. Warshaw, A. L.; Walker, W. A.; Cornell, R.; Isselbacher, K. J. *Lab. Invest.* **1971,** *25*, 675–684.

10. Gonnella, P. A.; Walker, W. A. *Adv. Drug Del. Rev.* **1987,** *1*, 235–248.

11. Saffran, M.; Bedra, C.; Kumar, G.S.; Neckers, D. C. *J. Pharm. Sci.* **1988,** *77*, 33–38.

12. Takaori, K.; Burton, J.; Donawitz, M. *Biochem. Biophys. Res. Commun.* **1986,** *137*, 682–687.

13. Friedman, D. I.; Amidon, G. L. *Pharm. Res.* **1989,** *6*, 1043–1047.

14. Kidron, M.; Bar-On, H.; Berry E. M.; Ziv, E. *Life Sci.* **1982,** *31*, 2937–2941.

15. Wood, A. J.; Maurer, G.; Niederberger, W.; Beveridge, T. *Transplant. Proc.* **1983,** *15*, 2409–2410.

16. Mathews, D. M.; Payne, J. W. *Curr. Top. Membr. Trans.* **1980,** *14*, 331–425.

17. Hu, M.; Amidon, G. L. *J. Pharm. Sci.* **1988,** *77*, 1007–1011.

18. Hu, M.; Sinko, P. J.; DeMeere, A. L. J.; Johnson, D. A.; Amidon, G. L. *J. Theoret. Biol.* **1988,** *131*, 107–114.

19. Tsuji, A.; Tamai, I.; Hirooka, H.; Terasaki, T. *Biochem. Pharmacol.* **1987,** *36*, 565–567.

20. Marcon-Genty, D.; Tome, D.; Kheroua, O.; Dumontier, A. M.; Heyman, M.; Desjeux, J. F. *Am. J. Physiol.* **1989,** *256*, G943–G948.

21. Schilling, R. J.; Mitra, R. K. *Int. J. Pharm.* **1990,** *62*, 53–64.

22. Rao, R. K.; Koldovsky, O.; Korc, M.; Pollack, P. F.; Wright, S.; Davis, T. P. *Peptide* **1990,** *11*, 1093–1102.

23. Fricker, G.; Bruns, C.; Munzer, J.; Briner, U.; Albert, R.; Kissel, T.; Vonderscher, J. *Gastroenterol* **1991,** *100*, 1544–1552.

24. Sinko, P. J.; Amidon, G. L. *J. Pharm. Sci.* **1989,** *78*, 723–726.

25. Oh, D. M.; Sinko, P. J.; Amidon, G. L. *Pharm. Res.* **1989,** *5*, S–91.

26. Oh, D. M.; Sinko, P. J.; Amidon, G. L. *Pharm. Res.* **1990,** *7*, S–119.

27. Yoshikawa, T.; Muranushi, N.; Yoshida, M.; Yamada, H.; Oguma, T.; Hirano, K. *Pharm. Res.* **1989,** *6*, 308–312.

28. Tamai, L.; Ling, H. Y.; Timbul, S. M.; Nishikido, J.; Tsuji, A. *J. Pharm. Pharmacol.* **1988,** *40*, 320–324.

29. Yee, S.; Amidon, G. L. In *Peptide-Based Drug Design*; Taylor, M. D.; Amidon, G. L., Eds.; American Chemical Society: Washington, DC, 1995; pp 135–148.

30. Friedman, D. I.; Amidon, G. L. *J. Pharm. Sci.* **1989,** *78*, 995–999.

31. Yee, S.; Amidon, G. L. *Pharm. Sci.* **1990,** *7*, S–155.

32. Stewart, B. H.; Dando, S. A.; Morrison, R. A. *Pharm. Res.* **1990,** *7*, S–516.

33. Dressman, J. B.; Yamada, K. In *Pharmaceutical Bioequivalence*; Welling, P.G.; Tse, F. L. S.; Dighe, S. V., Eds.; Marcel Dekker: New York, 1991; pp 235–266.

34. Aoyagi, N. Ph.D. Doctoral Thesis, *Comparative Studies of Griseofulvin Bioavailability Among Man and Animals*; Kyoto University, 1986.

35. Lapin, B. A.; Cherkovich, G. M. In *Pathology of Simian Primates*, Part I; Fiennes, R. N. T.-W., Ed.; Karger, S., Basel, Switzerland, 1972; pp 127.

36. Dreyfuss, J.; Shaw, J. M.; Ross, J. J. *Xenobiotica* **1978,** *8*, 503–508.

37. DeMiranda, P.; Krasny, H. C.; Page, D. A.; Elion, G. B. *J. Pharm. Exp. Ther.* **1981,** *219*, 309–315.

38. Kwan, K. C. *The Use of Animals as Substitutes for Humans in Oral Bioavailability Studies*; PMA/FDA Workshop: Washington, DC, July 1989.

39. Pond, W. G.; Houpt, K. A. *Biology of the Pig*; Comstock (Cornell University Press): Ithaca, NY, 1978.

40. Aoyagi, N.; Ogata, H.; Kaniwa, N.; Ejima, A.; Yasuda, Y.; Tanioka, Y. *J. Pharm. Dyn.* **1984,** 7, 7–14 (published by the Pharmaceutical Society of Japan, Tokyo).

41. Dressman, J. B. *Pharm. Res.* **1986,** 3, 123–131.

42. Ueda, Y.; Munechika, K.; Kikukawa, A.; Kanoh, Y.; Yamanouchi, K.; Yokoyama, K. *Chem. Pharm. Bull.* (Tokyo) **1989,** 37, 1639–1641.

43. Christensen, F. N.; Davis, S. S.; Hardy, J. G.; Taylor, M. J.; Whalley, D. R.; Wilson, C. G. *J. Pharm. Pharmacol.* **1985,** 37, 91–95.

44. Ogata H. *Applied Pharmacokinetics-Theory and Experimental;* Soft Science: Toyko, Japan, 1985; p 585.

45. Bialer, M.; Friedman, M.; Dubrovsky, J. *Biopharm. Drug. Dispos.* **1984,** 6, 401–411.

Factors Influencing Absorption and Bioavailability After Enteral Administration

5

I n Chapter 3 enteral absorption was defined as absorption from the gastrointestinal (GI) tract, which results in a compound entering the splanchnic circulation and passing through the liver before entering the general circulation. The ability of liver enzymes to metabolize compounds during this absorption stage, and also the unique location of the liver in the cardiovascular system, can result in considerable metabolism of compounds during their first-pass through this organ.

The liver is not the only site of drug degradation or metabolism affecting systemic bioavailability. Other sites include the GI lumen, gut wall, and to a lesser extent the lung. The impact of drug metabolism in these organs and tissues, in addition to chemical drug degradation in the GI tract, can have a major effect on systemic drug availability, depending on the extent and efficiency with which a drug is metabolized. Although presystemic or first-pass metabolism is generally associated with hepatic metabolism, the actual oral bioavailability of unchanged drug, F, is a product of the contribution of all systems, as shown in equation 5.1.

$$F = F_g \cdot F_h \cdot F_l \qquad (5.1)$$

where F_g, F_h, and F_1 represent fractions of drug that survive during initial passage through he GI tract, liver, and lung, respectively. The pharmacologic or therapeutic impact of these metabolic processes of course depends not only on the extent to which a compound is metabolized but also on the pharmacologic activity of any metabolites formed. If a compound is administered orally as a prodrug, presystemic metabolism to a pharmacologically active metabolite may be an essential component of its administration.

Metabolism in the Gut

Gut Lumen Metabolism

Metabolizing enzymes in the gut lumen originate from exocrine glands, cells that are shed from the gut mucosa, and from the gut flora. The gut flora, comprising aerobic and anaerobic organisms, is situated mainly in the distal intestine and colon. Enzymes produced by exocrine glands and by most mucosal cells are more active in the proximal small intestine and are inactivated by intestinal microorganisms. Pancreatic secretions containing enzymes involved in peptide and protein hydrolysis were described in Chapter 3. Although most intestinal secretions tend to be protective and lubricative in nature, they are also important for presystemic metabolism of some drugs. However, these enzymes are present in the gut wall, so that assignment to a location in the gut lumen is tenuous. Hydrolysis of pivampicillin to the parent drug ampicillin is an example of involvement of gut luminal esterases.

Gut Wall Metabolism

Drug metabolism in the gut wall is far more common than luminal metabolism, and this topic has been reviewed extensively (1–3). Gut wall metabolic enzyme activity is greatest in the mucosal epithelial cells of the duodenum and jejunum, and it decreases distally.

Many metabolic interactions occur in the gut wall, including Phase 1 reactions, such as oxidation, reduction, and hydrolysis, and Phase 2 conjugation reactions, such as glucuronidation, sulphation, N-acetylation, O-methylation, and glutathione and glycine conjugation (1) (see Chapter 9).

Recent work has shown that the stomach mucosa is the primary site for first-pass metabolism of ethanol by alcohol dehydrogenase (ADH). Blood levels of ethyl alcohol following oral administration are much lower than after intravenous administration of a small dose of ethanol (0.15 g/kg). Over 80% of an orally administered dose is metabolized by the stomach. However, when the same amount of ethanol is delivered to the duodenum by nasogastric tube, or is given to patients with partial gastrectomies in which the antrum is removed, the first-pass effect is completely abolished, and bioavailability is the same as after an IV dose. Therefore, neither the liver nor the intestine contributes significantly to the first-pass effect on ethyl alcohol (4).

Although gut wall metabolism of drugs is well established, and both induction and inhibition of these enzymes have been described, the quantitative contribution of gut wall presystemic metabolic clearance has not been established for most drugs. This lack of information is due in part to the difficulty of separating the contributions of gut wall, gut flora, and hepatic metabolism to observed bioavailability parameters, and to the generally far greater contribution of hepatic metabolic enzymes to presystemic metabolism.

Bacterial Metabolism

Some of the numerous enzymatic drug reactions catalyzed by intestinal bacterial enzymes are summarized in Table 5.1. These metabolic processes are a major factor in the metabolism of some drugs (5). They are particularly important for controlled-release products, which release significant amounts of drug in the distal small intestine and colon where bacterial counts are high.

There are about 60 types of bacteria in the colon, and the composition is fairly consistent between individuals, male and female, between different races, and between different animal species. For this reason, the results of studies in animals are often predictive of results in humans. The colon is a targeted site for drug delivery, for drugs that specifically act on the colon, and for drugs that may be better absorbed from the colon. 5-Aminosalicylic acid (5-ASA) is a commonly used compound for ulcerative colitis or large-bowel inflammatory disease. There are several drug delivery systems that utilize colonic bacteria to split 5-ASA from ester or azo prodrugs. There are also new dosage forms that have special coatings that are degraded in the colon by bacterial enzymes. Drug delivery systems that deliver drugs to the colon are likely to see increasing use with some poorly absorbed drugs such as polar antibiotics and with macromolecules such as peptides and proteins.

Metabolism in the Lung

Whether a drug is administered enterally or parenterally into the venous circulation, it has to pass through the lungs before entering the general circulation. Although passage through the lungs is not restricted to enteral absorption, it is nonetheless a potential site for presystemic drug metabolism after all enteral doses. It may therefore contribute to reduced drug bioavailability by this route. The possibility of lung metabolism casts some doubt on the claim made generally, and also in this book (*see* Chapter 3), that intravenous dosing leads to 100% systemic availability. If lung metabolism is extensive, this may not necessarily be the case. However, two factors make the assumption of 100% systemic availability reasonable. The first is that any first-pass lung metabolism is essentially common to both intravenous and enterally administered drugs, so that the effect will tend

Metabolism of Foreign Compounds Carried Out by Gastrointestinal Microflora	**TABLE 5.1**

1. Hydrolysis of glycosides	13. Reduction of double bonds
a. Glucuronides	14. Reduction of nitro groups
b. Other glycosides	15. Reduction of azo groups
2. Hydrolysis of sulfate esters	16. Reduction of aldehydes
3. Hydrolysis of amides	17. Reduction of ketones
4. Hydrolysis of esters	18. Reduction of alcohols
5. Hydrolysis of sulfamates	19. Reduction of N-oxides
6. Hydrolysis of nitrates	20. Reduction of sulfoxides
7. Dehydroxylation	21. Reduction of epoxides to olefins
a. C-Hydroxyl compounds	22. Aromatization
b. N-Hydroxyl compounds	23. Nitrosamine formation
8. Decarboxylation	24. Nitrosamine degradation
9. Dealkylation	25. Acetylation
a. O-Aklyl compounds	26. Esterification
b. N-Alkyl compounds	27. Methylation
10. Dehalogenation	28. Ketone formation
11. Deamination	
12. Heterocyclic ring fission	
a. O-Containing ring systems	
b. N-Containing ring systems	

Source: references 6–8.

to cancel out in any bioavailability calculation. The second is that the extent of lung metabolism is generally small relative to gut and liver metabolism, and it can therefore be disregarded. However, it cannot be totally ignored and is certainly important, as in the case of inhaled drugs (*see* Chapter 3). Although the lung may not metabolize drugs efficiently, it is nonetheless capable of accumulating highly lipophilic drugs such as lidocaine, propranolol, verapamil, and propafenone.

Hepatic Metabolism and the First-Pass Effect

The subject of hepatic metabolism will be addressed in detail in Chapter 9. Let it suffice here to say that the liver is the primary site for drug metabolism in the body. Although drug metabolism takes place in many tissues and organs, including the GI tract, bloodstream, spleen, lungs, kidneys, brain, and muscle tissue, the greatest concentration of drug metabolizing enzymes of all types occurs in the liver, and the greatest preponderance of metabolism occurs there.

The unique location of the liver in the cardiovascular system makes it extremely important, not only as a major eliminating organ in the body, but also as a major factor influencing systemic availability of enterally administered drugs. The two major eliminating organs in the body are the liver and kidneys, and the unique relationship between these organs, at least from a drug elimination point of view, is described in Chapter 11.

Despite this close relationship, there are important differences between the roles of the liver and the kidneys in the cardiovascular system. While the kidneys derive their blood supply exclusively from the renal arteries, the liver receives blood both from the hepatic arteries and the splanchnic portal circulation. It is this dual blood supply, one arising from the systemic circulation and the other carrying absorbed substances from the GI tract, that places the liver in a unique position as an eliminating organ. By receiving portal blood, the liver "sees" substances before they enter the systemic circulation, and it can metabolize drugs, with variable efficiency, thereby influencing their systemic availability. Blood reaching the liver via the hepatic arteries, on the other hand, carries substances that have already entered and been diluted in the systemic circulation so that, for those substances that reach the liver by this route and are metabolized there, this organ acts purely as an eliminating organ.

Regardless of the source of hepatic blood supply, vascular exchange within the liver occurs through the sinusoids to the hepatic vein. While red blood cells and other formed elements are confined to the inner axial core of the liver sinusoid, dissolved substances can freely exchange with the liver parenchymal cells via the fluids in the outer space of the sinusoid, the space of Disse.

The location of the liver between the splanchnic and systemic circulations is well suited to its many functions. These include synthesis of most plasma proteins, including albumin, α_1-acid glycoproteins, and clotting factors; regulation and synthesis of amino acids, fatty acids, cholesterol, and glucose; synthesis of bile acid and urea; removal from blood of ammonia, endotoxins, bilirubin, endogenous hormones, and other waste products; and metabolism of drugs to more hydrophilic substances that can subsequently be removed from the circulation by the kidneys.

The first-pass of an enterally absorbed drug through the liver may limit absorption of some drugs. This process is important for the following reasons:

1. From the time when the drug is being transported from the GI tract via the capillaries of the splanchnic circulation until it mixes with the general circulation at the inferior vena cava, the drug is confined to the volume of the splanchnic circulation and has not yet distributed into the rest of the vascular system, or to other body tissues and fluids into which it may eventually partition. Therefore, during the first-pass through the liver, a drug is at a higher concentration than it will be after it has mixed with other parts of the vascular system and other body tissues and fluids.

2. The liver is the principal organ for drug metabolism.

3. Drug metabolism is generally a first-order process. Thus, the higher the concentration of drug presented to the liver, the greater the quantity of drug metabolized.

Consider the combined effect of points 1–3. Drug molecules that are absorbed from the GI tract are confined to the splanchnic circulation and are presented to the metabolizing enzymes of the liver at a high concentration during the first-pass. This step is illustrated in Figure 5.1. If hepatic metabolism is assumed to be first-order in nature, then a certain proportion of drug will be removed by the liver during the first-pass, depending on the hepatic extraction ratio, resulting in reduced drug availability to the general circulation. The greater the extent and efficiency of hepatic extraction, the larger the proportion of absorbed drug removed by the liver, and the greater the first-pass effect.

The first-pass effect is likely to be important for any orally administered drug that is extensively metabolized and has a high hepatic extraction ratio. A drug may be efficiently absorbed from the GI tract and yet poorly absorbed into the general circulation because of this second line of defense. Little or no first-pass effect will occur if a drug undergoes little or no hepatic metabolism.

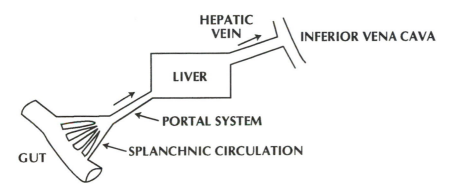

Absorption via the splanchnic circulation.

FIGURE 5.1

Quantitative Aspects of First-Pass Metabolism

Several methods have been described to predict the degree of first-pass metabolism of an orally administered compound. The two original methods were by Gibaldi et al. (6) and Rowland (7). One of several possible approaches based on the hepatic extraction ratio is described here. The oral availability of a drug that undergoes extensive hepatic metabolism can be predicted with this approach from equation 5.2.

$$\text{Oral availability} = 1 - E_h \qquad (5.2)$$

where E_h is the hepatic extraction ratio. If E_h is 0, the liver does not extract, no metabolism occurs, and all of the drug absorbed from the GI tract will likely reach the systemic circulation. On the other hand, if E_h is 1, that is, all of the drug is extracted during each pass through the liver; then the oral availability, however good the absorption, will be zero. Complete hepatic extraction of the drug may occur despite efficient absorption from the GI tract into the splanchnic circulation.

Equation 5.2 introduces the problem of determining the magnitude of the extraction ratio. One method by which it can be measured is by means of equation 5.3.

$$E_h = \frac{\text{Hepatic clearance}}{\text{Hepatic blood flow}} \qquad (5.3)$$

where hepatic clearance is the volume of blood that is cleared of drug by the liver per unit time, in milliliters per minute, and blood flow is equivalent to the splanchnic blood supply to the liver via the portal vein, which is approximately 1200 mL/min. Hepatic clearance can also be calculated from equation 5.4 after intravenous doses, where the dose is the intravenous dose, and $AUC^{0 \to \infty}$ is the area under the drug blood-concentration curve from zero to infinite time.

$$\text{Hepatic Clearance} = \frac{\text{Dose}}{AUC^{0 \to \infty}} \qquad (5.4)$$

A mathematical example will help to illustrate this method of predicting the hepatic extraction ratio. Suppose that a 100-mg dose of an extensively metabolized drug is administered by intravenous injection, and the area under the blood-concentration versus time curve is 280 µg/min/mL. To maintain consistency in units, the dose is multiplied by 1000 to express it in micrograms. Dividing the dose by the area under the curve as in equation 5.4 yields 357 mL/min. This value is the hepatic clearance. Hepatic clearance can then be divided by hepatic blood flow as in equation 5.3. The clearance is 357 mL/min, and if hepatic blood flow is 1200 mL/min, the hepatic extraction ratio is 0.298, or 0.3. From equation 5.2, oral bioavailability is 0.7, or 70%. This value represents the best systemic availability one might expect because the calculations assume that all of the drug is absorbed from the GI tract into the splanchnic circulation.

This approach is useful for predicting the systemic availability of oral drug doses, but it also provides additional information. Consider the example again, and recall that it predicts that drug availability to the general circulation should not exceed 70%, provided that the drug is efficiently absorbed from the GI tract into the splanchnic circulation. This hypothesis can be tested by giving the drug orally and measuring the systemic availability by comparing areas under drug blood-concentration curves from oral and intravenous doses. If the actual bioavailability is approximately 70%, then the prediction method is accurate (or at least appears to be). If the bioavailability value is less than 70%, then the drug probably was not efficiently absorbed from the GI tract because of a stability problem, poor solubility, an inappropriate formulation, or because the drug may not cross the GI epithelium very efficiently. For example, if the observed availability were actually 35%, one could conclude that only 50% of the drug was absorbed into the splanchnic circulation.

The other possible situation is when the observed oral bioavailability is greater than the predicted value. Although 70% bioavailability is predicted from the previous example, the actual availability may be 80% or 90% based on the relative area values from oral and intravenous doses. This situation is somewhat more difficult to interpret. Certainly, the lower predicted value is not related to GI absorption because the calculations assumed this value to be 100%. The only possible explanation relates to the degree of first-pass metabolism. The prediction method assumes first-order metabolism. However, if metabolism is not first-order but is saturable at high drug concentrations, which is more likely to occur during the first-pass then at any other time because of the high concentration of drug being presented to the liver, then the hepatic extraction ratio, E_h, will decrease and systemic availability will be greater than predicted. Some drugs with high, intermediate, and low hepatic extraction ratios are given in Table 5.2. These categories are further elaborated on in the discussion of liver failure in Chapter 10.

Enterohepatic Circulation

The human liver secretes between 250 and 1000 mL of bile every day into the duodenum via the gall bladder and the common bile duct. Bile contains bile salts that have a solubilizing effect and thus promote absorption of fats. Approximately 90% of bile salts and bile acids that are secreted in bile are reabsorbed from the intestine and returned to the liver where they are again available for secretion. A schematic of this system is shown in Figure 5.2. This closed-loop system is efficient because it enables bile salts and acids to be used many times. Enterohepatic cycling can also occur with drugs and other foreign compounds.

Some Drugs with High, Intermediate, and Low Hepatic Extraction Ratios **TABLE 5.2**

High	Intermediate	Low
Acebutolol	Acetaminophen	Aminopyrine
Alprenolol	Aspirin	Caffeine
Desipramine	Codeine	Carbamazepine
Isoproterenol	Erythromycin	Diazepam
Lidocaine	Metoprolol	Ibuprofen
Morphine	Nortriptyline	Isoniazid
Propoxyphene	Quinidine	Phenobarbital
Verapamil	Ranitidine	Theophylline

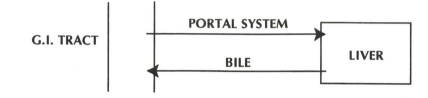

FIGURE 5.2   Schematic of Enterohepatic circulation, showing biliary excretion and reabsorption.

Biliary excretion is similar to renal tubular secretion (Chapter 11) because biliary excretion occurs by an active, energy-consuming process and can transport substances against a considerable concentration gradient. The ability of a drug to be excreted in bile is related to its chemical structure and molecular weight. Different species exhibit different molecular weight thresholds, or the minimum molecular weight necessary to be excreted in bile. The rat has a lower molecular weight threshold of approximately 200 for biliary excretion of quaternary ammonium compounds and about 300 for aromatic anions. Humans, on the other hand, have a lower molecular weight threshold of 300 for quaternary ammonium compounds and over 400 for most other compounds.

Drugs can be excreted in bile in unchanged form or as metabolized conjugates. The metabolized conjugate form is favored if conjugation brings the molecular weight of the compound above the minimum threshold value. In order to be excreted in unchanged form, drugs must meet the molecular weight criterion and also contain a polar functional group such as ammonium, carboxyl, or sulfate.

If, after being excreted in the bile into the duodenum, a compound or its metabolites are not reabsorbed, then they are voided in the feces or further degraded by bacterial microflora, and the bile becomes an efficient elimination route. However, the unchanged drug or deconjugated metabolites, which may exist because of the action of intestinal enzymes or intestinal bacterial degradation, can also be reabsorbed into the bloodstream and can thus set up a similar enterohepatic cycle to that of bile salts. Compounds such as cromolyn sodium, erythromycin, and rifampin are excreted unchanged in bile. Conversely, compounds such as carbenoxolone, estradiol, indomethacin, and morphine are excreted in the bile mainly as conjugates. The cephalosporin cefoperazone undergoes extensive biliary excretion in humans, this route of excretion accounting for 60–70% of total eliminated drug. Concentrations of cefoperazone in bile may be 10 to 20 times higher than those in serum (8, 9). The degree to which enterohepatic cycling occurs with many compounds in humans is not known, mainly because most human biliary excretion data are obtained during gall bladder surgery, and information is scanty for most compounds.

As noted previously, drugs can be extensively metabolized in the liver during the first-pass. Because bile formation also initiates in the liver, and because drugs are taken up by bile in that organ, drugs that are extensively excreted in bile will also have their systemic absorption reduced by this process in addition to any hepatic metabolism. Therefore, any drug that is extensively metabolized and also excreted in bile is in double jeopardy and is unlikely to exhibit good systemic availability regardless of possibly good GI absorption.

The exact nature of the enterohepatic cycle may be complex because hepatic extraction occurs from blood supplied by both the portal and hepatic veins. Thus, the recycling process involves drug that is just entering the body via the splanchnic circulation and also drug that is passing through the liver from both the portal and hepatic veins as part of the general systemic circulation. Whatever processes are involved, biliary recycling, or the enterohepatic cycle, can prolong the apparent elimination rate of a drug.

This effect has been elegantly demonstrated for the cardiac drug digitoxin. Oral maintenance doses of cholestyramine were administered orally 8 h after a dose of tritiated digitoxin. Cholestyramine is a nonabsorbable ion-exchange resin capable of binding digitoxin and some of its metabolites in the intestine and thereby preventing their absorption or reabsorption. Cholestyramine treatment caused a marked reduction in the elimination half-life of radioactivity in serum from 11.5 to 6.6 days, and this reduction was presumably the result of cholestyramine preventing reabsorption of biliary excreted digitoxin or its metabolites, with subsequent recycling.

Factors Influencing the First-Pass Effect

Age

Reduced first-pass metabolism in the elderly has been reported for several compounds including nalbuphine (10), verapamil (11), and clomethiazole (12). The decline in first-pass metabolism, resulting in increased systemic availability of unchanged drug, is often coupled with reduced systemic clearance so that plasma concentrations of unchanged drug can increase substantially. Unfortunately, as will be discussed in Chapter 10, changes in liver function and consequent changes in first-pass metabolism are notoriously difficult to predict from standard liver function tests. For example, no changes were observed in first-pass metabolism of amitriptyline, imipramine, metoprolol, morphine, pethidine, and propranolol despite observed age-related changes in liver function tests (1). Plasma levels of propranolol and morphine are increased in the aged as a result of reduced systemic clearance rather than increased systemic oral availability.

Liver Disease

The effect of liver disease on drug metabolism in general is discussed in Chapter 10. The relationships between liver disease and drug kinetics have been reviewed by Blaschke and Rubin (13).

Systemic bioavailability of propranolol is increased and systemic clearance is decreased in cirrhosis. Similar effects have been reported for lidocaine in cirrhosis and for propafenone in males with chronic alcoholic liver disease. In some cases, inconsistent results have been reported, and these reflect the complex relationship among liver function, drug metabolism, protein binding, and first-pass metabolism. For example, in one study the bioavailability of the beta-blocking agent labetolol was almost doubled in patients with chronic liver disease, whereas systemic clearance was unchanged. As labetolol is a highly extracted drug, it is likely that impaired liver function lowered the saturation metabolism threshold such that hepatic clearance was impaired at the high drug concentrations that occur during the first-pass, but not at the lower concentrations occurring subsequently.

Genetic Polymorphism

Data are continuing to accumulate in this area of research relevant to first-pass hepatic metabolism. Several distinct patterns have been documented. For example, hydroxylation of imipramine has been shown to be polymorphic, and oral bioavailability of this compound varies widely, from 27% to 80% (*14*). Extensive metabolizers demonstrated a high degree of first-pass metabolism and low bioavailability, whereas poor metabolizers exhibited high systemic bioavailability. Similarly, systemic concentrations of metoprolol, whose metabolism pattern is strongly associated with the debrisoquine phenotype, are six times higher, and its elimination half-life is three times longer, in poor metabolizers compared with extensive metabolizers. These types of studies show that any drug that is highly extracted by the liver and also undergoes polymorphic metabolism will exhibit polymorphic and hence variable first-pass metabolism.

Enzyme Induction and Inhibition

Any substance that changes the activity of hepatic metabolic enzymes is likely to affect the degree of hepatic first-pass metabolism. The effect will be greater for highly extracted drugs. Induction of metabolizing enzymes will likely reduce systemic availability of highly extracted drugs, whereas inhibition will have the opposite effect.

Enzyme Induction

Increased hepatic drug metabolism enzyme function due to the action of other drugs or chemicals will increase hepatic clearance and therefore reduce systemic availability of any drug that undergoes first-pass hepatic clearance. Enzyme induction has a somewhat greater effect on drug systemic availability than on systemic clearance.

Increased The antituberculosis drug rifampicin (rifampin) is a known enzyme inducer. The effect of rifampicin on verapamil pharmacokinetics has been evaluated by Barbarash et al. (*15*). The oral bioavailability of a single dose of verapamil after 15 days of rifampicin dosing was reduced from 26% to 0.02%. Systemic clearance of verapamil was increased by 24%.

Cigarette smoke contains a variety of organic and inorganic substances. Some of these compounds, including nicotine and certain polycyclic hydrocarbons, are known enzyme inducers. Steady-state propranolol concentrations in smokers have been reported to be one-half of those in nonsmokers (*16*). This reduction was not influenced so much by systemic clearance, as by an apparent increase in first-pass metabolism.

Enzyme Inhibition

Several drugs inhibit hepatic metabolism. Unlike enzyme induction, which may take a long period to develop, inhibition is an acute effect and is related to direct competition for metabolizing enzymes. Cimetidine inhibits the metabolism of other drugs by binding to the heme–protein of cytochrome P-450 (CYP) isozyme. Therefore, the bioavailability of highly extracted drugs that are metabolized by the same CYP enzymes that cimetidine inhibits would be expected to increase, and this is indeed the case. The systemic clearance of verapamil remained unchanged after preexposing individuals to cimetidine for 7 days (300 mg every 6 h). However, oral bioavailability increased from 26.3% to 49.3% (*17*).

An interesting example of a food component inhibiting first-pass hepatic metabolism was recently described for the calcium antagonist felodipine. In a study of six men with borderline hypertension, felodipine and dehydrofelodipine systemic availability increased 2.5- and 1.7-fold, respectively, when felodipine was taken with two servings of 250-mL double-strength grapefruit juice, relative to water (18). Under the same conditions, plasma levels of nifedipine and dehydronorfedipine increased 1.4- and 1.2-fold, respectively. The results with felodipine were reproduced in another study in nine healthy middle-aged men (19). The interaction with grapefruit juice, which is believed to be a class effect for the dihydropyridines, is thought to be due to inhibition of first-pass oxidative metabolism by flavonoids in the grapefruit juice. However, the precise mechanism of interaction has yet to be elucidated.

Summary

First-pass or presystemic clearance may be important for extensively metabolized drugs or drugs that are extensively cleared in the bile. First-pass metabolism occurs from all GI absorption sites except the mouth and the lower rectum. It may also occur after intraperitoneal dosage because much of the drug administered by this route is absorbed via the splanchnic circulation. Inhaled substances may undergo first-pass metabolism in the lungs. A method was described to predict the first-pass effect on the basis of intravenous data, and the implications of deviations from predicted values were discussed.

Biliary excretion and enterohepatic cycling can influence absorption and elimination of drugs that are subject to these phenomena.

Several factors may influence the degree of first-pass hepatic metabolism, including age, liver disease, genetic polymorphism, and enzyme induction and inhibition.

Problems

1. Following a 100-mg intravenous bolus dose of a drug that equilibrates evenly between plasma and blood red cells, the area under plasma curve ($AUC^{0 \to \infty}$) is calculated to be 200 μg/min/mL. Assuming that the hepatic blood flow rate is 1200 mL/min, predict the maximum systemic availability of the drug after an oral dose.

2. If the observed systemic availability of the oral dose is only 40%, what is the absorption efficiency of the drug from the GI tract into the splanchnic circulation?

References

1. Tam, Y. K. *Clin. Pharmacokinet.* **1993,** *25*, 300–328.
2. Caldwell, J.; Marsh, M. V. In *Clinical Pharmacology and Therapeutics: Presystemic Drug Elimination;* George, C. F.; et al., Eds.; Butterworth: London, 1982; pp 29–42.
3. Ilett, K. F.; Tee, L. B. G.; Reeves, P. T.; Minchin, R. F. *Pharmacol. Ther.* **1990,** *46*, 67–93.
4. Caballeria, J.; Baraona, E.; Rodamilans, M.; Lieber, C. S. *Gastroenterol* **1989,** *96*, 388–392.

5. Barr, W. H. In *Pharmaceutical Bioequivalence*; Welling, P. G.; Tse, F. L. S.; Dighe, S. V., Eds.; Marcel Dekker: New York, 1991; pp 149–167.

6. Gibaldi, M.; Boyes, R. N.; Feldman, S. *J. Pharm. Sci.* **1971**, *60*, 1338–1340.

7. Rowland, M. *J. Pharm. Sci.* **1972**, *61*, 71–74.

8. Craig, W. A. *Clin. Ther.* **1980**, *3*, 46–49.

9. Shimizy, K. *Clin. Ther.* **1980**, *3*, 60–79.

10. Jaillon, P.; Gardin, M. E.; Lecocq, B.; Richard, M. O.; Meignam, S.; et al. *Clin. Pharm. Ther.* **1989**, *46*, 226–233.

11. Storstein, L.; Larsen, A.; Midtbø, K.; Soevareid, L. *Acta Med. Scand.* **1984**, *681*(Suppl), 25–30.

12. Nation, R. L.; Vine, J.; Triggs, E. J. *Europ. J. Clin. Pharmacol.* **1977**, *12*, 37–145.

13. Blaschke, T. F.; Rubin, P. C. *Clin. Pharmacokinet.* **1979**, *4*, 423–432.

14. Brösen, K.; Gram, L. F. *Clin. Pharm. Ther.* **1988**, *43*, 400–406.

15. Barbarash, R. A.; Bauman, J. L.; Fischer, J. H.; Kondos, G. T. *Drug Intel. Clin. Pharm.* **1987**, *21*, 11A.

16. Vestal, R. E.; Wood, A. J. J.; Branch, R. A.; Shand, D. G.; Wilkinson, G. R. *Clin. Pharmacol. Ther.* **1979**, *26*, 8–15.

17. Smith, M. S.; Benyunes, M. C.; Bjornsson, T. D.; Shand, D. G.; Pritchett, E. L. *Clin. Pharmacol. Ther.* **1984**, *36*, 551–554.

18. Bailey, D. G.; Spence, J. D.; Munoz, C.; Arnold, J. M. O. *Lancet* **1991**, *337*, 268–269.

19. Edgar, B.; Bailey, D. G.; Bergstrand, R.; Johnson, G.; Regårdh, C. G. *Eur. J. Clin. Pharmacol.* **1992**, *42*, 313–317.

Physicochemical and Formulation Factors Affecting Drug Absorption

6

T he chemical and physical properties of a drug are of primary concern to the formulator because these characteristics can affect drug stability and absorption characteristics.

Chemical Factors

A variety of chemical options can be used to improve the stability and absorption of a drug without affecting its pharmacological properties. For example, erythromycin can be esterified to produce more acid-stable and fat-soluble derivatives for improved oral availability. The esters do not appear to be bacteriologically active but they become so when hydrolyzed to the free base. Another example of this type of drug is hetacillin, which is hydrolyzed to the active form ampicillin during or after absorption.

Both the stability and solubility of weak acids and bases tend to increase when they are in the form of water-soluble salts. Figure 6.1 shows how administration of soluble salts of penicillin V results in higher plasma levels compared to the free acid (1). Higher levels of the potassium salt than the calcium salt are consistent with their relative aqueous solubilities and dissolution rates in the gastrointestinal (GI) tract. The low levels obtained from penicillin G sodium are due to this penicillin being less stable than penicillin V in gastric juice. The rationale for better absorption of water-soluble salts of weak acids than the free acid form is illustrated in Figure 6.2. The pH of a solution of the salt of an acidic drug is given by equation 6.1.

$$pH = 0.5(pK_w + pK_a + \log C) \tag{6.1}$$

where pK_w is the negative logarithm of the dissociation constant for water, pK_a is the negative logarithm of the dissociation constant for the weak acid, and C is the

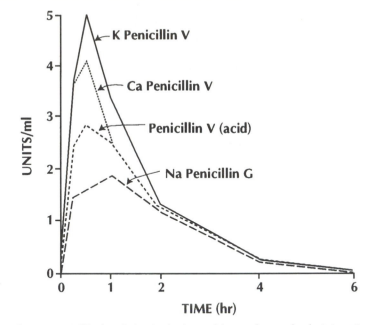

FIGURE 6.1
Mean plasma penicillin levels in 10 fasting subjects after oral administration of 105 units of penicillin in different forms. (Reproduced with permission from reference 1.)

molar concentration of total weak acid. For a weak acid with a pK_a of 4.0 and a concentration of 1 M, equation 6.1 becomes:

$$pH = 0.5(14 + 4 + 0) = 9.0 \qquad (6.2)$$

Equation 6.2 shows that the drug solution will have an alkaline reaction. Applying this concept to Figure 6.2 shows that as the salt of an acidic drug dissolves in the stomach, the salt generates a diffusion layer of relatively high pH that in turn promotes further dissolution of the weak acid. As the dissolved molecules move into the bulk stomach contents, the pH falls and may cause drug precipitation. If this does occur, the drug will likely precipitate in fine particles that tend to redissolve readily.

The same argument could be used for basic drugs, but the pH effect resulting from the use of salts of weak bases is far less important because this effect is swamped by the low pH of the stomach. Thus, salts of basic drugs are used primarily for handling and stability and are less important for improved dissolution.

Physical Factors

Different physical forms of a drug can be used to improve drug absorption. Typically, the crystal or polymorphic form of a drug, the state or form of hydration or solvation, and the physical size of drug particles can be varied, often with considerable impact on drug absorption.

Polymorphism and Amorphism

Many compounds can form crystals with different molecular arrangements, or *polymorphs*. Although these polymorphs have identical chemistry, they may have differ-

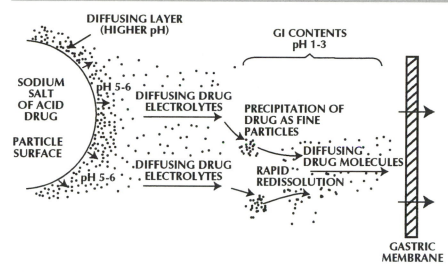

Schematic of the dissolution process in the stomach, for a water-soluble salt of a relatively insoluble weak acid. (Reproduced with permission from reference 1.)

FIGURE 6.2

ent physical properties such as melting point, dissolution rate, and solubility. For example, the vitamin riboflavin exists in several polymorphic forms, and these forms have a 20-fold range in aqueous solubility. Polymorphs that have no crystal structure, or amorphic forms, have different physical characteristics from the crystalline form.

Different polymorphs of organic compounds, including the amorphic form, may be produced depending on temperature, solvent of crystallization, and other factors during their preparation. One polymorph is often the most stable, and other forms tend to convert to that form at a rate and to an extent depending on the conditions.

The most important effect that polymorphism has in pharmacokinetics is on dissolution. Absorption of many orally administered drugs is controlled by dissolution rate. According to the Noyes–Whitney solution rate law, amorphous forms of a drug generally dissolve faster than crystalline forms (but there are exceptions) because no energy is required to break up the crystal lattice. Thus, from the standpoint of the rate of absorption, and often the extent of the absorption as well, the amorphous form is preferred over crystalline forms, and several drugs, including chloramphenicol palmitate, hydrocortisone, and prednisolone, are marketed in their amorphous forms. An interesting example of the use of combined polymorphs to provide optimal therapeutic effect is provided by insulin, where a mixture of amorphous and crystalline forms provides initial fast release from the amorphous component and sustained release from the crystalline component.

Solvation

During their preparation, drug crystals may incorporate one or more solvent molecules to form solvates. The type of solvate formed can influence drug dissolution rate. The most common solvate is water. If water molecules are already present in a crystal structure, the tendency of the crystal to attract additional water to initiate the dissolution process is reduced, and solvated (in this case hydrated) crystalline forms tend to dissolve more slowly than anhydrous forms. Although solvation was originally considered to be an important factor influencing absorption of some drugs such as ampicillin, observed differences in drug absorption from hydrated and anhydrous drug forms are not great and are generally not considered clinically significant. Significant differences have been observed in the dissolution of hydrated and anhydrous forms of caffeine, theophylline, glutethimide, and mercaptopurine. The significance of these differences regarding in vivo absorption has not been examined but is likely to be slight, as with ampicillin.

Particle Size

Unlike solvation, particle size can play a major role in the absorption of slowly dissolving drugs. The dissolution rate of solid particles is proportional to surface area, and surface area is related to the fineness of the particle. Particle size reduction has been used to increase the absorption of a large number of poorly soluble drugs, such as

- bishydroxycoumarin
- chloramphenicol
- digoxin
- griseofulvin
- medroxyprogesterone acetate
- nitrofurantoin
- phenobarbital
- phenacetin
- spironolactone
- tolbutamide

Griseofulvin is an interesting member of this group. This drug is a potent fungicide that is taken orally. Griseofulvin has extremely low aqueous solubility, and material of normal particle size results in poor and erratic absorption. Absorption was greatly improved when microsize particles were prepared, but it was improved even more when the drug was formulated in ultramicrosize particles as a monomolecular dispersion (solid solution) in polyethylene glycol.

Formulation Factors

Drug formulations are designed to provide a product that is attractive, distinctive, convenient to use, stable, and has the appropriate physicochemical characteristics to provide an optimal absorption profile. Currently available conventional dosage forms may be broadly categorized, in order of decreasing dissolution rate, as solutions, solid solutions, suspensions, capsules and tablets, coated capsules and tablets, and controlled-release formulations.

Solutions

Aqueous solutions in the form of elixirs, syrups, emulsions, or just simple solutions do not have a dissolution problem and generally result in faster and more complete absorption of passively absorbed drugs (i.e., the overwhelming majority of drugs) than other dosage forms. Some acidic drugs administered as water-soluble salts may precipitate in the acidic pH of stomach fluids, but such precipitates, as described earlier, are likely to be in finely divided form and should readily redissolve either within the stomach or as they pass into the relatively alkaline environment of the small intestine. Solution dosage forms are particularly useful for pediatric or geriatric patients and for patients who may have difficulty swallowing solid dosage forms. Because of their generally good oral bioavailability, solution dosage forms are frequently used as standards against which the bioavailability of other oral dosage forms is compared.

Although most oral solutions are aqueous, nonaqueous solutions or emulsions may be useful in particular cases. For example, a solution of indoxole in oil, as an oil-in-water emulsion, is absorbed three times more efficiently than an aqueous solution.

Solid Solutions

The solid solution is a novel formulation approach in which a drug is trapped as a solid solution, or molecular dispersion, in a water-soluble matrix, which leads to a large surface area to interact with GI fluids. Although the solid solution method is an attractive approach, particularly for lipophilic molecules that have dissolution and bioavailability problems, only one drug, griseofulvin, is currently marketed in this form. The marketed product Gris-PEG contains griseofulvin together with polyethylene glycol 400 and 800, and povidone.

Suspensions

A drug that is formulated as a suspension is in solid form, but the drug is finely divided, has a large surface area, and is freely available to interact with and dissolve in the GI fluids. The drug particles can also disperse and diffuse readily between the stomach and small intestine so that absorption of suspensions is likely to be less sensitive to stomach-emptying rate than other solid dosage forms.

Besides the esters of erythromycin that were discussed earlier in this chapter, many other antimicrobial agents are administered in suspension form. These include chloramphenicol palmitate, some penicillins, tetracyclines, nitrofurantoin, and sulfonamides.

Similar to solutions, suspensions are useful dosage forms for patients who experience difficulty taking solid medication. Adjusting the dose to a patient's need is easier with solutions and suspensions than with solid dosage forms. Giving one-half or two-thirds of a tablet or a fraction of a capsule may be difficult. Liquid dosage forms therefore have practical advantages besides simple dissolution effects. However, liquid dosage forms also have disadvantages; the principal ones are greater bulk, difficulty of handling, and possibly reduced stability compared to solid products.

Capsules and Tablets

Capsules and tablets are the most commonly used oral dosage forms. These two types of formulations differ from each other in that material in capsules is less impacted than in compressed tablets. Once a capsule dissolves, the encapsulated material can generally disperse quickly in a manner similar to a suspension. The capsule material, although water-soluble, can impede drug dissolution by interacting with drug material, but this interaction is uncommon.

The processes leading to tablet disintegration, dissolution, and absorption are described in Figure 6.3 (2). Tablets generally disintegrate in stages, first into granules and then into primary particles. As particle size decreases, dissolution rate increases because of increased surface area.

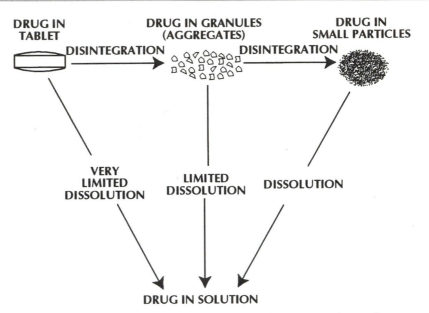

FIGURE 6.3   Tablet disintegration and dissolution. (Reproduced with permission from reference 2.)

Tablet disintegration into granules or into primary particles was once considered to be a sufficient criterion to predict in vivo absorption. However, now dissolution is recognized as a better criterion and bears a closer, albeit tenuous, relationship to in vivo drug absorption rates.

Regulatory agencies now require dissolution data for all new formulations in marketing submissions. The increasingly wide acceptance of dissolution as the best available in vitro parameter to establish drug uniformity and drug release rate and to predict in vivo absorption is reflected in the proliferation of such tests in official compendia. For example, the 1975 *United States Pharmacopeia* and *National Formulary* combined contained dissolution tests for 20 drug products. The 1980 *Pharmacopeia*, on the other hand, contained 53 such tests, and tests for many additional products have appeared in subsequent supplements to the *Pharmacopeia*.

In Vitro–In Vivo Correlations

There is currently considerable interest in the relationship between in vitro dissolution and in vivo bioavailability parameters for drugs and drug dosage forms. The U.S. Food and Drug Administration has spearheaded a research program to examine relationships between in vitro dissolution patterns and in vivo drug absorption or bioavailability characteristics. The intent of this research is to gain a better understanding of the relationships between these phenomena, but more particularly, to use these relationships to establish quantitative correlations between in vitro and in vivo parameters so that in vitro data can accurately predict and ultimately serve as a surrogate for in vivo performance. The cost savings of using a simple dissolution test to compare different drug formulations instead of a conventional in vivo bioequivalence or bioavailability study are significant, and such correlations, if they can be achieved, would likely find their way into official compendia. The main thrust of current research in this area is based on differentiation of drugs or formulations in terms of solubility and permeability, where solubility denotes intrinsic in vitro solubility and permeability denotes the ease with which the drug crosses GI membranes (3).

Drugs or drug products can be classified into four groups depending on solubility and permeability, namely high solubility, low permeability; high solubility, high permeability; low solubility, high permeability; and low solubility, low permeability. For drugs with low solubility and high permeability, dissolution is likely rate limiting for absorption, and establishing in vitro–in vivo relationships for these types of substances is feasible, and probably useful. For drugs with high solubility and low permeability, on the other hand, permeability is likely rate limiting for absorption, and in vitro dissolution has little or no effect on the rate or extent of absorption. These two limiting cases are relatively straightforward. However, with the two classes of high solubility, high permeability and low solubility, low permeability, the picture is less clear. Whether any clear-cut in vitro–in vivo relationships can be established in these cases is questionable and must be determined experimentally.

Because of the complex nature of this research, the conduct of appropriate experiments, which often involves prolonged GI intubation of subjects, is limited to a small number of laboratories. As a consequence, progress in achieving in vitro–in vivo correlations for particular drugs or formulations has been slow. It is likely that much more will be heard regarding this area of research, both for conventional and controlled-release dosage forms, in the future.

Excipients

Along with the active material contained in tablets and capsules, a variety of so-called inert ingredients are present. For example, tablets and capsules may contain magnesium stearate or talc to improve flow properties. Starch, magnesium aluminum silicate, methylcellulose, carboxymethylcellulose, or acacia may be present as binders or disintegrants. Lactose, kaolin, calcium sulfate, magnesium stearate, or other materials may be found as simple bulk agents and diluents. Tablets may also incorporate a variety of coatings to improve their stability, taste, appearance, and release characteristics. Although considered to be inert, these additives can affect drug dissolution and absorption. For example, changing the excipient (or diluent) from calcium sulfate to lactose and increasing the content of magnesium silicate increased the activity of oral phenytoin. The systemic availability of thiamine and riboflavin was reduced by the presence of fuller's earth, which adsorbs these drugs. Similarly, absorption of tetracycline from capsules is reduced by calcium phosphate because of complex formation.

Most of these types of interactions were reported some time ago, and the present level of sophistication associated with the pharmaceutical industry suggests that these interactions are unlikely to occur today. However, as recently as 1971, different formulations of digoxin yielded up to sevenfold differences in serum digoxin levels (4). This example should be taken as an object lesson never to underestimate the potential of excipient–drug interactions to increase or decrease drug absorption.

Coated Tablets

Tablets are often formulated with a coating, usually some form of acid-insoluble material such as shellac, resin, or styrene–maleic acid copolymer. These formulations are insoluble at acidic pH but dissolve readily at neutral or alkaline pH. Coated tablets are therefore ideally suited to prevent release of drug in the stomach, and yet permit release after the dosage form has left the stomach and has entered the relatively alkaline region of the small intestine.

Preventing release of drug in the stomach may be useful from two viewpoints. First, it protects acid-labile drugs from acid-catalyzed degradation. This phenomenon was discussed earlier for erythromycin and erythromycin stearate, but it is important for all acid-labile substances. In this case, the drug is protected from endogenous secretions.

The second viewpoint is not to protect the drug from the patient, but rather to protect the patient from the drug. Some drugs are irritating to the stomach and can cause local distress, nausea, and vomiting. Such drug substances include iron salts, diethylstilbestrol, and some nonsteroidal anti-inflammatory agents. Drug release can be delayed until the drug reaches the small intestine by using an acid-resistant coating, thus avoiding local toxicity in the stomach. Because release of the drug from these types of formulations is dependent on the stomach-emptying rate, absorption of the drug may be significantly affected by changes in stomach-emptying patterns. Bogentoft et al. (5) compared the effect of food on the absorption efficiency of enteric-coated aspirin tablets and a formulation of enteric-coated aspirin granules contained in conventional capsules. The results of this study are summarized in Figure 6.4. Plasma salicylate levels (salicylate is the major metabolite of aspirin formed by hydrolysis) from the two formulations are similar under fasting conditions. Food did not affect absorption from the granules but caused a marked delay and reduction in absorption from the coated tablets. Tablets yielded essentially zero absorption until 4 h after dosing, presumably as a result of the tablets being retained in the stomach and the acid-resistant coating remaining intact during that period. The lack of effect by food on the encapsulated granules is consistent with the more diffuse nature of this dosage form and the ease with which granules may move into the intestine after the capsule has dissolved.

Controlled-Release Formulations

There has recently been a marked increase in interest in this type of dosage form. The strong interest has been due to (1) greater appreciation of the advantages of controlled drug release, (2) a dramatic increase in the development of novel polymer systems and devices that are suitable for controlled release from oral dosage forms, and (3) a common strategy among pharmaceutical companies to develop controlled-release products in order to protect drug franchises in the face of generic competition, to maintain market share.

Some drugs currently available in controlled-release form are listed in Table 6.1. Several different commercial products are available for many drugs and drug combinations, and the number of drug substances available in controlled-release form is increasing rapidly.

Most of the oral controlled-release products currently available include diuretic and cardiovascular drugs, respiratory drugs, and compounds acting on the central nervous system. Little attention has been paid to antimicrobial agents.

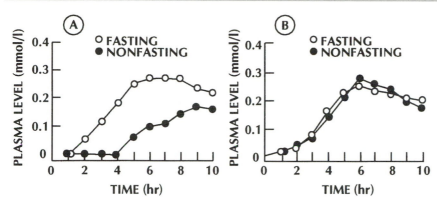

Mean plasma salicylate levels in eight subjects following single 1.0-g doses of aspirin as enteric-coated tablets (A) and enteric-coated granules in capsules (B) under fasting and nonfasting conditions. (Reproduced with permission from reference 5.)

FIGURE 6.4

Only one compound, tetracycline, is marketed in controlled-release form. Controlled release of antimicrobial agents that have appropriate pharmacokinetic properties still appears to represent an area of untapped potential. However, this situation may remain so until more information is available regarding the temporal relationships between circulating antibiotic levels and the efficacy of antibacterial effect.

Controlled-release dosage forms are invariably more expensive than conventional formulations and can be justified only when they offer one or more therapeutic advantages. Some of the possible advantages of controlled-release dosage forms are as follows:

Advantages of Controlled Drug Release

1. rapid onset and maintenance of therapeutic drug levels
2. reduced dosing frequency
3. reduced fluctuation in drug levels
4. reduced total amount of drug used
5. reduced inconvenience to the patient and increased compliance
6. reduced patient care time
7. less nighttime dosing
8. more uniform pharmacological response
9. reduced GI irritation
10. reduced side effects

Although each of these advantages is important, the only ones relevant to this chapter are items 1, 2, 3, and 8. Item 1 can be achieved only if the controlled-release formulation contains a fast-release component, or if a fast-release formulation is used to initiate therapy. Items 2 and 3 describe the essence of controlled release, which is to obtain prolonged circulating drug levels with less fluctuation compared to conventional dosage forms and to achieve these drug levels with less frequent drug administration. Item 2, which is often claimed as a "sufficient"

TABLE 6.1 **Drugs Available in Controlled-Release Form**

Vitamins, Minerals and Hormones
Ascorbic acid
Iron preparations
Methyltestosterone
Nicotinic acid
Potassium
Pyridoxine
Vitamin combinations

Diuretic and Cardiovascular Drugs
Acetazolamide
Ethaverine HCl
Isosorbide dinitrate
Nicotinyl alcohol
Nitroglycerin
Papaverine HCl
Pentaerythritol tetranitrate
Procainamide
Quinidine gluconate and sulfate
Reserpine

CNS Drugs
Amphetamine sulfate
Aspirin
Caffeine
Chlorpromazine
Dextroamphetamine sulfate
Diazepam
Diethylpropion HCl
Fluphenazine
Indomethacin
Lithium
Meprobamate
Methamphetamine HCl
Orphenadrine citrate
Pentobarbital
Pentylenetetrazole
Perphenazine
Phenmetrazine HCl
Phenobarbital
Phentermine HCl
Phenylpropanolamine HCl
Prochlorperazine

Respiratory Agents
Aminophylline
Brompheniramine maleate
Carbinoxamine maleate
Combination, antitussive
Combination, expectorant
Combination, upper respiratory
Dexchlorpheniramine maleate
Dimethindene maleate
Diphenylpraline HCl
Dyphylline
Phenylpropanolamine HCl
Pseudoephedrine HCl and sulfate
Theophylline
Trimeprazine
Tripelennamine HCl
Xanthine combinations

Antimicrobial
Tetracycline

Gastrointestinal Drugs
Belladonna alkaloids
Hexocyclium methylsufate
1-Hyoscyamine sulfate
Isopropamide iodide
Prochlorperazine maleate
Tridihexethyl chloride

Other
Pyridostigmine bromide

Source: Reproduced with permission from reference 5.

rationale for development of a controlled-release formulation, is becoming unacceptable as a sole criterion by regulatory agencies, in the absence of any other advantage. This is understandable given the increasing emphasis worldwide on cost containment in health care, particularly in the area of pharmaceuticals. Although Item 3 is often claimed as an advantage of controlled-release formulations, regulatory agencies generally insist that maximum and minimum plasma levels obtained during repeated dosing of the controlled-release formulation not differ significantly from those obtained from a conventional-release formulation. Thus, the advantage of reduced fluctuation refers principally to the overall fluctuation during a day in terms of the number of peaks and troughs in plasma levels rather than the absolute magnitude of peak and trough values. Item 8, which is also a primary goal of controlled-release dosage, may be predicted from theoretical drug level–response relationships, depending on the drug class, but it is generally difficult and expensive to prove clinically.

The major potential disadvantages of controlled-release oral dosage forms are:

Disadvantages of Controlled Drug Release

- possibility of dose-dumping
- reduced potential for accurate dose adjustment
- slow absorption may delay onset of activity
- increased potential for first-pass metabolism
- possible reduction in systemic availability
- drug release period restricted to residence

"Dose-dumping", a term used to describe the inadvertent rapid release of drug material due to faulty formulation or some other factor resulting in abnormally high levels of circulating drug, is particularly important for potent drugs that have a narrow therapeutic index. However, good manufacturing practice and the highly sophisticated dosage forms currently appearing on the market reduce the probability of this occurring.

Administering a fraction of a tablet or capsule to achieve fine dose adjustment is more difficult with some controlled-release dosage forms than with others. For example, controlled-release tablets that consist of granules in a tablet matrix can readily be subdivided to obtain a fraction of a dose. On the other hand, formulations such as repeat-action tablets or osmotic pump devices lose their sustained-release properties once the dosage form is fractured, so that fine dose adjustment for these dosage forms, apart from the use of available ranges of individual whole tablet or capsule strengths, is impractical.

Slow absorption inevitably delays the onset of drug activity from an initial dose, but this delay should be unimportant with repeated doses. Increased first-pass metabolism may occur with drugs that undergo extensive hepatic clearance, but only if hepatic clearance is saturable following rapid absorption from conventional doses. If saturation does not occur with conventional oral dosages, and if hepatic clearance is first-order in nature, then the same proportion of an oral dose will be cleared during the first-pass through the liver, regardless of the drug absorption rate.

Reduced and variable absorption from controlled-release formulations has been extensively documented. In most cases, drug bioavailability from controlled-release oral formulations is not greater than 80–85% of that from conventional formulations. The problem of reduced and variable absorption from oral controlled-release formulations is demonstrated in Figure 6.5 (6), which shows a wide range in theophylline absorption rate and extent from commercial controlled-release capsules.

Residence time within the GI tract is a potential disadvantage associated with oral controlled-release products, and this provides a major distinction between oral and parenteral controlled-release dosage forms. The actual time period available for an oral dosage form to effectively release drug for absorption is limited to the time period during which the drug is available for absorption within the GI tract. The residence time of the dosage form in the stomach is variable, depending on the activity of the patient, the presence of food in the stomach, and direct or indirect interactions with other drugs. After leaving the stomach, the dosage form, along with dissolved drug, passes into the optimal absorption region of the proximal small intestine. Distal to this region, absorption becomes less efficient, and the drug is furthermore exposed to the bacterial microflora. Because of variable GI transit time and also these other factors, estimating the optimum release period for an oral controlled-release dosage form becomes difficult.

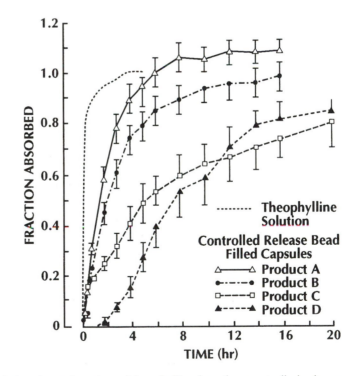

FIGURE 6.5   Cumulative absorption plots of theophylline from four controlled-release capsules and an aqueous solution. Error bars indicate 1 standard deviation. (Reproduced with permission from reference 6.)

Apart from the disadvantages just discussed, the pharmacokinetic characteristics of some drug types make them inherently unsuited for controlled-release formulations. Some typical characteristics are listed here.

Drugs that are Unsuited for Controlled Release

- short biological half-life
- long biological half-life
- potent drug with a narrow therapeutic index
- large doses
- poorly absorbed
- low or slow solubility
- active absorption
- time course of circulating drug levels does not agree with the pharmacological response
- extensive first-pass metabolism

A controlled-release dosage form of a drug that has a biological half-life of less than 2 h, or is administered in large doses, may need to contain a prohibitively large amount of drug. On the other hand, drugs with a long biological half-life of 8 h or more are sufficiently sustained in the body when administered in conventional doses, and prolonged release is not necessary.

Absorption of poorly water soluble compounds is often limited by their dissolution rate. Incorporation of such compounds into a controlled-release formulation is therefore unnecessary and is likely to markedly reduce absorption efficiency. Administering drugs such as warfarin, whose pharmacological effect is considerably prolonged relative to its blood profile, in controlled-release form is of no therapeutic advantage. Similarly, incorporating compounds such as fluorouracil, amino acids, and perhaps some β-lactam antibiotics and thiazide diuretics, which appear to exhibit reduced absorption efficiency at sites distal to the proximal small intestine, is likely to reduce absorption efficiency while achieving little or no prolongation of effect. As stated previously, if a drug undergoes extensive first-pass hepatic clearance that is saturable with conventional fast-release dosages, systemic availability may be decreased from a controlled-release dosage form because of nonsaturation of the presystemic clearance mechanisms. However, if hepatic clearance is not saturated with conventional doses, slower absorption of the drug should not affect first-pass presystemic clearance.

Although the arguments just given provide useful general rules regarding controlled-release dosage forms for a particular drug, there are the inevitable exceptions. Nitroglycerin has a short biological half-life of less than 0.5 h. It is rapidly metabolized by the liver and is generally considered to be poorly absorbed from oral doses. However, a large number of controlled-release oral nitroglycerin products are available in addition to an increasing number of topical and transdermal preparations. Low circulating levels of nitroglycerin obtained from these products are thought to provide adequate prophylaxis against angina attacks but would not be adequate to treat an acute angina episode. On the other hand, many of the drugs listed in Table 6.1 have biological half-lives greater than 8 h. Controlled release of these products may reduce toxic side effects simply by preventing sharp initial peaks in circulating drug levels that may occur with conventional doses.

However, these products are unlikely to provide more sustained blood levels or a more prolonged therapeutic effect than conventional dosage forms.

Renewed interest in controlled release has led to a variety of new formulations. Products that are representative of well-established release forms and also some more novel categories are summarized in Table 6.2. Of those listed, the osmotic pump has probably enjoyed the greatest success. This system, patented by Alza Corporation, is unique in that drug release from the system is highly controlled, utilizing osmotic pressure as the driving force for expulsion of drug from the dosage form, and is largely independent of the GI environment. This mechanism provides highly reproducible zero-order drug delivery, which is ideal for controlled release products (7, 8). The osmotic pump formulation has been particularly successful in a marketed form of nifedipine, Procardia XL.

Pharmacokinetics of Controlled Release

Despite the large and ever-expanding array of formulations devoted to oral controlled release, and the complex and varied physical properties involved in the release of drug from these formulations, the number of kinetic models necessary to describe overall drug release phenomena from existing dosage forms is relatively small (9). The major release patterns are summarized in simple graphical form in Figure 6.6. The patterns can be divided into two major categories, those that release drug at a slow zero- or first-order rate, and those that provide an initial rapid dose followed by slow zero- or first-order release of the sustained component.

The sustained nature of drug release from these dosage forms, even when a fast-release component is present, leads to considerable problems for in vitro dissolution testing. For conventional oral dosage products, in vitro dissolution crite-

TABLE 6.2	**Oral Controlled-Release Dosage Forms**	
Category	Product	Active Ingredient
Slow erosion with initial fast release dose	Tedral SA	Theophylline, ephedrine HCL, phenobarbital
Erosion core only	Tenuate Dospan	Diethylpropion HCl
Repeat action tablets	Chlor-Trimeton Repetabs	Pseudoephedrine sulfate, chlorpheniramine maleate
Pellets in capsules	Combid Spansule	Isopropamide iodide, prochlorperazine maleate
Pellets in tablets	Theo-Dur	Theophylline
Leaching	Desbutal Gradument	Methamphetamine HCl, pentobarbital sodium
Ion-exchange resins	Biphetamine	Amphetamine, dextroamphetamine
Complexation	Rynata	Chlorpheniramine, phenylephrine, and pyrilamine tannate
Microencapsulation	Nitrospan	Nitroglycerin
Flotation–diffusion	Valrelease	Diazepam
Osmotic pump	Acutrim	Phenylpropanolamine

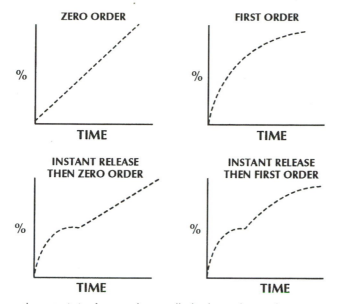

Drug release characteristics from oral controlled-release dosage forms.

FIGURE 6.6

ria are expressed in terms of the fastest possible dissolution rate; there is no upper limit. However, the situation is different for controlled-release products. The optimum dissolution rate for controlled-release products is not the fastest rate that can be obtained, but rather some intermediate value that will reflect prolonged release of drug in the GI tract. Thus, a dissolution window is required for these products, and deviation from the optimum rate can be too fast or too slow.

Because of the difficulty of establishing guidelines for these dosage forms, the large number of formulations, and the different release profiles as illustrated in Figure 6.6, no official guidelines for in vitro dissolution tests for oral controlled-release dosage forms exist at the present time. Also, relationships between in vitro drug release and in vivo bioavailability characteristics are not clearly established. The appropriateness of a controlled-release formulation and equivalence between different controlled-release products, or between controlled-release and conventional products, must currently be based on in vivo data.

Although oral dosing is easier and more convenient than any other dosage route, oral absorption of many drugs is poor. This is the case not only with polypeptide and protein drugs but also with the multitude of insoluble agents that enter the drug development pipeline. As molecules become larger and more complex in the quest for new or superior therapeutic efficacy, their absorption characteristics tend to decline.

In order to address this family of problems, a plethora of absorption enhancers and membrane-permeation enhancers continues to be developed and considered. Some classes of compounds that have been shown, largely in animal studies, to increase absorption of other drugs are listed in Table 6.3 (10).

Absorption Enhancers

Nonsteroidal Anti-Inflammatory Agents (NSAIDS)

Several studies have shown that NSAIDS, in particular indomethacin, diclofenac, mepirazole, phenylbutazone, and salicylate promote the absorption of a variety of other drugs including insulin, ampicillin, cephalothin, pepleomycin, cefoxitin, and cefmetazole. Most of these observations were made in the rat and frequently after rectal administration [10]. Several mechanisms by which NSAIDS promote drug absorption have been postulated, including reduction of the barrier function of the mucous layer, increased lipid bilayer fluidity, modification of membrane proteins, increased membrane permeability, and binding of mucosal calcium ions. However, the exact mechanism or mechanisms are not known. As NSAIDS are often irritating to the intestinal mucosa, and this effect may well relate to their absorption enhancing ability, the feasibility of using NSAIDS to promote drug delivery is uncertain.

Surfactants

Surfactants are used widely as additives in pharmaceuticals. In view of their solubilizing effects, and also their potential to change membrane permeability, they have been examined extensively as absorption enhancers. Most mechanistic studies have been carried out in animals. Typically, polyoxyethylene ethers have been shown to enhance gastric or rectal absorption of lincomycin [11], penicillin, cephalosporins [12], and fosfomycin [13] in rats and rabbits. In rats, colonic absorption of interferon-α was increased from 3% to 8% by polyoxyethylene esters of oleic acid and oleic acid glycerides [14].

Several studies have examined the effects of surfactants on intestinal absorption of insulin, with variable results. For example, both rectal and jejunal absorption of insulin in animals was promoted by anionic and cationic surfactants, including polyoxyethylene ethers. However, in humans, oral polyoxyethylene-20-oleyl ether resulted in poor and variable insulin absorption [15].

Any enhancing effect of surfactants on drug absorption, and there are little data to support this effect in humans, appears to be related to increased drug solubilization, modification of mucosal permeability, or reduction of resistance of the unstirred water layer at the GI membrane surface [16, 17]. In general, the un-ionic surfactants have benign effects on membrane structure, although cationic surfactants have been associated with reversible cell loss and loss of goblet cells, particularly after rectal delivery in animals. These effects severely limit consideration of surfactants as absorption promoters, particularly for long-term drug treatment.

Bile Salts

Bile contains glycine and taurine conjugates of cholic acid and chenodeoxycholic acid, which emulsify dietary fat, facilitate lipolysis, and transport lipid molecules through the unstirred water layer of the intestinal mucosa by micellar solubiliza-

TABLE 6.3	**Some Classes of Intestinal Drug Absorption Enchancers**	
Nonsteroidal anti-inflammatory agents	Liposomes	
Surfactants	Azone	
Bile salts	Cell permeation enhancers	
Medium-chain fatty acids	Nanoparticles	
Mixed micelles		

tion (*10*). The ability of bile salts to promote lipid absorption has prompted their investigation as absorption enhancers for drugs, again with only modest success. Studies in animals have demonstrated increased intestinal absorption of several compounds, including heparin (*18*) and interferon-α (*19*). Absorption of insulin has been shown to be increased by bile salts, both in experimental animals and in humans (*20*). Increased serum insulin, and also a hypoglycemic response, in humans due to coadministration of sodium cholate with insulin is shown in Figure 6.7 (*21*). Although bile salts have thus been shown to have potential as absorption enhancers, their effect on drug uptake appears to correlate well with mucosal damage. This phenomenon, together with possible co-carcinogenic and co-mutagenic properties of secondary bile salts (*22*), minimizes the attractiveness of these compounds as enhancers.

The presence of medium-chain fatty acids and glycerides in food products has stimulated research into their effectiveness as drug absorption enhancers. Representative compounds from both of these classes have been shown to increase drug absorption under a variety of conditions, almost exclusively in animals, and in most cases after rectal administration (*10*). However, some oral studies have been conducted with positive results. Oral insulin bioavailability, estimated by the

Medium-Chain Fatty Acids and Glycerides

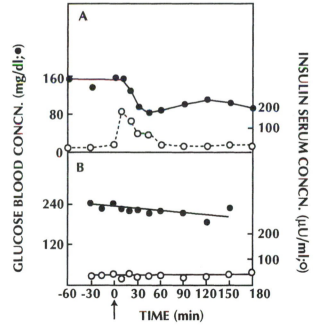

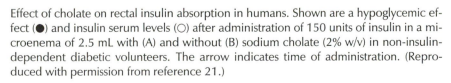

Effect of cholate on rectal insulin absorption in humans. Shown are a hypoglycemic effect (●) and insulin serum levels (○) after administration of 150 units of insulin in a microenema of 2.5 mL with (A) and without (B) sodium cholate (2% w/v) in non-insulin-dependent diabetic volunteers. The arrow indicates time of administration. (Reproduced with permission from reference 21.)

FIGURE 6.7

hypoglycemic effect, was increased to 9–13% relative to intramuscular administration by a mixture of sodium dodecanoate and cetyl alcohol (23). Aftriaxone absorption was enhanced by glyceryl-1-monooctanoate after oral, duodenal, and rectal administration in various animal species (24).

Despite the obvious potential of these classes of compounds as absorption enhancers, both have been shown to have negative effects on mucosal membrane integrity, so that additional research is needed to evaluate the relative risks and benefits.

Mixed Micelles

Mixed micelles consist of fatty acids solubilized by the addition of surfactants or bile salts. The effects of these solubilized lipids on intestinal drug absorption have been reviewed by Muranishi (25). Several studies have demonstrated that mixed micelles are effective absorption enhancers for compounds such as heparin, streptomycin, gentamycin, and insulin. Jejunal insulin absorption was increased from 0.4% to 31% relative to intramuscular administration by mixed micelles (26). Generally, the effect of mixed micelles on intestinal drug absorption tends to be greater in the distal region of the GI tract than in the proximal region. The mechanism for increased absorption is not known. Some publications claim that mixed micelles are safe to use, whereas others report a disordering effect on the epithelial cells of the small intestine (27).

Liposomes

Liposomes are vesicles composed of bilayers or multilayers containing phospholipids and cholesterol surrounding aqueous compartments (10). A drug is entrapped within the liposome and, at least in theory, is released from the liposome for absorption at the intestinal membrane surface. Although this dosage form received considerable attention during the 1970s and 1980s, and several animal studies have suggested some potential for absorption enhancement, the lack of absorption-promoting properties in other studies, as well as stability liabilities, has resulted in somewhat reduced interest in liposomes as oral-absorption enhancers.

Azone

Azone (1-dodecylazacycloheptan-2-one) and related compounds have been studied as transdermal drug-penetration enhancers. They have also been considered as possible oral-absorption enhancers. Although some efficacy has been shown, it appears that an emulsifying agent has to be present for azone to penetrate the intestinal mucosal membrane to promote drug absorption (28). Although one study reported the absence of gross morphological damage after exposure of mucosa to azone (29), there is little additional information on the overall effect of azone on mucosal structure. Additional safety studies are needed on azone as an intestinal absorption enhancer.

Cell Permeation Enhancers

Although most drug absorption enhancers avoid direct alteration of the mucosal membrane, cell permeation enhancers use this as a means of increasing the penetration and absorption of the drug into cells. One form of enhancer that is currently of some interest consists of glycosylated molecules, or facial amphiphiles, that interact with membranes, temporarily increasing their permeability. These molecules are designed to self-assemble in membranes to form transient pores that permit hydrophilic compounds such as polypeptides and oligonucleotides to cross the membrane. Although still in its infancy, this technology has considerable potential as an absorption enhancer. No adverse effects have been seen with these compounds to date (*30, 31*).

Nanoparticles

On the basis of known relationships between particle surface area and dissolution rate, it is reasonable to predict that ultrafine particles may increase dissolution of relatively insoluble molecules to the point where acceptable dissolution and absorption can be achieved. If these particles can then be stabilized in some way to avoid chemical breakdown, aggregation, and agglomeration, and also retain fluidity, then a stable and useful drug product may be achieved.

This concept has found expression in a proprietary nanoparticle technology that has considerable potential as a drug absorption enhancer. With this technology, a drug is reduced to nanometer-size particles in the presence of stabilizers. Originally investigated for intravenous computer imaging (*32*), this technology shows considerable promise for improving absorption of poorly water soluble compounds. As the nanoparticle system is purely "formulation" in nature, it is unlikely to have any effect on mucosal integrity or function. It is interesting that, as demonstrated by its use for intravenous imaging, nanoparticles are sufficiently small that they can be administered parenterally, apparently without ill effects. In this writer's opinion, much more will be heard of this novel type of absorption enhancer.

Summary

Solution dosages and suspensions generally give rise to more satisfactory bioavailability than capsules or tablets.

Coated tablets or capsules may be used to protect a drug from acid degradation in the stomach, or to prevent local irritation in the stomach.

In vitro dissolution may predict in vivo drug bioavailability in some cases, but in vitro–in vivo correlations are not yet sufficiently developed or understood for in vitro data to replace in vivo studies.

Controlled-release formulations are used to prolong drug activity, reduce fluctuation in blood concentrations, and increase patient convenience and compliance. Controlled-release formulations represent one of the most active research areas in the pharmaceutical industry. The potential advantages and disadvantages of controlled release should be considered before initiating a development program for a particular compound.

Many enhancers and enhancer technologies have had varied success as absorption promoters. Some of the more recent technologies, based on either formulation or membrane effects, show considerable promise for increasing oral absorption of poorly absorbed compounds.

References

1. Juncher, H.; Raaschou, F. *Antibiot. Med. Clin. Ther.* **1957,** *4*, 497–507.
2. Cadwallader, C. D. *BioPharmaceutics and Drug Interactions*; Hoffman-LaRoche: Nutley, NJ, 1971.

3. Amidon, G. L.; Lennernas, H.; Shah, V. P.; Crison, J. R. *Pharm. Res.* **1995**, *12*, 413–420.

4. Lindenbaum, J.; Mellow, M. H.; Blackstone, M. O.; Butles, V. P. *New Engl. J. Med.* **1971**, *285*, 1344–1347.

5. Bogentoft, C.; Carlsson, I.; Ekenved, G.; Magnusson, A. *Eur. J. Clin. Pharmacol.* **1978**, *14*, 315–355.

6. Weinberger, M.; Hendeles, L.; Bighley, L. *New. Engl. J. Med.* **1978**, *299*, 852–857.

7. Theeuwes, F. *Curr. Med. Res. Opin.* **1983**, *8*(Suppl), 220–225.

8. Theeuwes, F.; Swanson, D.; Wong, P.; Bonsen, P.; Place, V.; Heimlich, K.; Kwan, K. C. *J. Pharm. Sci.* **1983**, *72*, 253–258.

9. Welling, P. G. *Drug Dev. Ind. Pharm.* **1983**, *9*, 1185–1225.

10. van Hoogdalem, E. J.; de Boer, A. G.; Breimer, D. D. *Pharmacol. Ther.* **1989**, *44*, 407–433.

11. Brookes, L. G.; Marshall, R. C. *J. Pharm. Pharmacol.* **1981**, *33*, 43P.

12. Davis, W. W.; Pfeiffer, R. R.; Quay, J. F. *J. Pharm. Sci.* **1970**, *59*, 960–963.

13. Ishizawa, T.; Hayashi, M.; Awazu, S. *J. Pharm. Pharmacol.* **1987**, *39*, 892–895.

14. Bocci, V.; Corradeschi, F.; Naldini, A.; Lencioni, E. *Int J. Pharm.* **1986**, *34*, 111–114.

15. Galloway, J. A.; Root, M. A. *Diabetes* **1972**, *21*(Suppl 2), 637–648.

16. Florence, A. T. *Pure. Appl. Chem.* **1981**, *53*, 2057–2068.

17. Plá-Delfína, J. M.; Pérez Buendiá, M. D.; Casabó, M. D.; Peris-Ribera, V. G. *Int. J. Pharm.* **1987**, *37*, 49–64.

18. Guarini, S.; Ferrari, W. *Experientia* **1985**, *41*, 350–352.

19. Bocci, V.; Naldini, A.; Corradeschi, F.; Lencioni, F. *Int. J. Pharm.* **1985**, *24*, 109–114.

20. Kidron, M.; Bar-On, H.; Berry, E. M.; Ziv, E. *Life Sci.* **1982**, *31*, 2837–2841.

21. Raz, I.; Bar-On, H.; Kidron, M.; Ziv, E. *Isr. J. Med. Sci.* **1984**, *2*, 173–175.

22. Rainey, J. B.; Maeda, M.; Williamson, R. C. N. *Cell Tissue Kinet.* **1986**, *19*, 485–490.

23. Touitou, E.; Rubinstein, A. *J. Pharm. Pharmacol.* **1978**, *30*, 662–663.

24. Beskid, G.; Unewsky, J.; Behl, C. R.; Siebelist, J.; Tossounian, J. L.; McGarry, C. M.; Shah, N. H.; Cleeland, R. *Chemotherapy* **1988**, *34*, 77–84.

25. Muranishi, S. *Pharm. Res.* **1985**, *2*, 108–118.

26. Shichiri, M.; Kawamori, R.; Goriya, Y.; Kikuchi, M.; Yamasaki, Y.; Shigeta, Y.; Abe, H. *Acta Diabetol.* **1978**, *15*, 175–183.

27. Taniguchi, K.; Muraneshi, S.; Sezaki, H. *Int. J. Pharm.* **1980**, *4*, 219–228.

28. van Hoogdalem, E. J.; de Best, M. A.; de Boer, A. G.; Breimer, D. D. *J. Pharm. Sci.* **1989**, *78*, 691–692.

29. Murakami, M.; Takada, K.; Muranishi, S. *Int. J. Pharm.* **1986**, *31*, 231–238.

30. *A New Paradigm for Carbohydrate Techologies: Fact Sheet*; Transcell Technologies: Princcton, NJ, Spring, 1994.

31. Cheng, Y.; Ho, D. M.; Gottlieb, C. R.; Kahne, D. *J. Am. Chem. Soc.* **1992**, *114*, 7319–7320.

32. Gazelle, G. S.; Wolf, G. L.; Bacon, E. R.; McIntire, G. L.; Cooper, E. R.; Toner, J. L. *Invest. Radiol.* **1994**, *29*, S268–S288.

Clinical Factors and Interactions Affecting Drug Absorption

7

uring drug development, the efficiency with which an orally administered drug is absorbed and also its absorption rate (i.e., its overall bioavailability) are usually determined in panels of healthy individuals under controlled and generally fasting conditions in the absence of other drugs. This type of testing procedure is required by regulatory agencies so that drug bioavailability can be established under controlled conditions without interference from other substances.

In clinical practice, however, drugs are seldom taken under such ideal conditions. Patients who are receiving medication may be suffering from a variety of illnesses, particularly those involving the gastrointestinal (GI) tract, that could affect drug absorption. Similarly, patients often receive more than one drug at the same time, particularly in hospitals and in geriatric therapy, where it is common practice for patients to receive several drugs simultaneously. Medication may also be taken under varying conditions relative to meals. In fact, mealtimes are often used as practical reminders for drug dosage. Remembering to take two tablets at breakfast time is easier than remembering to take them at 10 a.m. or 1 hour before breakfast.

These conditions—diseases of the GI tract, drug–drug interactions, and drug–food interactions—are often present in varying degrees, individually or collectively, when drugs are administered. This chapter will show that these conditions can cause substantial and often unpredictable changes in drug absorption.

Diseases of the GI tract are potentially important conditions affecting drug absorption. A survey conducted in 1968 showed that digestive diseases were responsible for 10–15% of hospital admissions, 30% of major operations, and 9% of all deaths in the United States (1). The term GI disease is difficult to define because it can include such conditions as diseases of the liver, pancreas, and other

Influence of GI Disease on Drug Absorption

organs and tissues closely but indirectly related to the GI tract. The discussion in this chapter is limited to those conditions and surgical procedures directly involving the GI tract, including diseases of the stomach, diseases of the small and large intestine, and intestinal infections.

Diseases of the Stomach

Despite the large number of diseases affecting the stomach, little is known regarding their influence on drug absorption. Many conditions that might be expected to have a profound effect on drug absorption, such as carcinoma or peptic stricture, have not been studied.

One condition that has been studied to a small extent and is currently of interest in some laboratories is achlorhydria, which occurs when diminished secretion of hydrochloric acid results in relatively alkaline stomach contents. The results of the few studies that have been done do not suggest an altered drug absorption pattern that might be expected from principles based on pH dependency of ionization and lipophilicity. Achlorhydria has no apparent effect on the absorption of phenoxymethylpenicillin, but it has been shown to cause an increase in the absorption of aspirin (2, 3). Increased absorption of aspirin in this case was presumably due to faster tablet dissolution. Tablet dissolution was favored over drug absorption as an explanation because the increased ratio of ionized to un-ionized drug in the relatively alkaline conditions due to achlorhydria should have decreased, rather than increased, aspirin absorption rate.

Surgery

Surgical procedures that involve removal of part of the GI tract might be expected to influence drug absorption as a consequence of reduction in the epithelial surface area or changes in motility or secretory patterns. Knowing how these procedures actually influence drug absorption would be useful, but again little information is available. Generating information from these patient populations is difficult, and the present rate of progress suggests that it may be some time before the relationships among GI surgery, drug absorption, and therapeutic consequences are understood. Some interactions that have been reported for stomach surgery are listed in Table 7.1 (4).

Partial gastrectomy has caused reduced absorption of some drugs, but the reduction is attributed to different factors. For example, reduced iron absorption was attributed to loss of a gastric factor necessary for inorganic iron absorption, and reduced ethionamide absorption was attributed to slow dissolution, causing unreliable absorption in postresection patients. Absorption of ethambutol, quinidine, and sulfisoxazole was reduced in cases of gastric surgery with vagotomy, but absorption of these and other compounds was not affected by gastric surgery alone. This observation suggests that delayed stomach emptying due to loss of vagal control caused the actual reduction in absorption. Mean serum levels of quinidine, ethambutol, and sulfisoxazole in patients before and after gastric surgery with vagotomy are shown in Figure 7.1 (5).

Resection of the small bowel is a surgical procedure for reducing nutrient absorption in obese individuals. This type of procedure might reasonably be expected to decrease drug absorption. However, as with gastric surgery, a decrease does not necessarily occur. Absorption of hydrochlorothiazide, levonorgestrol, and norethindrone is reduced following intestinal shunt surgery, but absorption of ampicillin, digoxin, phenazone, and propylthiouracil appears to be unaffected.

Loss of sulfasalazine activity for treatment of Crohn's disease following colonic resection is an interesting example of an indirect effect by surgery affecting drug absorption and action. Sulfasalazine is partially absorbed in the intact form into the systemic circulation, but most of the drug is cleaved at the azo linkage by intestinal bacteria to release the active moiety 5-aminosalicylate. Reduced effectiveness of sulfasalazine in postsurgery patients appears to result from loss of colonic bacteria following resection. Some effects of intestinal surgery on drug absorption are summarized in Table 7.2 (6).

A large number of diseases can afflict the small intestine. Again, information on most of these diseases is fragmentary and provides little substantive insight into their effect on drug absorption. However, two conditions, celiac disease and Crohn's disease, have been studied extensively. The reported apparent effects of these conditions on drug absorption are summarized in Table 7.3.

Celiac disease is an inflammatory condition of the proximal small intestine that is caused by ingestion of gluten, a viscous protein in cereals. The condition is generally kept in remission by a gluten-free diet.

Crohn's disease is also an inflammatory condition, but it differs from celiac disease because it tends to occur in the distal small intestine and proximal large intestine, is of largely unknown etiology, and is usually treated with steroids and sulfasalazine, although it may require surgical resection in some cases.

Both celiac disease and Crohn's disease cause malabsorption of nutrients, and therefore they might be expected to have a similar effect on drug absorption. However, studies conducted in patients have produced a variety of results (7). Both

Diseases of the Small Intestine

Effects of Stomach Surgery on Drug Absorption

TABLE 7.1

Procedure	Effect on Drug Absorption		
	Increased	Unchanged	Decreased
Partial gastrectomy	Ethanol p-Aminosalicylate	Ampicillin Digoxin Isoniazid	Cephalexin Ethionamide Folate Iron Nitrofurantoin Sulfamethoxazole
Antrectomy, gastroduodenostomy		Ethambutol Isoniazid Quinidine Sulfisoxazole Tetracycline	
Antrectomy, gastroduodenostomy, vagotomy			Ethambutol Quinidine Sulfisoxazole

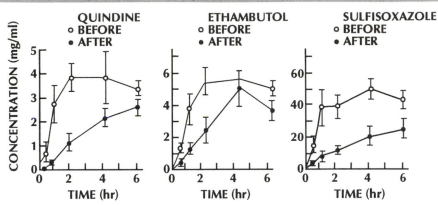

FIGURE 7.1 Mean serum concentrations of quinidine, ethambutol, and sulfisoxazole in nine patients before and after antrectomy and gastroduodenostomy with selective vagotomy. (Reproduced with permission from reference 5.)

TABLE 7.2 **Effects of Intestinal Surgery on Drug Absorption**

	Effect on Drug Absorption	
Procedure	Unchanged	Decreased
Intestinal shunt surgery	Ampicillin	Hydrochlorothiazide
	Digoxin	Levonorgestrol
	Phenazone	Norethindrone
	Propylthiouracil	
Colonic resection		Sulfasalazine

TABLE 7.3 **Apparent Effects of Intestinal Diseases on Drug Absorption**

	Effect on Drug Absorption		
Condition	Increased	Unchanged	Decreased
Celiac disease	Aspirin	Ampicillin	Acetaminophen
	Cephalexin	Indomethacin	Amoxicillin
	Clindamycin	Lincomycin	Penicillin V
	Erythromycin	Methyldopa	Pivampicillin
	Fusidate	Pivmecillinam	Practolol
	Propranolol	Rifampin	
	Sulfamethoxazole		
	Trimethoprim		
Crohn's disease	Clindamycin	Erythromycin	Acetaminophen
	Fusidate	Hydrocortisone	Cephalexin
	Oxprenolol	Rifampin	Lincomycin
	Propranolol		Methyldopa
	Sulfamethoxazole		Metronidazole
	Trimethoprim		

diseases may cause decreased absorption, as shown in Table 7.3, but both may also cause increased absorption, or have no effect. Some of the effects are quite dramatic; for example, circulating levels of fusidate, propranolol, and trimethoprim are approximately doubled in celiac patients compared to normal controls.

The unexpected increase in apparent absorption of some drugs in patients with celiac and Crohn's disease has led to attempts to understand the underlying mechanism. Increased absorption of propranolol in celiac disease has been attributed to altered drug diffusion from the proximal small intestine, and also to saturable first-pass metabolism. Present evidence seems to favor the latter explanation, but this is not the complete story.

One factor that is common to celiac and Crohn's disease is an increased erythrocyte sedimentation rate. This phenomenon occurs in many inflammatory conditions, including rheumatoid arthritis. As indicated in Figure 7.2, propranolol levels are markedly increased in patients with raised erythrocyte sedimentation rates who are suffering from Crohn's disease, and also from rheumatoid arthritis (8). The latter condition is unlikely to affect drug absorption, so that increased drug levels may not result entirely from drug absorption effects. Further research has established that erythrocyte sedimentation rates are associated with elevated plasma levels of the acute-phase protein α_1-acid glycoprotein (AAG). Propranolol and many other basic drugs bind avidly to this protein, so that when levels of AAG increase in conditions of stress, fever, or other inflammatory conditions including celiac and Crohn's disease and rheumatoid arthritis, more of the circulating drug binds to the protein. This binding causes a shift in the distribution of drug from tissue into plasma. Thus, part of the increase in circulating propranolol levels in celiac and Crohn's disease, and possibly other conditions, appears to result from

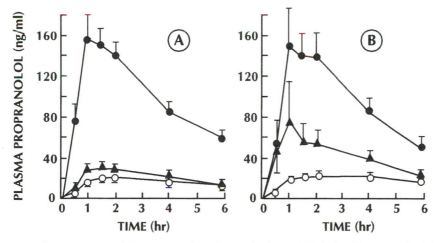

Mean plasma propranolol concentrations (± standard error) (A) in healthy controls (○) and in patients with rheumatoid arthritis with ESR ≤ 20 mm/h (▲) and with ESR >20 mm/h (●), and (B) in healthy controls (○) and in patients with Crohn's disease with ESR ≤ 20 mm/h (▲) and with ESR > 20mm/h (●). (Reproduced with permission from reference 8.)

FIGURE 7.2

redistribution of drug into the plasma at the expense of tissues as a consequence of increased AAG levels, as well as an absorption effect (8). Studies in rats have shown that adjuvant-induced arthritis can cause a threefold increase in circulating propranolol levels compared to normal controls after intravenous injection, where no absorption is involved (9).

Diseases of the Large Intestine

Diseases of the large intestine are less likely to affect drug absorption than those of the small intestine because most absorption is complete by the time drug has reached this region of the GI tract. However, diseases of the large intestine are a potential problem for drugs that are absorbed distally, particularly from controlled-release formulations. As with most other regions of the GI tract, the influence of diseases of the large intestine on drug absorption is largely unknown.

Loss of sulfasalazine activity for the treatment of Crohn's disease following colonic resection was described in the section "Surgery". Loss of activity in that instance is due largely to loss of bacterial flora following surgery, thereby preventing cleavage of the molecule to release 5-aminosalicylate. Another study reported no alteration in hydrocortisone absorption in patients suffering from regional enteritis and active ulcerative colitis. The problem of large bowel disease does not appear to have been addressed for rectal administration, where varying proportions of administered drug may be absorbed from the distal large intestine.

Intestinal Infections

The GI tract is susceptible to a variety of infections such as shigellosis, gastroenteritis, cholera, food poisoning, and infestations by worms and protozoa. Although these conditions present their own particular problems for the patient and call for prompt treatment, the main problem as far as drug absorption is concerned is that intestinal infections frequently cause diarrhea, which can affect drug absorption. This effect has been demonstrated for ampicillin and nalidixic acid. Absorption of both compounds was reduced in children with acute shigellosis, and the extent of reduction in absorption was related to the severity of the condition. Pregnancies have occurred after use of oral contraceptives during periods of diarrhea, where poor GI absorption is presumably due to rapid intestinal transit of the contraceptive pill.

Drug–Drug Interactions Affecting Absorption

Drug–drug interactions leading to improved or impaired absorption of one or both of the interacting substances are a major problem in therapy, particularly in severely ill or geriatric patients who may be on multiple therapy. Some compounds that have been shown to affect the absorption of other drugs will be discussed in the next two sections. Many interactions have been reported in detail (10), and the underlying mechanisms of some interactions have recently been reviewed (11).

Indirect Action

One of the primary factors affecting drug absorption is the stomach-emptying rate. Any drug that can alter this rate is likely to influence the absorption of other drugs or itself.

Propantheline, an anticholinergic drug, reduces GI motility, the stomach-emptying rate, and gastric acid secretion. It is used as adjuvant therapy for treatment of peptic ulcer. Metoclopramide, on the other hand, increases the stomach-emptying rate and is used to prevent gastric reflux. These two agents have opposite effects on absorption of other drugs, presumably because of their opposing influences on the stomach-emptying rate. Typically, metoclopramide increases the absorption rate of acetaminophen, ethanol, levodopa, and lithium, whereas propantheline decreases the absorption rate of acetaminophen, ethanol, lithium, and hydrochlorothiazide. Although these results suggest the possibility of the rarest and most elusive of all goals, a general rule regarding drug absorption, exceptions to the rule have already been reported. Specifically, the absorption of chlorothiazide and digoxin has been reported to be increased by propantheline and reduced by metoclopramide.

Increased absorption efficiency of chlorothiazide and digoxin in the presence of propantheline, and the opposite effect with metoclopramide, is probably due to dissolution-rate-limited absorption. Slower stomach emptying permits a greater percentage of the drug to pass into solution before it passes into the small intestine. An alternative explanation involving saturable and site-specific absorption has been proposed for chlorothiazide but has yet to be confirmed.

Neomycin is administered orally for bowel sterilization and also for treatment of enterocolitis. In addition to its antibiotic activity, neomycin has a variety of effects on the GI tract, including binding and precipitation of bile acids and fatty acids, interference with the micellar phase, and consequent inhibition of fat absorption. In addition to affecting fat absorption, neomycin impairs absorption of vitamins A and B_{12}, digoxin, and penicillin V.

Direct Action

Antacids, kaolin–pectin, charcoal, cholestyramine, and metal ions exert their effects on other drugs directly. By far the most common among these agents are the antacids. By raising gastrointestinal pH, antacid preparations may increase the solubility of acids but decrease the solubility of bases. Antacids may also increase the degree of ionization of acids and decrease that of bases. The solubility effect should favor absorption of acids, but the ionization effect should favor absorption of bases. Antacids that contain heavy metal ions may also chelate with other drugs.

Regardless of the mechanism or mechanisms involved, antacids reduce the absorption of most drugs, including digoxin, isoniazid, penicillamine, most tetracyclines, chlorpromazine, and vitamin A. Most of these interactions result in marked reduction in absorption, ranging from 10–45% for chlorpromazine to 80% for some tetracyclines. The effect of sodium bicarbonate on tetracycline absorption is illustrated in Figure 7.3 (*12*).

It is interesting that increased gastric pH due to antacids has caused a marked increase in absorption of aspirin from an enteric-coated tablet. Although this increase appears to be the only reported interaction of this type, absorption of most drugs from enteric-coated formulations could be accelerated by antacids because of faster dissolution of the acid-resistant, but alkaline-sensitive, enteric coating.

Three other substances, kaolin–pectin, charcoal, and cholestyramine resin, inhibit absorption of other drugs by adsorbing compounds to their surface. Interest in activated charcoal has increased recently because this material not only

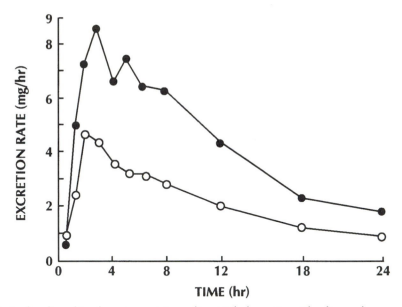

FIGURE 7.3 Effect of sodium bicarbonate on tetracycline HCl absorption. The figure shows mean urinary excretion rates in subjects receiving a single 250-mg tetracycline HCl capsule with 200 mL of water (●) or with 200 mL of water containing 2 g of sodium bicarbonate (○). (Reproduced with permission from reference 12.)

reduces drug absorption but also increases the elimination rate of substances by trapping them after diffusion or excretion into the GI tract. This method is potentially useful for treatment of drug overdose and poisoning.

Metal ions interact with other drugs by chelation and can substantially reduce drug absorption. Dramatic reduction in serum levels of four different tetracyclines by moderate doses of ferrous sulfate is shown in Figure 7.4 (*13*). These types of interactions have been recognized for some time, and tetracycline and iron salts should not be administered together. Iron tablets have been shown to inhibit penicillamine absorption (penicillamine is a potent metal chelator) by 75% and also to significantly decrease penicillamine-induced urinary excretion of copper. The clinical significance of this study is uncertain, but it is likely to be important because of the magnitude of the effect.

Interactions of Drugs that Share Common Absorption Mechanisms

Passive Absorption.   Compounds with proposed enhancing effects on passive transcellular and pericellular absorption include oleic acid, sodium lauryl sulfate, and bile salts (*14*). Kimura et al. summarized in a detailed report the effects of di- and trihydroxy bile salts on the in situ absorption of a broad range of compounds (*15*). Sodium taurodeoxycholic acid (STDC) interacts most likely directly with the intestinal membrane by forming mixed aggregates with phospholipids in the lipid bilayer, thereby altering membrane structure. Compounds that are poorly absorbed by the unperturbed membrane usually exhibit an increased rate of

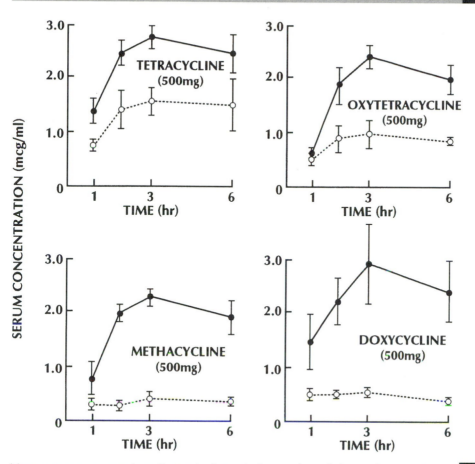

Mean serum concentrations (± standard error) after single oral doses of tetracycline, oxytetracycline, methacycline, and doxycycline taken alone (●) or simultaneously with 200 mg of ferrous sulfate (40 mg of Fe^{2+}) (○). (Reproduced with permission from reference 13.)

FIGURE 7.4

absorption in the presence of bile salts. Both STDC and sodium taurocholic acid (STC) also appear to deplete calcium at tight junctions, resulting in opening of cell junctions, facilitating pericellular diffusion. Absorption of well absorbed compounds, in contrast, may decrease when administered with bile salts. This decrease may be due, at least in part, to reduction in thermodynamic activity as a consequence of micelle formulation, and to depletion of some mucosal components (lipids and/or proteins) to which the drug may have high affinity.

Another mechanism whereby opening of the intercellular space facilitates passive drug uptake is solvent drag. Electrolytes and glucose induce convective transport of other molecules by increasing the osmolarity in aqueous channels between the cells, resulting in increased net water uptake (16). Solvent drag has also been utilized to increase oral bioavailability of small peptides and peptide analogs. Permeability of the dipeptide D-kytrophin and also of the tripeptide analog cephradine increased in the presence of glucose. However, no effect of glucose was observed on uptake of a growth-hormone-releasing hexapeptide or octapeptide somatostatin analog, which may reflect a size limit for the pericellular route.

Carrier-Mediated Absorption.   The amino acid and small-peptide carriers, which transport di- and tripeptides, are two well-characterized active transport systems in the small intestine (17, 18). Although the former exhibits high substrate specificity, the latter facilitates transport of a diverse group of peptides and peptide analogs. Compounds thought to be transported via the small-peptide carrier include amino-β-lactam antibiotics and angiotensin converting enzyme (ACE) inhibitors.

Drugs that are thought to be absorbed by the amino acid pathway include L-dopa and L-methyldopa. Intestinal wall permeability of L-dopa in rats was decreased 10-fold in the presence of the inhibitor L-leucine (19). In humans, competitive uptake by L-leucine caused a reduction in L-dopa intestinal absorption from 40% to 21% (17). Cefatrizine absorption was significantly inhibited by the dipeptide Phe–Phe, whereas L-Phe had no effect on its absorption, suggesting that cefatrizine uptake occurs via the small-peptide transporter (18). Mutual absorption inhibition has been demonstrated among some amino-β-lactam antibiotics but not among others. Absorption of amoxicillin was significantly inhibited by cyclacillin, cephadrine, and cephalexin, but these compounds had no effect on ampicillin absorption. Similarly, cephalexin absorption was reduced by L-carnosine but not by glycine–glycine. More work is needed to determine the substrate specificity and selectivity of the small-peptide carrier. Inhibition of the absorption of the ACE inhibitor captopril by dipeptides has been demonstrated from in situ perfusion studies in rats (20). However, as captopril is also absorbed passively, it is not possible to assess the clinical importance of interactions involving the dipeptide carrier.

P-glycoprotein (Gp 170) was first described in cancer tissue, where it serves as an efflux transporter for chemotherapeutic agents leading to drug resistance. P-glycoprotein is present also in normal tissue such as the blood–brain barrier, kidneys, liver, and jejunum (21). Several drugs, including vinblastine, vincristine, and verapamil, are substrates for this secretory carrier.

Secretion with this carrier may contribute to the observed nonlinear oral absorption kinetics of some drugs, such as the hydrophilic β-blocking agents celiprolol, pafenolol, and talinolol, which exhibit a dose-dependent increase in their oral pharmacokinetics. Celiprolol was shown to be secreted into the intestinal lumen of rats (22). Studies in Caco-2 cells indicated that celiprolol secretion could be inhibited by vinblastine and verapamil, further supporting involvement of P-glycoprotein in the secretory mechanism (23). Intestinal secretion of pafenolol has also been linked to observed nonlinear oral bioavailability. A recent study performed with talinolol in humans, as well as inhibition studies in Caco-2 cells, also attributed its variable and dose-dependent oral absorption to a potential interaction with P-glycoprotein (24). It is not known to what extent this transport protein may be involved in clinically significant drug interactions in vivo.

Drug–food interactions that alter drug absorption are influenced not only by the drug but also by the nature of the food and the formulation in which the drug is administered.

The type and size of a meal may have a marked effect on the nature of a drug–food interaction. Liquid meals may have a totally different effect than solid meals on drug absorption. The time interval between eating and medication also affects the nature and extent of drug–food interaction.

The formulation in which a drug is administered may have a dramatic effect on the nature of drug–food interactions. Generally, the more disperse a formulation, the smaller the drug–food interaction. Typically, drugs administered in solution are less affected by food than drugs formulated in a compressed tablet.

Fasting gastric motility undergoes cycles of migrating motor complexes (MMCs). Each cycle lasts 2–3 h and comprises 4 phases, of which Phase 3, the "housekeeper wave", is the strongest. Non-nutrient liquids pass through the stomach largely independent of MMCs, but solids empty from the stomach into the duodenum mainly during Phase 3. Depending on when a solid meal is ingested relative to the MMCs, gastric residence time may vary from a few minutes to 2–3 h (25). Ingested food changes gastric motility to a postprandial pattern, during which time gastric residence time is increased, particularly by solid or hot meals, high-fat meals, and chyme of low pH or high osmolality. Solid foods tend to delay gastric emptying, whereas non-nutrient liquid meals may have the opposite effect. Distension of the stomach is the only known natural stimulus to stomach emptying. Faster stomach emptying with increasing fluid volume can be rationalized in terms of varying tension at receptors in the stomach wall (26). Ingested food also promotes gastric secretion of hydrochloric acid, which will tend to decrease gastric pH.

Once food passes from the stomach into the small intestine, it has a stimulating effect on intestinal motility and also on digestive enzyme and bile secretion. Delayed gastric emptying will delay absorption of drugs that are absorbed predominantly from the small intestine, but not of drugs that are absorbed from the stomach. It will delay absorption of acidic compounds or drugs in enteric-coated formulations by delaying drug transit from the stomach to the relatively alkaline region of the small intestine. On the other hand, delayed gastric emptying might increase systemic availability of compounds by permitting more material to dissolve in the stomach before passing into the small intestine.

It has already been noted that large fluid volumes empty from the stomach at a faster rate than small volumes. Once in the intestine, the presence of large fluid volumes might increase drug absorption by providing a more liquid environment and also by a solvent drag effect across the intestinal epithelium. On the other hand, large fluid volumes might delay absorption because of a reduced serosal–mucosal drug concentration gradient.

Ingested food may affect splanchnic blood flow, but the direction and extent of change may vary with the type of food. Protein has been shown to increase splanchnic blood flow, whereas a liquid glucose meal has been shown to cause a decrease. In most cases, after solid meals one would expect splanchnic blood flow to increase and to promote drug absorption via this route.

Drug–Food Interactions Affecting Absorption

Influence of Food on the Gastrointestinal Tract

How much drug enters the systemic circulation is also a function of presystemic clearance. Food may exert a direct influence on intrinsic drug metabolism, both presystemic and systemic. For example, oxidative metabolism may be increased by high-protein meals or diets high in cruciferous vegetables. Grapefruit juice has been shown to inhibit the metabolism of dihydropyridine calcium channel blockers. Systemic availability of orally administered felodipine was increased more than twofold by coadministered grapefruit juice (27). A less pronounced effect was observed with nifedipine.

Direct Effect of Food on Drug Absorption

In addition to its indirect effects on GI physiology, food may affect drug absorption directly. Food may act as a physical barrier inhibiting drug dissolution and preventing drug access to the mucosal surface of the GI tract. Specific ions or other substances in food may interact with drugs, resulting in reduced drug absorption, as with the chelation of tetracycline and penicillamine by metal ions, and complex formation by proteins. For drugs that are actively absorbed, direct competition for active carriers may occur between protein fragments and drug molecules, giving rise to decreased drug systemic availability.

Drug–food interactions are a particular problem for oral controlled-release products. These delivery devices present a greater quantity of drug to the patient per single dose unit than conventional dosage forms and are designed to deliver a drug at a controlled rate over a prolonged period. With these formulations, a marked effect by administered food on systemic availability may have a serious and prolonged effect on circulating drug levels.

The following sections discuss recent reports of concomitant food giving rise to reduced, delayed, and increased drug absorption, and also reports of food having no significant effect on drug absorption. The material is based principally on a recent review on this topic (28).

Interactions Causing Reduced Drug Absorption

Drugs whose absorption is reduced by food are given in Table 7.4. Systemic bioavailability of the biophosphonate alendronate sodium is approximately 0.75%. Absorption is further reduced by approximately 40% when alendronate is taken before food, and by as much as 90% when it is taken up to 2 h after food. These observations have led to a recommendation that alendronate be taken at least 30 min before food.

Serum levels of the cholinesterase inhibitor ambenonium chloride are dramatically reduced by food in patients with myasthenia gravis. Peak serum concentrations (C_{max}) and areas under the serum-concentration profiles [AUC(0–3h)] were both reduced by 70% after a breakfast compared to the fasting state. Mean serum-concentration time curves following single 10-mg doses of ambenonium chloride are shown in Figure 7.5 (29).

The very hydrophilic atenolol molecule differs from the lipophilic compounds of this class, metoprolol and propranolol, in that its absorption is reduced by food. The mechanism by which food decreases atenolol absorption was studied in healthy volunteers who received single doses of a commercial tablet, or a capsule containing bile acids (30). The bile acid formula reduced mean atenolol plasma C_{max} by 28% and AUC by 30% relative to commercial tablets. Coadministration of

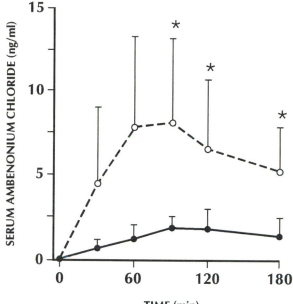

Mean serum concentrations (± standard error) of ambenonium chloride after oral administration of 10 mg of ambenonium chloride to six myasthenia gravis patients under fasting (○) and nonfasting (●) conditions. *P < 0.05. (Reproduced with permission from reference 29.)

FIGURE 7.5

Drugs Whose Absorption Is Reduced by Food

TABLE 7.4

Alendronate sodium	Naproxen
Ambenonium chloride	Navelbine
Atenolol	Nitrendipine
Azithromycin	Norfloxacin
Cefprozil	Paracetamol
Ceftibuten	Phenytoin
2-Choro-2′-deoxyadenosine	Pravastatin
Cicaprost	Ro-42-5892
Ciprofloxacin	Rufloxacin
Didanosine	SKF-106203
Dideoxycytidine	Sotalol
Doxazosin	Sulpiride
Flecainide	Tacrine
Hydralazine	Tetracycline
Levodopa, Carbidopa	Verapamil
Metformin	Zidovudine
Methotrexate	882-C
MK-679	

Source: Adapted with permission from reference 28.

azithromycin with a large meal may reduce absorption by up to 50%. The mechanism of this interaction is unknown, but this basic, highly lipid soluble compound is degraded by acid-catalyzed hydrolysis of the ether bond to the neutral cladinose sugar, which may contribute to its poor oral bioavailability.

Two studies have demonstrated reduced absorption of new cephalosporins in the presence of food (*31*) and an "elemental diet" (*32*). Administration of cefprozil to children after a meal caused a 22% reduction in mean C_{max} values and a 16% reduction in urinary excretion. Simultaneous administration of ceftibuten and an elemental diet composed of egg white hydrolysate, an elemental diet composed of amino acids, and also a mineral solution had a similar effect.

Food moderately reduced the systemic availability of 2-chloro-2'-deoxyadenosine (CdA) in patients with leukemia. Reduced systemic availability may be related to slower absorption resulting in greater presystemic metabolism. Although neither a standard breakfast nor a high-fat, high-calcium breakfast altered ciprofloxacin absorption, absorption was markedly impaired by co-administered milk and yogurt. Systemic availability was reduced 35% by coadministered milk and yogurt compared to water. Thus, ciprofloxacin is affected by some diary products to a similar extent as tetracycline derivatives. Ciprofloxacin, together with other fluoroquinolones, is known to bind heavy metals, including calcium, to form insoluble chelates.

Food ingestion had a marked effect on the absorption of the purine nucleoside analog didanosine in men seropositive for human immunodeficiency virus (HIV) but free of AIDS symptoms (*33*). Decreased bioavailability after postprandial administration may have been due to prolonged gastric retention leading to increased drug degradation. In a separate study, the bioavailability of didanosine from a chewable tablet was reduced approximately twofold by a standard breakfast.

A milder food effect was reported for the novel quinazoline antihypertensive agent doxazosin. After a standard light breakfast, mean C_{max} in hypertensive subjects was reduced by 18%, and the AUC by 12%, compared to fasting. Toxicity in the form of ventricular tachycardia occurred in a baby boy when dextrose was substituted for milk foods during flecainide therapy. Serum flecainide levels, based on single-point determinations, were doubled following dextrose substitution. This effect justifies close flecainide blood monitoring in infants who may be on intermittent milk diets.

Both a standard breakfast and bolus enteral nutrient treatments reduced blood levels of hydralazine compared to fasting. Typically, the fasting treatment yielded a C_{max} value of 87 ng/mL as compared to 11 ng/mL and 15 ng/mL after standard breakfast and enteral bolus treatments, respectively.

Plasma levels of levodopa, 3-*O*-methyldopa, and carbidopa were compared in patients with Parkinson's disease after they took a controlled-release formulation containing levodopa and carbidopa (*34*). Co-administration with a high-protein meal reduced plasma levels of both levodopa and carbidopa. However, concentrations of the levodopa metabolite 3-*O*-methyldopa increased. Although levodopa and carbidopa levels were generally reduced by high-protein meals, the plasma levels exhibited a flattened concentration–effect profile.

As noted earlier with ciprofloxacin, co-administration with milk or yogurt markedly reduces the systemic availability of norfloxacin. C_{max} and the AUC values were reduced approximately 50% by milk or yogurt compared to water. Plasma profiles obtained in this study are shown in Figure 7.6 (35). The results of this study provide further evidence that, although absorption of oral fluoroquinolones is not affected to a significant extent by solid foods, dairy products containing solubilized calcium may reduce circulating drug levels to an extent that may be clinically significant.

In a study in hypercholesterolemic men, pravastatin C_{max} dropped by 49% and AUC by 31% in the presence of food. However, reduction in mean total cholesterol and low-density lipoproteins was identical when pravastatin was taken under fasting and nonfasting conditions. The lack of effect on blood lipids may have been due to food causing increasing extraction of pravastatin by the liver, the primary site of cholesterol synthesis and clearance, and also pravastatin activity.

Systemic availability of the cognition agent tacrine was reduced by approximately 20%, and peak plasma levels by 40%, when it was administered following a standard breakfast. However, administration with meals also reduced the incidence and severity of gastrointestinal side-effects, thereby improving patient tolerance.

Several studies have examined the effect of food on the absorption of the HIV agent zidovudine. All studies reported significantly delayed and reduced absorption. Although there is no known direct relationship between circulating levels

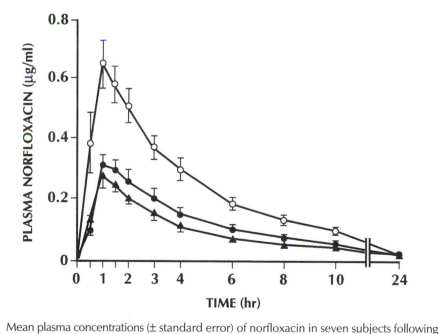

Mean plasma concentrations (± standard error) of norfloxacin in seven subjects following a single 500-mg oral dose of norfloxacin with 300 mL of milk (●), yogurt (▲), or water (○). (Reproduced with permission from reference 35.)   **FIGURE 7.6**

and efficacy of zidovudine, doses found to be effective in AIDS patients generally achieve peak serum levels of 0.27 µg/mL. However, lower and more prolonged levels obtained after meals may maintain efficacy while reducing toxicity.

Interactions Causing Delayed Drug Absorption

Compounds whose absorption is delayed by food are listed in Table 7.5. Delayed absorption may be unimportant for most drugs. However, the importance of an interaction depends on its extent and also the drug therapeutic index. Peak circulating levels of some drugs listed in Table 7.5 are reduced by up to 60% with food ingestion, and changes of this magnitude are likely to be therapeutically important. For example, a 50% reduction in aniracetam C_{max} by food is likely to have clinical consequences. Possible clinical significance is also claimed for the delay in β-methyldigoxin absorption by food. Mean serum β-methyldigoxin C_{max} levels were reduced from 1.79 ng/mL fasting to 0.96 ng/mL after food.

Several studies have reported delayed cephalosporin absorption by food. Absorption of cefaclor was delayed following rice and bread meals, the effect increasing with meal size. Similar delays in absorption were observed when cefaclor was taken immediately after a standard breakfast, causing a 51% reduction in mean C_{max} values. Other studies have shown cefadroxyl absorption to be unaffected by food whereas absorption of cephalexin is delayed in a similar manner to cefaclor. Peak plasma levels of cefdinir were reduced approximately 50% by food in children who received the drug as a fine granule suspension.

TABLE 7.5 — Drugs Whose Absorption Is Delayed by Food

Acetorphan	Loracarbef
Albuterol	Methotrexate
5-Aminosalicylic acid	Monofluorophosphate
Aniracetam	Moricizine
β-methyldigoxin	Nicorandil
Cefaclor	Nifedipine
Cefdinir	Ofloxacin
Cefprozil	Paracetamol
CL-275838	Penciclovir (Famciclovir prodrug)
Diclofenac	Rifabutin
Diltiazem	Salsalate
Doxycycline	Terazocin
Erythromycin acistrate	Terfenadine
Fadrozole	Theophylline
Famotidine	Tiagabine
Flurbiprofen	Topiramate
Fluvastatin	Trazodone
Fusidate sodium	Valproic Acid
Hydroxychloroquine	Vigabatrin
Isosorbide-5-mononitrate	Zalospirone
Lomefloxacin	Zidovudine

Source: Adapted with permission from reference 28.

Co-administered food had a marked effect on circulating levels of diclofenac. Absorption from a hydrogel bead capsule in male subjects was delayed, yielding a mean C_{max} value of 312 ng/mL at 6.25 h as compared to 502 ng/mL at 2 h in the fasting state. The results obtained using diclofenac capsules are in contrast to those obtained using diclofenac enteric-coated tablets, in which peak circulating levels were delayed but not reduced by food. The results obtained in a food-effect study with erythromycin acistrate (2'-acetylerythromycin stearate) are consistent with results reported on food interactions with erythromycin and its derivatives. Food generally reduces systemic availability of erythromycin and erythromycin stearate while absorption of erythromycin esters is either unchanged or increased. The rate of erythromycin acistrate absorption was delayed to a small extent following a light meal and to a greater extent following a heavy meal (36). Although the effect was attenuated after repeated doses, food significantly delayed absorption in some individuals, resulting in undetectable levels up to 24 h postdosing. Studies in healthy male subjects showed that food can markedly affect circulating levels of the 3-hydroxy-3-methylglutaric acid (HMG)-CoA reductase inhibitor fluvastatin. Peak plasma levels were reduced by 70%, and the time of peak drug concentration, t_{max}, was increased fourfold by a standard low-fat breakfast following a solution dose of fluvastatin; whereas peak levels were reduced by 55%, and t_{max} values were increased up to fourfold from a capsule dose. A similar but somewhat attenuated effect was observed following a low-fat evening meal. A substantial food effect occurs with the oral β-lactam antibiotic loracarbef. Although systemic availability was similar in subjects who received 200-mg doses in fasting and nonfasting states, peak plasma levels were reduced over 50%, and the time of peak was doubled by food.

Two studies have examined the effect of food on bioavailability of methotrexate. One study demonstrated a 16% reduction in C_{max} and a 30-min delay in t_{max} after a standard breakfast. In another study, in which methotrexate was given during fasting or after a standard French breakfast in patients with rheumatoid arthritis, C_{max} was reduced by 31%, and t_{max} was increased by 40 min. Absolute systemic availability was not influenced by food. A more dramatic effect was obtained when fluoride in the form of monofluorophosphate was administered after a standard meal and during fasting. Food caused a 67% reduction in peak plasma fluoride levels, while AUC and urinary excretion values were essentially unchanged.

A high-fat meal markedly reduced the absorption rate of the calcium channel blocking agent nifedipine, reducing peak plasma levels by 47%. The related compounds nicardipine and nitrendipine were unaffected. Milk and yogurt had only a small effect on ofloxacin absorption, whereas a standard meal caused a 20% reduction in peak plasma levels and a moderate delay in t_{max}. Ofloxacin is thus similar to other fluoroquinolone antibiotics in that it is only modestly affected by food.

Two studies have contributed further to the wide spectrum of food effects reported for theophylline. Absorption of theophylline from Theo-Dur tablets was delayed to a small extent, but mean peak levels increased from 4.7 to 6.3 µg/mL following a postprandial evening dose as compared to the fasting state. In another study, the rate of theophylline absorption from a multiparticulate controlled-release formulation was delayed. Following both eating and fasting treatments,

37–39% of the absorbed dose was absorbed from the colon, indicating that this is an important absorption site for sustained-release theophylline products.

Different meals influenced the extent of drug–food interaction with the serotonin agonist–antagonist zalospirone. Administration of a single zalospirone dose after a meal containing 43% fat resulted in a 31% reduction in C_{max} values compared to the fasting state. Administration after a meal containing 19% fat resulted in only a 13% reduction. Peak drug levels were delayed twice as long after the high-fat meal compared to the low-fat meal.

In a study in asymptomatic HIV-infected patients, zidovudine C_{max} was reduced 57% when zidovudine was administered 30 min after a high-fat breakfast. In another study in symptomatic HIV-infected men, serum zidovudine C_{max} was reduced 32% when drug was administered after a liquid protein meal relative to the fasting state. Absolute absorption values were unaffected.

Interactions Causing Increased Drug Absorption

The studies cited in Table 7.6 represent a substantial portion of the many reports on drug–food interactions and reflect both the broad spectrum of drug–food interactions and their unpredictability. Compounds discussed in this section tend to be poorly water soluble, but not always so. Some interactions are trivial, and others are potentially clinically important.

Conflicting food-interaction results were reported with the onchocerciasis agent amocarzine. In male Guatemalan patients, systemic availability increased 20% when the drug was taken with a "copious" breakfast compared to fasting. When the dose was increased, both the peak plasma levels and systemic availabil-

TABLE 7.6	**Drugs Whose Absorption Is Increased by Food**
Alprazolam	Itraconazole and fluconazole
Amiodarone	Levodopa
Amocarzine	5-Methoxypsoralen
Astemizole and pseudoephedrine	Moclobemide
Atovaquone	Nifedipine
Bay-X-1005	Oxcarbazine
Brofaromine	Oxybutinin
Buflomedil	Phenytoin
Cefetamet pivoxil	Progesterone
Cefuroxime	Repirinast
CGP-43371	Sparfloxacin
Clarithromycin	S-1108
Cyclosporine	Theophylline
Danazol	Ticlopidine
Diltiazem	Tramadol
Encainide	Vanoxerine
Felodipine	Vinpocetine
Fenretinide	Zalospirone
Gepirone	566-C-80
Itraconazole	

Source: Adapted with permission from reference 28.

ity of amocarzine were increased approximately threefold when the drug was given after a standard breakfast, relative to fasting. The remarkable increase in absorption due to food after the high dose of amocarzine may be related to the greater degree of solubilization by the meal or to decreased presystemic metabolism. Substantially increased absorption due to food was reported for the lipophilic antiprotozoal agent atovaquone. Peak atovaquone plasma levels increased over fivefold, and systemic bioavailability over threefold, when the drug was given after a high-fat breakfast as compared to fasting.

A remarkable food effect involved the lipophilic hypolipidemic compound CGP-43371 (37). Administration of single 800-mg capsule doses of CGP-43371 after breakfast caused an 11-fold increase in peak plasma drug levels and a 13-fold increase in overall bioavailability relative to the fasting state. Plasma levels from this study are shown in Figure 7.7. As CGP-43371 is absorbed mainly from the ileum, delayed gastric emptying would enable more compound to disintegrate and dissolve before reaching this absorption site.

Absorption of the heterocyclic steroid derivative danazol, and also the retinoid fenretinide, is substantially increased by food. Systemic availability of danazol from a capsule dose was increased over threefold by food in healthy female subjects, while bioavailability and peak plasma levels of fenretinide increased three-fold following a high-fat meal as compared to fasting. Further examination of the effect of meal composition showed that a high-fat meal increased fenretinide bioavailability threefold relative to a carbohydrate meal, and a high-protein meal yielded intermediate results.

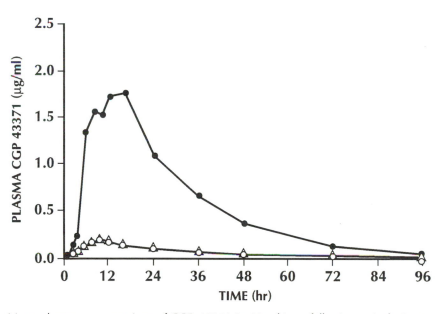

Mean plasma concentrations of CGP 43371 in 12 subjects following a single 800-mg oral dose of CGP-43371 as a dispersion ($\triangle$) or capsule ($\bigcirc$) under fasting conditions or as a capsule after a standard meal ($\bullet$). (Reproduced with permission from reference 37.)   **FIGURE 7.7**

Itraconazole systemic availability increased two- to threefold following a standard breakfast as compared to fasting. In contrast to itraconazole, absorption of fluconazole was relatively insensitive to food, both C_{max} and AUC being slightly reduced or unchanged by meals. Although these divergent results are consistent with previous data on these agents, there is no mechanistic explanation for their different behavior.

A dramatic food effect occurred with oral micronized progesterone. Repeated doses of micronized progesterone were administered in capsules for 5 days to 15 healthy postmenopausal women, either 2 h before or immediately after a standard breakfast. Peak day 1 and day 5 plasma levels of progesterone increased fivefold, and systemic availability twofold, with food. Increased progesterone absorption is attributed to a direct drug–food interaction in the gastrointestinal tract, or to increased blood flow to the liver causing decreased presystemic clearance.

A similar food interaction was observed with the piperazine derivative vanoxerine, a dopamine reuptake inhibitor. Administration of 100 mg of vanoxerine to healthy men after low-fat and fatty breakfasts increased systemic availability 1.8-fold and 3.6-fold, respectively. Despite the considerable increase in systemic availability after the high-fat meal, C_{max} was increased less than twofold because of delayed absorption. One subject who was not markedly affected by food intake was a "poor metabolizer" of debrisoquine, which suggests that decreased first-pass metabolism, possibly related to increased splanchnic blood flow, may have contributed to the food effect in the other subjects.

The last drug listed in this category reflects, again, the dramatic effect that food can have on circulating drug profiles. Systemic availability of a novel antiprotozoal agent, 566 C80, was increased 3.3-fold, and C_{max} 5.4-fold, when the drug was administered after food (38). In attempts to elucidate the mechanism of this interaction, 566 C80 was given during fasting, with meals of varying fat content, as an aqueous suspension, as an oily emulsion, and after infusion of cholecystokinin octapeptide (CCK-OP). Results from these studies led to the conclusion that increased absorption of 566 C80 could be accounted for by dietary fat.

Cases in which Food Has No Effect on Drug Absorption

The reports summarized in Table 7.7 describe studies in which food had little or no effect on drug absorption. Some compounds in this table have been cited in previous tables, which reflects the varied results that may be obtained under different study conditions or with different formulations of the same drug. As the examples included in Table 7.7 cannot really be considered interactions, comment is restricted only to the more interesting cases.

Although absorption of hydrochlorothiazide has previously been reported to be both increased and decreased by food from conventional single-drug formulations, absorption of both hydrochlorothiazide and bisoprolol was unaffected by food when they were administered in a combination tablet. No significant differences were observed in plasma pharmacokinetic parameters or in the percentage of hydrochlorothiazide excreted in urine.

Ingestion of the selective monoamine oxidase A (MAOA) inhibitor brofaromine together with cheese resulted in no change in blood pressure in one subject, and a maximum change of only 20 mm Hg in three subjects. The mean increase in blood pressure was only 11 mm Hg as compared to 40 mm Hg from an equivalent dose of tyramine. The lack of interaction with tyramine-rich foods may greatly increase the benefit–risk ratio of these MAOA inhibitors.

Although other studies have reported increased absorption of the cephalosporin prodrug cefetamet pivoxil from tablets, food had no effect on absorption from an oral syrup formulation. Mean C_{max} values in plasma were 2.7 µg/mL with food and 2.9 µg/mL fasting, and absolute bioavailability was 38% and 34% after fed and fasting doses, respectively. It is interesting that the syrup yielded significantly lower absolute systemic bioavailability than a tablet under fed conditions.

Significant increases in cyclosporin absorption with food have been reported in renal transplant patients, but two other further studies have reported a minimal effect of food on cyclosporin absorption. Cyclosporin was administered to 14 renal transplant patients immediately following a moderate or trace fat breakfast (39). Neither meal had any significant effect on cyclosporin pharmacokinetic parameters. Mean C_{max} values were 410, 346, and 365 ng/mL, and mean AUC values were 2115, 2085, and 2145 ng h/mL following fasting, moderate-fat, and high-fat treatments at an average cyclosporin dose of 3.51 mg/kg/day. In another study, conducted in healthy male subjects, a standard light breakfast caused a 17% reduction in mean peak plasma cyclosporin levels but had no effect on areas under plasma profiles.

Food had no effect on the rate and extent of absorption of a new cholinesterase inhibitor, E 2020. Following single 2-mg oral doses of E 2020 in healthy male volunteers, mean peak plasma E 2020 levels of 3.3 and 3.2 ng/mL, and AUC values of 166.5 and 172.8 ng h/mL, were obtained under fasting and fed conditions.

Drugs Whose Absorption Is Not Affected by Food　　　　　**TABLE 7.7**

Alprazolam	Metoprolol succinate
Amlodipine	Morphine sulfate
Bambuterol	Mosapride citrate
Bisoprolol	Moxonidine
Brofaramine	Nefiracetam
Brompheniramine	Paroxetine
Bromocriptine	Piroximone
Carbamazepine	Procainamide
Cardizem	Propranolol
Cefetamet pivoxil	Pseudoephedrine
Cimetidine	Ranitidine
Cyclosporine	Rifabutin
Diazepam	Sparfloxacin
E-2020	Temafloxacin
Fluvoxamine	Theophylline
Hydrochlorothiazide	Tiaprofenic acid
Ibuprofen	Trimetazidine
Levodopa	Verapamil
Methotrexate	

Source: Adapted with permission from reference 28.

A recent report on the lack of effect of food on levodopa absorption illustrates, again, the unpredictability of drug–food interactions. In a study performed on healthy volunteers, a meal containing 30.5 g of protein had no effect on levodopa absorption from a 150-mg solution dose, whereas absorption was reduced by 10%, and peak plasma levodopa levels by 26%, when the drug was ingested following a meal containing only 10.5 g of protein (40). Poor bioavailability of levodopa following the low-protein meal relative to the fasting state suggests that low-protein diets do not increase levodopa absorption, and any beneficial effects of a low-protein diet on the efficacy of levodopa may be related to reduced competition for transport across the blood–brain barrier rather than increased systemic availability. As with other quinolone antibiotics, absorption of sparfloxacin appears to be essentially unaffected by food. Following single doses to healthy subjects, the time of peak sparfloxacin plasma levels was increased from 3.1 to 4.7 h by food, but peak levels and systemic availability were unaffected. In another study, plasma sparfloxacin levels under fed and fasted conditions were almost super imposable.

The last compound to be discussed in this section is of interest largely because of the divergent food-interaction results often reported with controlled-release preparations. Verapamil absorption was reduced by 30% and peak serum levels by 48% when a sustained release tablet was taken with food (41). However, when verapamil was administered in a different sustained-release formulation, there was minimal food effect. Healthy male volunteers received single verapamil doses in a newly marketed sustained-release formulation either while fasting or 10 min after a standard breakfast. Plasma profiles of verapamil following the sustained-release capsule were superimposable in the fed and fasting states. Plasma profiles of the metabolite norverapamil were similarly unaffected. The divergent results obtained in these studies are most likely related to the formulations used. Release of drug from membrane controlled-release formulations, including those using osmotic pump technology, is generally less sensitive to changes in the gastrointestinal environment than other controlled-release formulations.

Conclusions

The number of articles on food–drug interactions illustrates the high level of interest in this topic. Drugs and drug formulations fall into four major categories: those whose absorption is reduced, delayed, increased, or not affected by food. Many of the drugs whose absorption was reduced or increased also exhibited delayed absorption. The nature of a drug–food interaction may be at least partially predictable from a physicochemical perspective, but accurate predictability is plagued by many exceptions, which are often spectacular in nature.

The clinical importance of a drug–food interaction depends on the extent of change in circulating drug levels, the margin of safety, and the slope of the drug concentration–response curve. A small change in circulating drug levels for a drug with a relatively flat dose–response curve may be of little clinical consequence. However, a large change in circulating drug levels of a drug with a steep dose–response curve, and a narrow safety margin, may have severe clinical consequences.

Summary

Disease states have been shown to alter drug absorption, but information on this subject is fragmentary. Predicting the effect of a particular condition on drug absorption is not possible using current information in most cases. Consideration of possible changes in drug–protein binding and distribution states is important when examining disease-state pharmacokinetics.

Many drugs interact with other drugs in the GI tract to affect absorption. These interactions may be direct or indirect. The direction of change is largely predictable, but there are enough exceptions for drugs to be considered individually.

Substances in foods can interact with drugs and lead to reduced, delayed, or increased drug absorption. The nature and extent of interaction, again only partially predictable, may be therapeutically important.

References

1. Bank, S.; Saunders, S. J.; Marks, I. N.; Novis, B. H.; Barbezat, G. O. In *Drug Treatment Principles and Practice of Clinical Pharmacology and Therapeutics*; Avery, G. S., Ed.; ADIS: Sydney, Australia, 1980; pp 193–194.
2. Davies, J. A.; Holt, J. M.; Mullinger, B. J. *Antimicrob. Chemother.* **1975,** *1* (Suppl.), 69–70.
3. Pottage, A.; Nimmo, J.; Prescott, L. F. *J. Pharm. Pharmacol.* **1974,** *26,* 144–145.
4. Welling, P. G. In *Pharmacokinetic Basis for Drug Treatment*; Benet, L. Z. et al, Eds.; Raven: New York, 1984; pp 29–47.
5. Venho, V. M. K.; Aukee, S.; Jussila, J.; Mattila, M. J.; Sand, J. *Gastroenterol* **1975,** *10,* 43–47.
6. Welling, P. G.; Tse, F. L. S. *J. Clin. Hosp. Pharm.* **1984,** *9,* 163–179.
7. Parsons, R. L.; Hossack, G.; Paddock, G. *J. Antimicrob. Chemother.* **1975,** *1,* 39–50.
8. Schneider, R. E.; Bishop, H.; Hawkins. C. F. *Br. J. Clin. Pharmacol.* **1979,** *8,* 43–47.
9. Bishop, H.; Schneider, R. E.; Welling, P. G. *Biopharm. Drug Dispos.* **1981,** *2,* 291–297.
10. Welling, P. G. *Clin. Pharmacokinet.* **1984,** *9,* 404–434.
11. Lipka, E.; Crison, J. R.; Schug, B.; Blume, H.; Amidon, G. G. In *Mechanisms of Drug Interactions*; D'Arcy, P.; McElnay, J. C.; Welling, P. G., Eds.; Springer-Verlag: Heidelberg, Germany, 1996; pp 13–43.
12. Barr, W. H.; Adir, J.; Garrettson, L. *Clin. Pharm. Ther.* **1971,** *12,* 779–784.
13. Neuvonen, P. J.; Gothini, G.; Hackman, R.; Bjorksten, K. *Br. Med. J.* **1970,** *4,* 532–534.
14. Muranishi, S. *Crit. Rev. Ther. Drug Carrier Syst.* **1990,** *7,* 1–33.
15. Kimura, H.; Inui, K-I.; Sezaki, H. *J. Pharmacobio-Dyn.* **1985,** *8,* 578–585.
16. Fleisher, D. In *Peptide–Based Drug Design*; Taylor M. D.; Amidon, G. L., Eds.; American Chemical Society: Washington, DC, 1995; 501–525.
17. Mailliard, M. E.; Stevens, B. R.; Mann, G. E. *Gastroenterol* **1995,** *108,* 888–910.
18. Sinko, P. J.; Hu, M.; Amidon, G. L. *J. Contr. Rel.* **1987,** *6,* 115–121.
19. Lennernas, H.; Nilsson, D.; Aquilonius, S. M.; Ahrenstedt, O.; Knutson, L.; Paalzow, L. K. *Br. J. Clin. Pharmacol.* **1993,** *35*(3), 243–250.
20. Hu, M.; Amidon, G. L. *J. Pharm. Sci.* **1988,** *77*(2), 1007–1011.
21. Cornwell, M. M. In *Molecular and Clinical Advances in Anticancer Drugs*; Ozols, R., Ed.; Kluwer Academic: Boston, 1991; pp 37–56.
22. Kuo, S. M.; Whitby, B. R.; Artursson, P.; Ziemniak, J. A. *Pharm. Res.* **1994,** *11*(5), 648–653.
23. Karlson, J.; Kuo, S. M.; Ziemniak, J.; Artursson, P. *Br. J. Pharmacol.* **1993,** *110*(3), 1009–1016.
24. Wetterich, U.; Mutschler, E.; Spahn-Langguth, H.; Langguth, P. *Naunyn-Schmiedenberg's Arch. Pharmacol.* **1995,** *351* Suppl. R1.

25. Ewe, K.; Press, A. G.; Dederer, W. *Eur. J. Clin. Invest.* **1989**, *19*, 291–297.

26. Hopkins, A. *J. Physiol. London* **1966**, *182*, 144–149.

27. Bailey, D. G.; Spence, J. D.; Munoz, C.; Arnold, J. M. O. *Lancet* **1991**, *337*, 268–269.

28. Welling, P. G. In *Mechanisms of Drug Interactions*; D'Arcy, P.; McElnay, J. C.; Welling, P. G., Eds.; Springer-Verlag: Heidelberg, Germany, 1996; pp 45–123.

29. Ohtsubo, K.; Fujii, N.; Higuchi, S.; Aoyama, T.; Goto, I.; Tatsuhara, T. *Eur. J. Clin. Pharmacol.* **1992**, *42*, 371–374.

30. Barnwell, S. G.; Laudanski, T.; Dwyer, M.; Story, M. J.; Guard, P.; Cole, S.; Attwood, D. *Int. J. Pharm.* **1993**, *89*, 245–250.

31. Nakamura, H.; Iwai, N. *Jpn. J. Antibiot.* **1992**, *45*, 1489–1504.

32. Iseki, K.; Satoh, Y.; Sugawara, M.; Miyazaki, K. *J. Pharm. Soc. Jpn.* **1994**, *114*, 233–240.

33. Knupp, C. A.; Milbrath, R.; Barbhaiya, R. H. *J. Clin. Pharmacol.* **1993**, *33*, 568–573.

34. Roos, R. A. C.; Tijssen, M. A. J.; Van der Velde, E. A.; Breimer, D. D. *Clin. Neurol. Surg.* **1993**, *95*, 215–219.

35. Kivistö, K. T.; Ojala-Karlsson, P.; Neuvonen, P. J. *Antimicrob. Agents Chemother.* **1992**, *36*, 489–491.

36. Järvinen, A.; Nykänen, S.; Mattila, J.; Haataja, H. *Arzneim–Forsch* **1992**, *42*, 73–76.

37. Sun, J. X.; Cipriano, A.; Chan, K.; Klibaner, M.; John, V. A. *J. Pharm. Sci.* **1994**, *83*, 264–266.

38. Hughes, W. T.; Kennedy, W.; Shenep, J. L.; Flynn, P. M.; Hetherington, S. V.; Fullen, G.; Lancaster, D. J.; Stein, D. S.; Palte, S.; Rosenbaum, D.; Liao, S. H. T.; Blum, M. R.; Rugers, M. *J. Infect. Dis.* **1991**, *163*, 843–848.

39. Honcharik, N.; Yatscoff, R. W.; Jeffery, J. R.; Rush, D. N. *Transplantation* **1991**, *52*, 1087–1089.

40. Robertson, D. R. C.; Higginson, I.; MacKlin, B. S.; Renwick, A. G.; Waller, D. G.; George, C. F. *Br. J. Clin. Pharmacol.* **1991**, *31*, 413–417.

41. Hoon, T. J.; McCollam, P. L.; Beckman, K. J.; Hariman, R. J.; Bauman, J. L. *Am. J. Cardiol.* **1992**, *70*, 1072–1076.

Drug Distribution

8

fter entering the general circulation, a drug is carried throughout the body and distributed to various tissues. The rate and extent to which a drug penetrates into tissues are controlled by the rate of delivery to the tissues by the circulation, the ease with which the drug can leave the circulation and enter the tissues, and the affinity of the drug for particular tissues (1).

Before considering various aspects of drug distribution in detail, familiarity with some body fluid volumes would be useful. Table 8.1 presents the average values of some body fluid volumes. If a drug is distributed only within plasma and does not penetrate into any other tissues including red blood cells (an unlikely event), then it is contained in a small volume of 3 L, or 4% of body weight. If, on the other hand, a drug freely enters circulating red blood cells but enters no other tissues, it is contained in a volume of approximately 5 L, or 7% of body weight. If a drug is distributed into extracellular water but has difficulty crossing cell membranes to reach intracellular water, then the drug will be contained in a volume of about 15 L, or 21% of body weight. If the drug can also penetrate into intracellular water, then it will be contained in a volume of approximately 42 L, or 60% of body weight. Calculation of drug distribution would be simple if these were the only distribution options available. In reality, distribution is more complex, but these simple examples illustrate the effect that drug distribution may have on plasma (or serum) drug concentrations.

Before discussing drug distribution further, plasma and serum need to be defined. If a blood sample is drawn in the absence of an anticoagulant, the clear supernatant obtained after the red blood cells have been allowed to clot is serum. On the other hand, if a blood sample is obtained in the presence of an anticoagulant, then the clear supernatant obtained after the red blood cells have been removed by centrifugation is plasma. Thus, plasma differs from serum in that plasma still contains coagulating factors, whereas serum has lost these during the coagulation process. From a pharmacokinetic viewpoint, the two fluids are generally regarded as identical.

However, data obtained from serum and plasma may be influenced by several factors. For example, if a drug is unstable to enzymatic or chemical degradation, then different drug concentrations may be obtained in serum that has been standing at room temperature for clot formation compared to plasma, which is obtained by immediate centrifugation after a blood sample is obtained. Clot formation to obtain serum may also occlude certain drugs.

Distribution–Concentration Relationships

Suppose a single-bolus intravenous dose of 100 mg of three different drugs was administered to normal individuals, and that the drugs are distributed, respectively, into whole blood, extracellular water, or total body water; these fluids represent volumes of 5, 15, and 42 L, respectively. Suppose also that the drugs are distributed homogeneously into these volumes, and that plasma concentrations of unchanged drug can be determined before any significant elimination has occurred. What plasma concentrations would be obtained?

When the 100 mg of drug is in whole blood, and provided the drug equilibrates freely between red blood cells and plasma, the drug concentration in plasma is 20 mg/L, or 20 µg/mL. Similarly, when the drug is distributed into extracellular water, the drug concentration in plasma is 6.7 µg/mL; and when the drug is distributed into total body water, the drug concentration is 2.4 µg/mL. Thus, the same quantity of each of the three different drugs in the body will yield plasma concentrations of 20, 6.7, and 2.4 µg/mL, depending on distribution characteristics, an eight-fold range in plasma concentrations for the same quantity of drug.

This example shows that drug distribution can have a marked influence on plasma drug concentrations and, unless the distribution characteristics of a drug are known, false conclusions can be drawn regarding drug absorption or systemic availability from plasma data alone. In this example, one might naively assume, from the measured plasma levels, that the drug that is distributed into extracellular water is absorbed three times more efficiently from an oral dose than the drug that is distributed into total body water. In fact, in this case both drugs are absorbed to exactly the same extent.

The actual distribution of a drug may of course be more complex than this. Many drugs bind to plasma proteins, and this binding may prevent them from

TABLE 8.1	**Volumes of Some Body Fluids**	
Fluid	Volume (L)	Percent of Body Weight
Plasma	3	4
Blood	5	7
Lymph	10	14
Intracellular water	27	39
Extracellular water	15	21
Total body water	42	60

Note: Volumes are approximate for a 70-kg male person.

leaving the plasma volume, depending in part on the relative affinity of drug for plasma and tissue proteins. Drugs may concentrate in red blood cells or may not enter these cells efficiently, so that the blood concentration might differ from the plasma concentration. An example of a drug that concentrates in red blood cells is the anticancer agent tubercidin. Approximately 80–98% of this drug concentrates in red blood cells when added to whole blood (2).

Drugs may have a high affinity for particular organs and tissues, particularly fatty tissues, so that the "apparent" distribution volume of a drug may be larger than the total body volume. Pentothal is a fat-soluble drug that rapidly enters the brain after it is administered. Subsequently, this drug is partially redistributed into body fat and then slowly released into the general circulation. The initial rapid concentration in brain tissues causes rapid onset of action, whereas redistribution into fatty tissue causes short duration of action. Digoxin is another drug that is extensively taken up by extravascular tissues. Its apparent distribution volume is about 500 L, or seven times larger than total body volume. This large apparent distribution volume is not possible if distribution is homogeneous, and the drug is clearly concentrated in, or being sequestered by, specific organs and tissues. High organ affinity may differ in different species. The anticonvulsant drug gabapentin has high affinity for pancreas tissue in rodents, but not in monkeys or humans.

In order to pass from intravascular to extravascular fluids, that is, to leave the general circulation, drugs must cross the capillary membrane. These membranes are "normal" in that they are more permeable to lipophilic than hydrophilic compounds, but "abnormal" in that they are more permeable than other membranes to most compounds (3). Fat-soluble substances tend to cross capillary membranes rapidly. The anesthetic gases cross essentially as if there were no membrane at all, with rapid exchange. Water-soluble molecules cross the membrane more slowly and at a rate that is inversely proportional to molecular size. The permeability of muscle capillary to some water-soluble compounds is demonstrated in Table 8.2. Small molecules appear to pass through the capillary walls by simple diffusion, whereas large molecules have difficulty getting across. The low membrane permeability of albumin (MW 69,000) is important for drugs that are bound to this macromolecule.

Capillary Permeability

Permeability of Muscle Capillary to Water-Soluble Molecules — TABLE 8.2

Molecule	Molecular Weight	Diffusion Coefficient Across Capillary Membrane ($cm^3/s \cdot 100\ g$)
Water	18	3.7
Urea	60	1.8
Glucose	180	0.64
Raffinose	594	0.24
Inulin	5,500	0.036
Myoglobin	17,000	0.005
Serum albumin	69,000	<0.001

Source: Adapted with permission from reference 3.

The permeability of capillary membranes varies in different regions of the body in accord with their particular purposes. For example, permeability is greatly increased in renal capillaries by pores in the membranes of renal endothelial cells. Membrane permeability is also increased in the liver by the lack of a complete membrane lining hepatic sinusoids. Both factors, porous membranes and lack of membrane lining, facilitate easy and rapid transfer of substances into vital organs of elimination and metabolism.

Perfusion and Diffusion Effects

Distribution of a drug between blood and tissue may be perfusion or diffusion *rate limited*. The distribution of most lipid-soluble drugs is *perfusion rate limited*, that is, transfer between blood and tissue depends on the rate and extent to which a particular tissue is perfused by blood. On the other hand, distribution of most water-soluble substances is *diffusion rate limited*, that is, it is controlled by the rate at which the molecules diffuse across the tissue membrane(s).

The rates at which blood perfuses some tissues are shown in Table 8.3 (4). These rates vary enormously, from 10 mL/min · g in lung to only 0.03 mL/min · g in resting muscle or fat. The rate at which a drug is presented to tissue can be described by equation 8.1

$$\text{Rate} = QC_A \tag{8.1}$$

where Q is the blood flow rate and C_A is the arterial drug concentration entering the tissue. If drug is taken up by tissue, then the net rate of tissue uptake is described by equation 8.2.

$$\text{Net rate of uptake} = Q(C_A - C_V) \tag{8.2}$$

where C_V is the concentration of drug in venous blood leaving the tissue. The tissue–blood distribution ratio can be described by a constant, K_p, so that when equilibrium between drug in tissue and arterial blood is reached (if it is reached), the amount of drug in tissue is given by equation 8.3.

TABLE 8.3   **Blood Perfusion Rates in Certain Tissues**

Tissue	Blood Perfusion Rate (mL/min · g of tissue)
Lung	10
Kidney	4
Thyroid gland	2.4
Adrenal gland	1.2
Liver	0.8
Heart	0.6
Brain	0.5
Muscle	0.03
Fat	0.03

Source: Adapted with permission from reference 4.

$$\text{Amount of drug in tissue} = K_p V_T C_A \tag{8.3}$$

where V_T is the tissue volume. Although equation 8.3 indicates how much drug gets into tissue, it does not indicate the rate of tissue uptake. This rate can be determined if the rate of tissue perfusion is known and if the distribution is perfusion rate limited. This rate is determined simply by dividing the amount of drug in tissue at equilibrium (equation 8.3) by the rate at which drug is introduced to the tissue (equation 8.1), as in equation 8.4. Therefore, the time of tissue uptake depends on the distribution ratio, K_P, and the perfusion rate to that tissue, Q/V_T.

$$\text{Time taken for equilibration} = \frac{K_p V_T C_A}{Q C_A} = \frac{K_p V_T}{Q} = \frac{K_p}{Q / V_T} \tag{8.4}$$

Consider how equation 8.4 works for distribution of a fat-soluble drug into the kidney and liver, and into muscle and fat. Actual perfusion rates and hypothetical distribution ratios are given in Table 8.4 (4). The results given in the table show that the time taken for drug to equilibrate into the kidney from a constant blood concentration is 0.25 min, whereas equilibration into liver would take 1.25 min, into muscle 33 min, and into fat 1666.7 min, or nearly 28 h. The longer time taken for a drug to equilibrate into those tissues where it tends to concentrate is due simply to the longer perfusion process needed to introduce drug into that tissue. An example of a drug of this type is pentothal. This lipophilic drug has high affinity for both brain and fat tissue. Because the brain is perfused approximately 16 times faster than fat tissue, drug levels in the brain increase faster. Drug accumulates in fat tissue more slowly because of slower perfusion. However, drug is eventually redistributed from brain to fat tissue, so that pentothal levels in the brain and fat may eventually be equal. The largest proportion of the body load would then be located in general fat because of its greater overall mass.

Perfusion-limited drug transport assumes that drug diffusion between blood and tissues is fast. Perfusion-limited transport is more common with nonpolar than polar compounds, whereas diffusion-limited transport is more common with polar than nonpolar compounds. The best examples of diffusion-limited transport are provided by drug entry into the central nervous system. However, penetration of drugs into other tissues may also be diffusion rate limited. Drugs tend to enter the prostate gland slowly, and adequate levels of antibiotics are difficult to obtain in this organ. Conversely, avid uptake of environmental contaminants such as

Time Required for a Perfusion-Rate-Limited Drug to Equilibrate in Tissue   **TABLE 8.4**

Tissue	Perfusion Rate (Q/V_T) (mL/min · mg of tissue)	K_p	Time to Equilibrate $K_p/(Q/V_T)$ (min)
Kidney	4	1	0.25
Liver	0.8	1	1.25
Muscle	0.03	1	33.3
Fat	0.03	50	1,666.7

Source: Adapted with permission from reference 4.

pesticide residues into fat tissue is due to the lack of a diffusion barrier for these fat-soluble compounds.

The arguments developed so far in this chapter are based on the equilibrium concept. In many cases, however, circulating drug levels are not constant but rise and fall with each dose. In these situations, equilibrium between blood and tissue may not be reached. For example, many antibiotics have short biological half-lives, and maintaining blood levels for a sufficient time for therapeutic levels to be achieved in infected tissues may be difficult.

Binding of Drugs to Tissues

Many drugs bind to plasma proteins. The most important protein as far as drug binding is concerned is albumin, although binding to other proteins, particularly the orosomucoid α_1-acid-glycoprotein (AAG), may also occur.

Although basic drugs often bind to AAG, acidic drugs bind predominantly to albumin. Albumin is a water-soluble protein with a molecular weight of 69,000. It circulates at a concentration of 4 g/100 mL in blood plasma, but exists also at lower concentrations in extravascular fluids. Albumin carries about 100 negative and 100 positive charges and has an isoelectric point of about pH 5. Thus, at a plasma pH of 7.4, albumin has a net negative charge, but it nonetheless attracts both anions and cations. Binding of drugs to albumin is usually rapidly reversible. The extent to which a drug binds depends on the affinity of the drug to albumin, the number of binding sites on the albumin molecule for that particular drug, and the concentrations of both drug and albumin.

Because plasma proteins cross the capillary membrane slowly, drugs bound to these proteins are essentially confined to the same volume as the protein, which is about 3 L. Drugs also bind to extravascular macromolecules, and this binding will be considered later in this chapter. Intravascular binding means that protein-bound drugs cannot reach their site of action, particularly if the site is in some tissue outside the bloodstream. For example, protein-bound drugs cannot cross the blood–brain barrier. They also generally cannot passively enter the liver, kidney, or other organs to be excreted, although there are some important exceptions.

The literature contains many values for the percentage of drug bound to plasma proteins. For example, probenecid is approximately 80% bound. This means that 80% of the concentration of total probenecid circulating in plasma is bound to proteins. Practically all assay methods used to measure drug or metabolite concentrations in plasma measure total (bound and unbound) drug. For example, if the total drug concentration in plasma is 10 μg/mL and the drug is 80% bound to plasma proteins, then the concentration of bound drug is 8 μg/mL and the concentration of free or unbound drug is only 2 μg/mL.

Failure to take this type of protein binding into account can cause major errors when calculating the distribution volume of a drug. For example, suppose 100 mg of a drug is administered by bolus intravenous injection, and the plasma concentration of total drug, determined before a significant quantity of drug has been eliminated, is 8 μg/mL. If protein binding were ignored, the distribution volume would be calculated as 100/8, which is 12.5 L. Thus, one might conclude that the drug has only modest distribution characteristics and is unlikely to penetrate significantly into extravascular tissues and fluids.

However, this drug is actually 80% bound to plasma proteins. A conventional drug assay would not detect this, but a protein-binding determination by equilibrium dialysis or by ultrafiltration would. If the drug is 80% bound to plasma proteins, and the total concentration in plasma is 8 μg/mL, then the bound concentration is 80% of 8, or 6.4 μg/mL. As the plasma albumin to which the drug is bound is confined essentially to the plasma volume of 3 L, the total quantity of drug bound (and therefore not free to leave the plasma volume and participate in any type of free drug equilibrium) is obtained by multiplying 6.4 by 3000 to yield 19,200 μg, or 19.2 mg. Because this quantity of drug is bound, albeit reversibly, to protein, the quantity must be subtracted from the 100 mg of total drug in the body. This leaves 80.8 mg of drug unbound to protein and therefore free to equilibrate between plasma water and other fluids. To find the distribution volume, the quantity 80.8 mg must be divided by the free drug concentration of 1.6 μg/mL, which yields a volume of 50.5 L. This volume is four times greater than the volume calculated from the total drug concentration, and is similar to the volume of total body water, suggesting that, instead of having restricted distribution, this drug is well distributed into tissues, undoubtedly entering both extracellular and intracellular water.

Unfortunately, the influence of protein binding on drug distribution is not as simple as these calculations imply. The calculations take into account binding in plasma, but not binding and other phenomena that may occur in extravascular fluids and tissues. Drugs may bind to extravascular proteins and other tissues, as previously noted for digoxin. Such binding tends to pull the drug out of plasma and thereby reduce the plasma drug concentration. Reduced drug concentration in plasma resulting from extravascular binding may imply a larger distribution space than actually exists. A drug may thus be subject to both intravascular and extravascular binding, these factors having opposing effects on apparent distribution volume. Table 8.5 provides a list of the percentage binding of some drugs to plasma proteins at therapeutic drug concentrations. Binding can vary from zero to almost 100%.

Percent Binding of Drugs to Plasma Proteins at Therapeutic Drug Concentrations   TABLE 8.5

Drug	Percent Bound	Drug	Percent Bound
Diazoxide	99	Methotrexate	45
Dicoumarol	98	Methadone	40
Diazepam	96	Meperidine	40
Digitoxin	95	Phenacetin	30
Prednisone	90	Acetaminophen	25
Phenytoin	87	Ampicillin	25
Riafampin	85	Digoxin	23
Chlorpropamide	80	Cephalexin	22
Pentothal	75	Barbital	10
Carbamazepine	72	Promethazine	8
p-Aminosalicylate	65	Antipyrine	4
Glutethimide	54	Isoniazid	0
Carbenicillin	47		

As plasma protein binding of drugs is generally reversible, it is important to consider the equilibrium between bound and unbound drug, as in equation 8.5 (3).

$$C + P_F \leftrightarrows P_F C \tag{8.5}$$

where C is free drug concentration and P_F is the concentration of free receptors. This equation, after taking into account the number of binding sites on the protein molecule for a particular drug, is rearranged to equation 8.6.

$$\frac{C(nP - P_F C)}{P_F C} = K \tag{8.6}$$

where n is the number of binding sites per protein molecule, K is the dissociation constant of the drug–protein complex, and nP is the total concentration of receptors ($P_F + P_F C$). Equation 8.6 is then rearranged to give equation 8.7.

$$(K + C) P_F C = nPC \tag{8.7}$$

However, if the relationship in equation 8.8 is used,

$$r = \frac{P_F}{P} = \frac{nC}{K + C} \tag{8.8}$$

where r equals the number of moles of drug bound per mole of protein, then equation 8.9 is obtained.

$$rK = nC - rC \tag{8.9}$$

If both sides of equation 8.9 are divided by CK, equation 8.10 is obtained.

$$\frac{r}{C} = \frac{n}{K} - \frac{r}{K} \tag{8.10}$$

If r is plotted against C from equation 8.8, a hyperbolic curve is obtained, as in Figure 8.1. This figure shows that as the concentration of drug increases, binding sites on the protein molecule start to fill, and the number of molecules of drug per molecule of protein increases. Finally, at saturation, the value of r becomes a constant of value n, the number of binding sites.

If r/C is plotted against r from equation 8.10, a linear plot is obtained, as in Figure 8.2. This figure is the familiar Scatchard plot, which is also used to obtain the value of n. The Scatchard plot may exhibit more than one linear segment, each relating to a different group of binding sites with common affinities.

Of somewhat more interest to pharmacokineticists is the percentage of circulating drug that is bound to plasma proteins, β. The value of β can be defined in terms of the concentrations of free and bound drug as in equation 8.11.

$$\beta = \frac{P_F C}{P_F C + C} = \frac{1}{1 + (C/P_F C)} \tag{8.11}$$

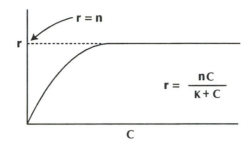

Plot of r vs. C from equation 8.8.

FIGURE 8.1

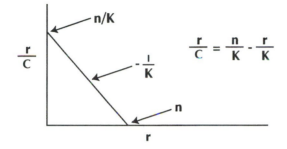

Plot of r/C vs. r from equation 8.10.

FIGURE 8.2

From the relationship in equation 8.8, the right side of equation 8.11 is rearranged to give equation 8.12.

$$\beta = \frac{1}{1 + K/nP + C/nP} \tag{8.12}$$

If β is then plotted against C/nP or, more conveniently, against the logarithm of that value, then for different values of K/nP a series of curves is obtained, as in Figure 8.3. The information in this figure shows that at low drug concentrations (the left side of the figure), drugs are maximally bound to protein to an extent dictated by the constant K/nP. As n and P can be considered to be constant for a particular drug, each of the curves in the figure can be considered to be a function of K. As the drug concentration increases, free drug eventually far exceeds bound drug, and the percentage binding, regardless of the value of the binding constant, approaches zero.

This figure implies that the percentage of drug bound to plasma proteins continuously changes with changing drug concentration. Because pharmacokinetic calculations are frequently based on total drug concentrations, a continuously changing relationship between concentrations of bound and unbound drug, and hence between the pharmacokinetics of total and free drug, would be expected.

In practice, the percentage binding of most drugs is less concentration-dependent than Figure 8.3 implies. The concentration scale on the abscissa of this figure

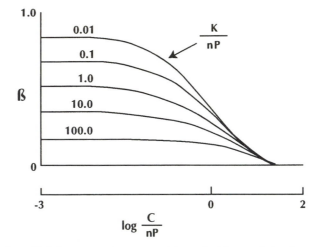

FIGURE 8.3   Plot of β vs. log (C/nP) from equation 8.12. (Adapted with permission from reference 3.)

is logarithmic so that, assuming that *n* and *P* are constant, the total concentration range is approximately 10^5, which is far greater than the normal therapeutic range for most drugs. The curves also tend to be horizontal and linear over quite a large proportion of the concentration range, depending on the value of *K*, so that drug binding to plasma proteins is usually fairly constant and independent of drug concentration, at least within the therapeutic range. Thus, although protein binding should still be taken into account when calculating drug distribution and clearance values, protein binding generally does not affect such parameters as rate constants because they are identical for both the free and total drug.

Dispyramide and phenytoin are exceptions to this general rule. The percentage binding of these drugs decreases with increasing drug concentration in plasma within the therapeutic range (5). For drugs of this type it is important to measure free-drug concentrations in plasma before attempting to examine their pharmacokinetics.

Intravascular and Extravascular Drug Binding

Regardless of the type of plasma protein to which a drug is bound, the effect of binding is always to draw drug into plasma at the expense of extravascular tissues and fluids. If a drug binds to extravascular tissues or proteins, exactly the opposite effect will occur, and drug will be pulled out of plasma. Most drugs that bind to macromolecules are probably influenced by both intravascular and extravascular binding. Many different kinds of extravascular molecules are capable of binding drugs, including albumin, fatty tissues, and cell membranes. Extravascular or tissue binding can significantly modulate the effect of drug binding to plasma proteins (6, 7).

In order to obtain greater appreciation of the opposing effects of both intravascular and extravascular binding, their influence on four major pharmacokinetic parameters will be considered. These parameters are as follows:

1. The percentage of total drug that is free, or unbound, in the body.

2. The concentration of free drug in plasma.

3. The concentration of total drug in plasma.

4. The apparent distribution volume.

The concentration of free drug in plasma, C_F, in equilibrium with free drug in other body fluids can be approximated by equation 8.13.

$$C_F = \frac{A_T}{V_F + 3\gamma + (V_F - 3)\varepsilon} \tag{8.13}$$

where A_T is the total amount of drug in the body in milligrams, V_F is the actual distribution volume of free drug in liters, the value 3 is the plasma volume in liters, and γ and ε represent the ratios of bound to unbound drug in plasma and in tissue, respectively.

From equation 8.13, the amount of free drug in the body, A_F, can readily be calculated with equation 8.14.

$$A_F = C_F V_F \tag{8.14}$$

The concentration of bound drug in plasma, C_B, is obtained from equation 8.15, the concentration of total drug in plasma, C_T, from equation 8.16, and the percentage of free drug in the body from equation 8.17.

$$C_B = C_F \gamma \tag{8.15}$$

$$C_T = C_F + C_B \tag{8.16}$$

$$\text{Percent free drug} = \frac{A_F}{A_T} \times 100 \tag{8.17}$$

The apparent distribution volume, V_{app}, is given by equation 8.18.

$$V_{app} = \frac{A_T}{C_T} \tag{8.18}$$

A simple mathematical example demonstrates how these equations can be used. Consider two drugs, A and B. Assume that both drugs are distributed into an actual body volume of 20 L but that they differ in that at therapeutic concentrations Drug A is 90% bound to plasma proteins and 5% bound to tissue proteins, whereas Drug B is only 5% bound to plasma proteins and 90% bound to tissue proteins. Both drugs are administered so that the total quantity of drug in the body at the time of measurement is 100 mg.

The numerical values for the parameters defined in equations 8.13–8.18 are shown in Table 8.6 and demonstrate the remarkable effects that binding can have on both real and apparent pharmacokinetic parameters.

For example, the greater overall binding of Drug B, due to the large tissue volume, relative to plasma reduces the concentration of free drug that is in equilibrium and at the same concentration in both plasma and tissue fluids to about one-fourth of the concentration of Drug A. The amount of free Drug B in the circulation is also only one-fourth of the amount of Drug A, as is the percentage of drug that is free in the body. Although these differences appear large, they are smaller than some other parameters. For example, the concentration of bound drug and, perhaps more important for pharmacokinetic calculations, the concentration of total drug in plasma, are many times higher for Drug A than for Drug B. The concentration of total Drug A in plasma is 20.9 µg/mL, compared to only 0.6 µg/mL for Drug B. Similarly, the apparent distribution volume for Drug A is only 4.8 L (slightly larger than plasma volume), but the apparent distribution volume of Drug B is 164 L. The true distribution volume for both drugs is 20 L.

These drugs are, of course, two extremes in the possible spectrum of protein-binding combinations, but this example illustrates the profound effect that binding can have on drug distribution, observed plasma levels, and pharmacokinetic parameters.

Consider now the percentage of free drug in the body, and how it is affected by plasma and tissue binding. Figure 8.4 describes three hypothetical drugs that have actual distribution volumes of 12, 20, and 42 L. Consider first the effect of variable drug binding to plasma proteins, represented by the solid curves. When plasma binding is zero, the drug is 100% free in the body. When plasma binding is 100%, the drug is 100% bound. However, between these limits the change in the percentage of free drug is not linearly related to protein binding but is curved, and the most rapid changes occur when protein binding exceeds 80–85%. Consider

TABLE 8.6   **Influence of Intravascular and Extravascular Binding on the Pharmacokinetic Parameters of Two Drugs, A and B, That Have Different Intravascular and Extravascular Binding Characteristics**

Parameter	Drug A	Drug B
V (actual)	20 L	20 L
A_T	100 mg	100 mg
Percent bound in plasma	90%	5%
Percent bound in tissues (µg/mL)	5%	90%
C_F (µg/mL) (Eq. 8.13)	100/[20 + 3(9) + 17(5/95)] = 2.09	100/[20 + 3(5/95) + 17(9)] = 0.58
A_F (mg) (Eq. 8.14)	2.09 × 20 = 41.8	0.58 × 20 = 11.6
C_B (µg/mL) (Eq. 8.15)	2.09 × 9 = 18.8	0.58 × 5/95 = 0.030
C_T (µg/mL) (Eq. 8.16)	2.09 + 18.8 = 20.9	0.58 + 0.030 = 0.61
Percent free (%) (Eq. 8.17)	(41.8 × 100)/100 = 41.8	(11.6 × 100)/100 = 11.6
V_{app} (L) (Eq. 8.18)	100/20.9 = 4.8	100/0.61 = 163.9

Note: For details see text.

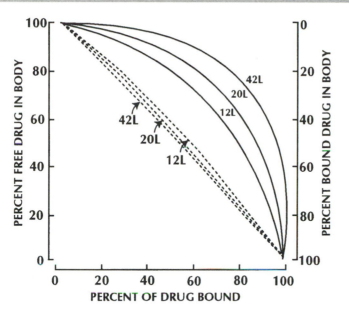

Changes in the percentage of drug that is free (unbound) in the body vs. changes in the percentage of circulating drug bound to serum proteins (—) and the percentage of extravascular drug bound to tissues (---). When considering binding at one site (serum protein or tissue), the binding at the other site is assumed to be 0. Curves were generated from equations 8.13–8.15, with A_T = 100 mg. The volumes 12, 20, and 42 L represent true distribution volumes of the free drug. (Reproduced with permission from reference 6.)

FIGURE 8.4

next the dashed curves representing tissue binding. As might be expected, the two extremes for these curves are identical to those for the solid lines. Thus, when all the drug is bound, whether in plasma or tissue, no free drug is present in the body. However, in the case of tissue binding, because of the relatively large volume of tissue fluids compared to plasma (the tissue–plasma volume ratio varies from 4:1 to 14:1 for the three drugs considered here), the relationship between the percentage of drug that is free and bound is almost linear throughout the entire tissue-binding range, and this relationship essentially becomes linear for the drug that has the largest distribution volume of 42 L.

In Figure 8.5, when binding to plasma proteins is zero, the concentration of free drug (which in this instance will equal the concentration of total drug) in plasma is obtained by dividing the total body load of 100 mg by the actual distribution volume to yield concentrations of 8.3, 5, and 2.4 µg/mL for the drugs with 12-, 20-, and 42-L volumes, respectively. On the other hand, when binding to plasma proteins is 100%, the concentration of free drug is zero; no free drug is present in the body. Between these extremes, the solid curves show that the concentration of free drug changes only slowly until plasma-protein binding again reaches 80–85%, at which point the concentration of free drug drops sharply toward zero as protein binding approaches 100%.

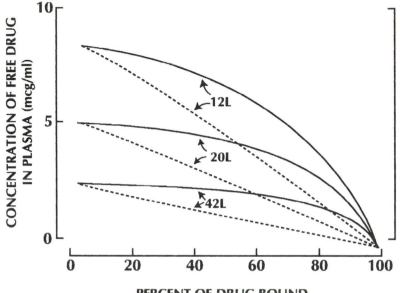

FIGURE 8.5 Changes in the concentration of free drug, C_F, vs. changes in the percentage of circulating drug bound to plasma proteins (—) and the percentage of extravascular drug bound to tissue (---). Curves were generated from equation 8.13. Conditions are as in Figure 8.4. (Reproduced with permission from reference 6.)

Tissue binding has the same effect in Figure 8.5 as in Figure 8.4; a virtually linear relationship exists between the free-drug concentration and tissue binding, and identical values are obtained when binding to tissue is 0 and 100% compared to those values obtained when plasma binding is 0 and 100%.

In Figure 8.6, the concentration of total (free and bound) drug in plasma is examined under the same conditions. However, the curves are quite different from those in the previous figures. At zero binding to plasma protein (left side of the figure), the concentration of total drug in plasma is the same as the concentration of free drug in Figure 8.5. This similarity is reasonable because in both cases there is no bound drug. However, as the degree of binding to plasma proteins increases, the concentration of total drug in plasma also increases until the three solid curves converge to a value of 33.3 µg/mL. This concentration results from all three drugs being totally confined to plasma volume, and the concentration value is determined simply by dividing the total drug load of 100 mg by 3 L. Therefore, although the concentration of free drug in plasma decreases with increased plasma-protein binding, the concentration of total drug increases. The changes in total drug concentration (note the alteration in the ordinate values at 10 µg/mL in this figure) increase dramatically after binding exceeds 80%. Tissue binding, on the other hand, has a similar effect on the concentration of total drug to that of free drug. A linear decline in total drug in plasma occurs as drug is bound more extensively to tissues and is drawn out of plasma.

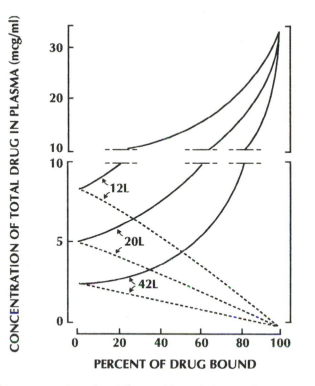

Changes in the concentration of total (free and bound) drug in plasma, C_T, vs. changes in the percentage of circulating drug bound to plasma proteins (—) and the percentage of extravascular drug bound to tissue (---). Curves were generated from equations 8.13–8.18. Conditions are as in Figure 8.4. (Reproduced with permission from reference 6.)

FIGURE 8.6

Figure 8.7 describes binding-related changes in the drug apparent distribution volume. Values of apparent distribution volume are often cited in the literature but are poorly understood. In this figure, the situation differs dramatically from that seen in the previous figures.

The calculated apparent volumes in Figure 8.7 are obtained from equation 8.18. Apparent volumes are the types of volumes that are generally calculated from total drug levels in plasma. At zero plasma-protein binding, and also at zero tissue binding, the apparent volumes are identical to the true volumes because all of the drug is free. However, as plasma-protein binding increases, the apparent volume decreases linearly to reach a value of 3 L when all of the drug is bound. So for any degree of drug binding to plasma proteins, the apparent volume of distribution will always underestimate the true volume, and the degree of underestimation will increase as the extent of plasma-protein binding increases.

Tissue binding has the opposite effect. As the drug binds more avidly to tissues (follow the dashed curves from left to right in Figure 8.7), the drug is drawn out of plasma, and the drug apparent distribution volume increases to large values as the concentration of total drug in plasma is reduced. This situation is typical for many

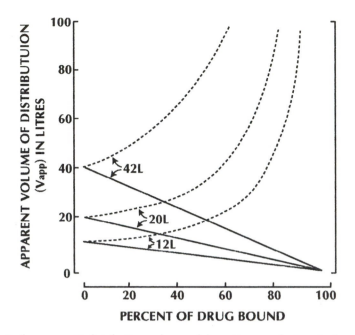

FIGURE 8.7

Changes in the apparent distribution volume of drug, V_{app}, vs. changes in the percentage of circulating drug bound to plasma proteins (—) and the percentage of extravascular drug bound to tissue (---). Curves were generated from equations 8.13–8.18. Conditions are as in Figure 8.4. (Reproduced with permission from reference 6.)

drugs. For example, digoxin has an apparent distribution volume many times greater than body volume, and trimethoprim has an apparent distribution volume approximately double that of body volume. These large apparent volumes of distribution are due to binding or sequestration of drug by tissue, which leads to extensive and heterogeneous tissue distribution at the expense of plasma levels, and occurs regardless of drug binding to plasma proteins. The large apparent distribution volume of trimethoprim relative to that of sulfamethoxazole is the reason why these two drugs are administered in a combination ratio of 5:1 sulfamethoxazole to trimethoprim: this ratio results in a plasma ratio of 20:1 sulfamethoxazole to trimethoprim, which is the optimum ratio for therapeutic effect.

A common misconception is that if a drug binds to plasma proteins, it cannot penetrate tissues and a small distribution volume results. Certainly, if plasma binding is high and tissue binding is low or zero, then the drug will concentrate in plasma. However, if a drug has affinity for macromolecules generally, the drug will tend to bind both in plasma and tissues. Thus, binding of drugs to plasma and tissue proteins should be related, and a significant correlation between the apparent distribution volume and plasma-protein binding might be expected. Such a correlation has been demonstrated with the four cephalosporin drugs shown in Figure 8.8. Therefore, just because a drug is highly bound to plasma proteins does not mean that the drug cannot penetrate into tissues. Erythromycin is 90% bound to plasma

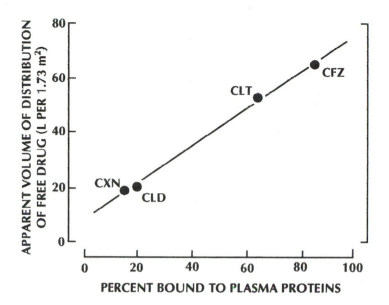

Relationship between plasma protein binding and the volume of distribution of free **FIGURE 8.8**
drug for four cephalosporins: CFZ, cefazolin; CLT, cephalothin; CLD, cephaloridine;
and CXN, cephalexin. Correlation coefficient = +0.998. (Reproduced with permission
from reference 6.)

proteins, but only 1–3% of the total body load of this antibiotic is contained within
the plasma volume. The remainder of the drug is in extravascular tissues.

Figures 8.5 and 8.6 show that changes in the degree of protein binding significantly affect circulating levels of free and total drug only when binding is about
80% or greater. Many drugs and endogenous substances are capable of displacing
other drugs from plasma binding sites and thus altering the unbound–bound equilibrium. However, for all drugs except highly bound molecules, these changes in
the unbound–bound equilibrium are likely to be of little pharmacokinetic or clinical significance. Even with highly bound drugs, increases in free-drug concentration due to displacement are likely to be rapidly compensated for by tissue distribution and elimination of free drug. Thus, changes in drug concentrations due to
altered binding are, for the most part, of little clinical significance whether the
drug is highly bound or not.

Penetration of Drug into the Central Nervous System

The central nervous system (CNS) comprises approximately 2% of body weight,
but CNS circulation receives 16% of cardiac output. Blood flow varies to different
regions of the brain. For example, flow to the cortex is 1–2 mL/min per gram,
whereas flow to cerebral white matter is only 0.24 mL/min per gram.

Surrounding the entire CNS is cerebrospinal fluid (CSF). The CSF has a total
volume of about 120 mL, and it has a turnover rate of 10–25% per hour. The pathways of CSF flow are shown in Figure 8.9 (8). CSF is formed mainly at the ventri-

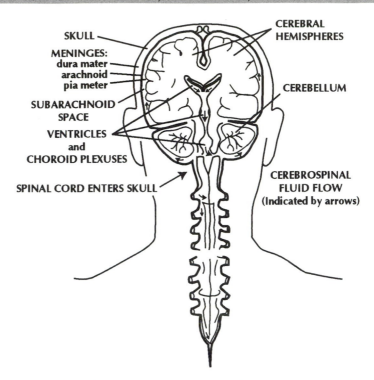

SKULL

MENINGES:
dura mater
arachnoid
pia meter

SUBARACHNOID
SPACE

VENTRICLES
and
CHOROID PLEXUSES

SPINAL CORD ENTERS SKULL

CEREBRAL
HEMISPHERES

CEREBELLUM

CEREBROSPINAL
FLUID FLOW
(Indicated by arrows)

FIGURE 8.9

Pathways of CSF flow. Flow is from the ventricles to the cerebral and spinal subarachnoid space. (Adapted with permission from reference 8).

cles and choroid plexus in the brain stem, and it reenters the general circulation mainly at the arachnoid villi. CSF has a much lower protein content than plasma. CSF serves many purposes:

1. It acts as a hydraulic cushion against brain injury.

2. It maintains control over ion concentrations in the CNS.

3. It maintains the CNS at a slightly acidic pH of about 7.2 as compared to blood pH of 7.4.

Figure 8.10 shows the physical–anatomical relationships among blood, CSF, and CNS tissue (8). Drugs can enter the CNS by two different routes, and each route contains a barrier to drug entry. The consequence of these barriers is that fat-soluble compounds can readily enter the CNS but water-soluble compounds generally cannot.

Drugs enter the brain mainly by the indirect route via the CSF. They enter the CSF at the epithelium of the choroid plexus. To enter the CSF from blood, a drug has to pass through the capillary endothelial cells, basement membranes, and stroma. These barriers do not appear to hinder the drug. However, drugs then have to pass through the choroid epithelial cells, which appear to contain some type of tight junction that prevents intercellular transport and thus inhibits pas-

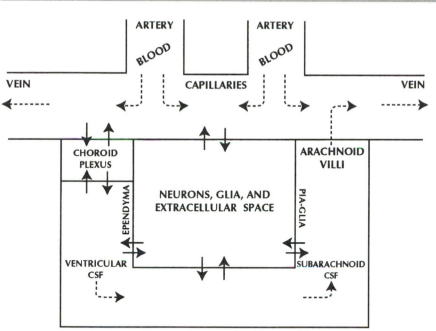

Routes by which compounds can enter and leave the CNS. (Reproduced with permission from reference 8.)

FIGURE 8.10

sage of water-soluble substances. Once within the CSF, there is no particular barrier preventing drugs from entering the cells of the CNS.

The second route of drug entry into the CNS is directly from the capillaries that constitute the CNS blood supply. In this case, no defense region of the CSF as described for the first route exists. The cells of the capillaries are therefore adapted to provide an alternative and similar defense in the form of glial connective tissue cells (or astrocytes) associated with the basement membrane of the capillary endothelium. Glial connective tissue cells serve a similar purpose as the choroid epithelial cells in restricting entry into the CNS to fat-soluble substances.

Thus two routes exist for drug entry into the CNS, but each route incorporates a defense structure to prevent, or a least minimize, penetration of water-soluble substances into the brain. These defense structures constitute the blood–brain barrier.

Drugs generally exit from the CNS either directly into the capillaries or, perhaps more commonly, at the arachnoid villi along with CSF that is reentering the general circulation. Some compounds, including weak acids and bases, can be removed from CSF by an active process at the choroid plexus.

The efficiency with which the blood–brain barrier works, and the presence of diffusion-limited entry into the CNS, is demonstrated in the classic studies of Brodie et al. (9). Some results obtained in these studies are shown in Figure 8.11. The ionization constants, partition coefficients, other physical characteristics, and the entry rate of some drugs into the CSF are listed in Table 8.7 (10). Compar-

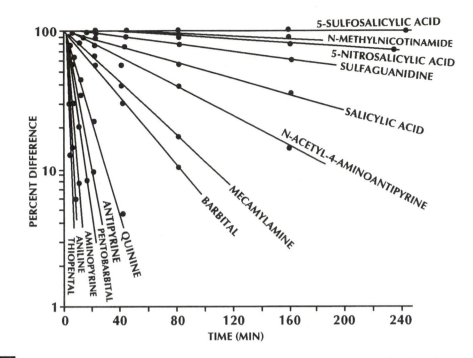

FIGURE 8.11   Rate of equilibration of various drugs between plasma and CSF. (Reproduced with permission from reference 9.)

Properties of Drugs and Times to Reach 50% Equilibrium Concentration Values Between Plasma and CSF   **TABLE 8.7**

Drug	Fraction Bound to Protein at pH 7.4	pK_a	Fraction Un-ionized at pH 7.4	Partition Coefficient (n-Heptane-Water) of Un-ionized Form	Effective Partition Coefficient	Time to Reach 50% Equilibrium Between Plasma and CSF (min)
Thiopental	0.75	7.6	0.61	3.3	2.0	1.4
Aminopyrine	0.20	5.0	1.0	0.21	0.21	2.8
Pentobarbital	0.40	8.1	0.83	0.05	0.042	4.0
Antipyrine	0.08	1.4	1.0	0.005	0.005	5.8
Barbital	<0.02	7.5	0.56	0.002	0.001	27.0
N-Acetyl-4-aminoanti-pyrine	<0.03	0.5	1.0	0.001	0.001	56.0
Salicylic acid	0.4	3.0	0.004	0.12	0.0005	115.0
Sulfaguanidine	0.06	>10.0	1.0	<0.001	<0.001	231.0

Source: Adapted from reference 10.

ison of these data with Figure 8.11 shows that the rate at which drugs enter the CSF is a function of ionization and lipid–water partitioning. For example, un-ionized salicylic acid (partition coefficient 0.12) partitions into n-heptane from water to a greater extent than pentobarbital (partition coefficient 0.05). However, pentobarbital is less ionized at pH 7.4, so that pentobarbital has an effective partition coefficient, and hence diffusion rate into the CSF from plasma, greater than that of salicylic acid.

Many antibiotics are water soluble and do not enter the brain efficiently. However, in some infections such as meningitis, the permeability of the CNS and CSF membranes increases, allowing more compound to enter the CNS. This phenomenon has been noted for such drugs as ampicillin, penicillin G, some cephalosporins, and lincomycin. Increased drug penetration into the CNS has also been noted in some experimental studies concerning brain tumors. The development of compounds, prodrugs, and drug delivery techniques that will penetrate the blood–brain barrier and provide therapeutic drug levels in the brain continues to challenge the ingenuity of pharmaceutical chemists and formulators.

Methods of Determining Drug Distribution

The ability to reach the site(s) of action at sufficient concentration and for sufficient time to elicit a required therapeutic effect is a vital characteristic of any drug. Although the material presented so far in this chapter helps the reader understand some factors that influence drug distribution, and its impact on drug pharmacokinetics, it does not describe how to accurately determine the precise distribution characteristics of any drug. The reason for this deliberate omission is that it is extremely time consuming and expensive, and sometimes virtually impossible, to obtain this type of information for any drug entity. The difficulty of obtaining this type of information, which involves developing specific assays for a drug in a variety of organs and tissues, is one of the reasons that physiological pharmacokinetic modeling (Chapter 18) is not commonly used in drug disposition studies.

However, methods do exist to obtain approximations of drug distribution characteristics based on the distribution of a radiolabel. With these methods, radioactivity is introduced into a drug molecule and the disposition of radioactivity in the body is determined. In this way, distribution of radioactivity can be determined as a function of dose, dosage route, and time. These methods have become increasingly sophisticated, having developed from simple "cut and burn" technology, in which portions of organs or tissues are oxidized and released radioactivity is counted, to advanced imaging technologies. However sophisticated these methods are, it must be remembered that they measure some form of radioactive label or tag that has been incorporated into a drug molecule. If a drug is not metabolized and the radiolabel is stable within the molecule, these technologies provide an accurate picture of drug distribution. On the other hand, if a radiolabel is unstable (for example, ^3H may undergo hydrogen exchange) or if a drug molecule is extensively metabolized, then these technologies indicate only the distribution of radioactivity, which may or may not bear a relationship to the distribution of the parent drug, depending on the position of the radiolabel relative to the drug metabolism profile. One of the principal technologies in this area is whole-body autoradiography (WBA). Several emerging technologies will also be discussed briefly.

In 1954, Ullberg first described whole-body autoradiography (*11*). With this technology, animals are typically dosed with radioactively labeled drug, sacrificed, and rapidly frozen. Freezing preserves the anatomical relationship of organs and tissues, and immobilizes radioactivity in body fluids and cavities. Carcasses are embedded in an ice matrix and sectioned in a cryostat microtome. Whole-body sections are captured on adhesive tape, dried, and apposed to X-ray film. The degree of exposure (film blackening) is related to the concentration of nuclide in the region of interest. Subjective evaluation of films provides a rapid description of the relative distribution of a compound and its metabolites in tissues and organs, substructures of organs (e.g., brain nuclei), and body cavities, and allows evaluation of the degree to which material is transported across such tissue barriers as brain, eye, testis, and placenta. Quantitative data were initially obtained from film by application of densitometric analysis (*12*). More recently, advances in computer technology have led to image analysis systems that employ a video camera or some other scanning device to render film autoradiograms into digital images (*13*). These images may also be enhanced for detailed organ structure identification.

Film autoradiography is a powerful tool for quantitatively determining the distribution of labeled compounds. However, because of the nonlinear response of photographic material to ionizing radiation, and also the time necessary for exposure of low levels of radioactivity, considerable experience is needed to accurately quantitate the wide range of concentrations that may be found in a typical drug time course study. There has therefore been considerable interest in two alternative systems, phosphor imaging (PI) and multiwire proportional counting (MWPC)–microchannel array detection, technologies that reduce the difficulties associated with film methodology. Output from both methods is in electronic form, and high-quality hard copies can be produced with high-resolution printers. There is a substantial time savings with PI in that images that may require months of exposure on film can be obtained in a few days with phosphor plates. Unlike autoradiography, PI detector response is linear. PI has a lower resolution than film autoradiography and is adversely affected by very high and low levels of radioactivity in samples. High activity produces a streaking phenomenon caused by light carryover in the detector, whereas with low activity, longer exposure is necessary and background exposure resulting from cosmic radiation increases. This problem can be resolved to some extent by carrying out exposure in a lead-protected environment.

MWPC relies on detection of the ionized track of a decay particle by a multiwire array (*14*). Perhaps the most attractive characteristic of this technique is that visualization occurs in real time on the computer screen and improves as more counts are accumulated by the array. Thus, determination is rapid, some 10 times faster than PI, and response is similarly linear. Resolution with MWPC is relatively low, so that images represent an average of counts over a region. This may limit the utility of the technique for drug distribution studies in that only gross organ substructure can be obtained.

Both PI and MWPC are inadequate for imaging tritium-labeled compounds. Protective shielding in the MWPC instruments effectively prevents detection of the low-energy beta emission. Similarly, standard phosphor plates have an antiscratch layer that acts as a barrier. Plates have been made without this coating, but they become contaminated and unusable in a short time. Because of their high cost, they do not represent a practical solution to low-energy beta autoradiography.

Other technologies have been developed with the express purpose of monitoring drug (or radioactivity) disposition in a single individual over time. Positron emission tomography (PET), single photon emission tomography (SPECT), gamma scintigraphy, and magnetic resonance imaging (MRI) are tools used primarily in the clinical setting, but they have considerable potential for the study of drugs and other xenobiotics in animals. The gamma camera provides a relatively low resolution image, but it can be used with several readily available nuclides. This technology has been used to evaluate the fate of dosage forms as they pass through the GI tract. MRI techniques are interesting because they are noninvasive and do not rely on radioactive tags. However, sensitivity is relatively low, and successful imaging relies on high concentrations of compounds in regions of interest. PET has become an important tool in diagnosis as well as a research tool for evaluating metabolic function and drug distribution. Nuclides used include ^{18}F, ^{11}C, ^{15}O, and ^{13}N. These are characterized by short decay half-lives (20.4 min for ^{11}C, 2.1 min for ^{15}O), and they must be prepared and incorporated into a molecule for study immediately before use (15). Because of the need for a cyclotron to make the nuclides, PET facilities are somewhat limited, and drug distribution studies are usually done in collaboration with a major hospital or university research center. Because of the short half-lives of the nuclides, pharmacokinetic data can be gathered only over a relatively short period of time.

For the discussion on methods of determining drug distribution, the author is indebted to William P. McNally for advice on this topic (16).

Summary

Capillary membranes are more permeable than most other physiological membranes to most compounds. Lipophilic molecules tend to cross the membranes readily. The ease with which hydrophilic compounds cross capillary membranes is inversely proportional to their molecular size.

Drug distribution into tissues may be perfusion or diffusion rate limited.

Drug binding to plasma proteins can be characterized mathematically in terms of drug concentration, affinity, and the number of binding sites on the protein molecule.

The free drug concentration in the body, the concentration of free drug in serum (or plasma), and the concentration of total drug in serum are related in a nonlinear fashion to plasma-protein binding.

Changes in plasma-protein binding have a significant effect on drug distribution and drug levels in plasma only when the drug is highly bound.

The apparent distribution volume of a drug is markedly dependent on plasma protein and tissue binding. Plasma protein binding tends to reduce the apparent distribution volume, whereas tissue binding tends to increase the apparent distribution volume relative to the true volume.

Both the degree of ionization and the lipid–water partition coefficient are important determinants of drug entry into the CNS. For many drugs, passage from the blood to the brain is diffusion rate limited. The low permeability of membranes separating the circulation from the CSF or CNS to water-soluble compounds represents the blood–brain barrier.

Imaging methods may be used to determine the distribution of radioactivity from radiolabeled compounds. The usefulness of this type of technology is related to the stability of the radioactive label on the drug molecule and to the degree to which the drug is metabolized.

Problems

1. A dose of 250 mg of a drug is administered to a 70-kg male patient by rapid intravenous injection. The concentration of total drug in plasma 5 min after injection is 16.6 µg/mL. Calculate the apparent distribution volume of the drug, assuming that the drug does not bind to plasma proteins and loss of drug from the body during the first 5 min is negligible.

2. Recalculate the distribution volume in Problem 1, assuming that:

    (i)   the drug is 70% bound to plasma proteins, and also that

    (ii) the drug is eliminated at such a rate that 20% of the dose is excreted from the body during the initial 5 min after dosing.

3. A drug has a true distribution volume in the body of 42 L. The drug is 90% bound to plasma proteins and 0% bound in tissue. Calculate the values of C_F, C_B, C_T, A_F, and V_{app} for the drug when the total body load is 100 mg.

4. Recalculate the values in Problem 3 assuming that the drug is 90% bound to plasma proteins and 80% bound in tissue.

5. Recalculate the values in Problem 3 assuming that plasma-protein binding is reduced to 60% and tissue binding remains unchanged at 80%.

References

1. Creasey, W. A. *Drug Disposition in Humans: The Basis of Clinical Pharmacology*; Oxford University: New York, 1979; pp 33–54.

2. Smith, C. G.; Reineke, L. M.; Burch, M. R.; Shefner, A. M.; Muirhead, E. E. *Cancer Res.* **1970**, *30*, 69–75.

3. Goldstein, A.; Aranow, L.; Kalman, S. M. *Principles of Drug Action: The Basis of Pharmacology*, 2nd ed.; John Wiley: New York, 1974; pp 130–138.

4. Rowland, M.; Tozer, T. N. *Clinical Pharmacokinetics: Concepts and Applications*; Lea Febiger: Philadelphia, PA, 1980.

5. Meffin, P. J.; Robert, E. W.; Winkle, R. A.; Harapat, S.; Peters, F. A.; Harrison, D. C. *J. Pharmacokinet. Biopharm.* **1979**, *7*, 29–42.

6. Craig, W. A.; Welling, P. G. *Clin. Pharmacokinet.* **1977**, *2*, 252–268.

7. Craig, W. A.; Kunin, C. M. *Ann. Rev. Med.* **1976,** *27*, 287–300.

8. Rall, D. In *Fundamentals of Drug Metabolism and Drug Disposition*; LaDu, B. N.; Mandel, H. G.; Way, E. L., Eds.; Williams and Wilkins: Baltimore, MD, 1981; pp 76–87.

9. Brodie, B. B.; Kurtz, H.; Schanker, L. S. *J. Pharmacol. Exp. Ther.* **1960,** *130*, 20–25.

10. Rowland, M.; Tozer, T. N. *Clinical Pharmacokinetics: Concepts and Application*; Lea Febiger: Philadelphia, PA, 1980; pp 34–47.

11. Ullberg, S. *Acta Radiol. Suppl.* **1954,** *118*,1–110.

12. Waddell, W. J.; Marlowe, C. *Xenobiot. Metab. Dispos.* **1994,** *9*, 408–416.

13. Som, P.; Yonekura, Y.; Oster, Z. H.; Meyer, M. A.; Pelletieri, M. L.; Fowler, J. S.; MacGregor, R. R.; Russell, J.; Wolf, A. P.; Fand, I.; McNally, W. P.; Bril, A. B. *J. Nucl. Med.* **1983,** *24*, 238–244.

14. Sweitzer, A. A.; Englert, D. F. *Quart. J. Nucl. Med.* **1995,** *39*, 42–43.

15. Fowler, J. S.; Wolf, A. P. In *Positron Emission Tomography and Autoradiography: Principles and Applications for the Brain and Heart*; Philps, N.; Mazziotta, J.; Schelbert, H., Eds.; Raven: New York, 1986; pp 391–450.

16. McNally, W. P. Personal Communication; Parke-Davis Pharmaceutical Research Division, Ann Arbor, MI, 1996.

Drug Metabolism

he metabolism of drugs and other foreign substances can be considered to be part of the body's natural defense mechanisms that defend the homeostatic balance of the system against invasion of foreign substances.

Although drug metabolism has been recognized for a considerable time, the study of drug metabolism first became an organized and well-defined scientific endeavor through the pioneering work and dedication of the late R. Tecwyn Williams. His book *Detoxication Mechanisms*, published in 1959, was the first authoritative text on metabolism of drugs and other chemicals (*1*). The title indicates one of the major natural functions of drug metabolism, which is the formation of more water soluble derivatives of parent compounds. This derivatization usually results in loss of pharmacological activity and more rapid excretion. Bernard B. Brodie once said, "If there were no such process as drug metabolism, it would take the body about 100 years to terminate the action of pentobarbital, which is lipid-soluble and cannot be excreted without being metabolized."

All of the reactions involved in drug metabolism are adaptations of biochemical, enzymatic processes that were in existence long before drugs were synthesized. The enzymes that oxidize phenobarbital or demethylate codeine, for example, were in existence in the body long before these drugs were discovered. The biochemical systems thus exist to metabolize these new compounds, but were not created for this purpose.

Drug metabolism takes place in many parts of the body. Metabolism can occur in the bloodstream, spleen, kidneys, brain, muscle tissue, the contents (bacteria) and wall of the GI tract, and many other tissues. The prostaglandins, for example, are extensively metabolized in lung tissue. However, the major organ in which drug metabolism takes place is the liver. The liver is the primary organ of waste dis-

Sites of Drug Metabolism

posal and works in close collaboration with the kidneys to eliminate unwanted substances from the body.

The liver is the largest gland in the body, weighing 1.2–1.6 kg. The liver receives blood from the region of the gastrointestinal (GI) tract, spleen, and gall bladder via the splanchnic circulation and hence the portal vein; it receives blood from the aorta via the hepatic artery. Approximately 80% of the total hepatic blood flow of 1.5 L/min is supplied by the portal vein. After entering the liver, the portal vein divides into a rich capillary network of hepatic sinusoids. Kupffer cells are found in the walls of the sinusoids. Blood penetrates into the liver parenchymal cells from the sinusoids through fine intracellular canaliculi.

The sinusoids converge toward the center of the liver lobule to form sublobular veins. Sublobular veins combine to form hepatic veins, which rejoin the general circulation at the inferior vena cava. This unique hepatic vascular architecture provides intimate contact between substances being transported in portal blood and the hepatic parenchymal cells.

Around each portal vein branch is a plexus of bile capillaries. Bile capillaries enter into interlobular bile ducts, which eventually combine to form the hepatic duct. In humans and higher mammals, bile is stored in the gall bladder before entering the duodenum via the common bile duct.

The hepatic artery is primarily responsible for oxygenation of the liver. Although this artery supplies only 20% of total hepatic blood flow, the greater part of the oxygen supplied to the liver is provided by the hepatic artery.

The site within the liver parenchymal cell where most metabolism takes place is the microsomal fraction, or *microsomes*. Microsomes is a name given to a liver fraction obtained by sequential centrifugation of homogenized liver. More specifically, liver is homogenized at 9000 g, the precipitation discarded, and the supernate centrifuged at 100,000 g. The resulting precipitate is the microsomal fraction.

The microsomal fraction is derived from a subcellular component, the endoplasmic reticulum. Two types of endoplasmic reticulum are designated according to their microscopic appearance and activity: rough-surfaced and smooth-surfaced. The rough-surfaced form contains the ribosomes and ribonucleoproteins involved in protein synthesis. The drug-metabolizing enzymes are associated primarily with the smooth-surfaced endoplasmic reticulum, although there are other sites of metabolism. For example, alcohol dehydrogenase is located primarily in the soluble fraction of the hepatocyte.

Alteration of drugs by the biochemical processes of the body tends to reduce their pharmacological activity and increase elimination rate. Loss of pharmacological activity is relatively simple to understand because the metabolite in most cases is not only more water-soluble than the parent drug, and therefore less capable of crossing membranes (for example, the blood–brain barrier), but the metabolite has also lost most of the original intrinsic pharmacological activity.

The most important consequence of drug metabolism is increased aqueous solubility of metabolites compared to the parent drug. As will be described in Chapter 11, drugs that reach the kidneys in the bloodstream are subject to three major processes: filtration from blood into the kidney tubules at the glomerulus, active secretion from blood into the tubules of the kidney, and passive reabsorption from the distal convoluting tubules back into the bloodstream. The filtration and active-secretion mechanisms have their own particular characteristics, but they are not sensitive to fat solubility, unless fat solubility affects some other parameter. Passive reabsorption, on the other hand, is restricted to lipophilic substances capable of crossing the membranes that line the distal tubules and thereby reenter the general circulation.

By converting a drug into a more water-soluble form (a metabolite), drug metabolism prevents distal tubular reabsorption from occurring and thereby promotes renal excretion. Metabolites may also be cleared by other routes including the lungs, bile, saliva, and sweat, but renal elimination is the most important route in most cases.

Although drug metabolism usually produces more water-soluble compounds, in some cases the metabolites are less water-soluble than the parent drug. Good examples of this phenomenon can be found among the sulfonamides. This group of antimicrobial agents has a common basic structure, and sulfonamides are metabolized primarily by acetylation at the free amino group. N-acetylation may give rise to decreased water solubility as, for example, with sulfathiazole and sulfacetamide. This reduced solubility may cause deposition of metabolites in the kidneys, with possible crystal urea.

Mechanisms of Drug Metabolism

Drug metabolism reactions occur in two distinct phases, Phase I and Phase II. Some drugs participate in both phases, and other drugs undergo either Phase I or Phase II metabolism. Some examples are given in Scheme 9.1.

Phase I Mechanisms

During Phase I, compounds are chemically activated, mainly by oxidation but also by reduction, or hydrolysis. This process could be considered as putting a chemical handle on a compound to prepare it for a possible Phase II reaction. These chemical activation reactions may result in pharmacological activation as, for

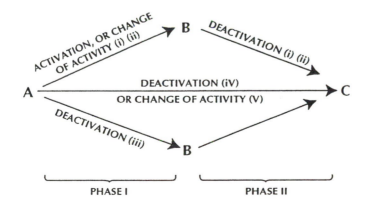

The two major phases of drug metabolism.

SCHEME 9.1

example, in chloral hydrate oxidation; change of activity, as in codeine demethylation to form morphine; or loss of pharmacologic activity, as in hydroxylation of phenobarbital. The last classification, loss of pharmacologic activity, is the most common.

Phase II reactions involve chemical conjugation or synthesis. Very few conjugation reactions exist for Phase II reactions, but these reactions act upon an enormous variety of substrates. Phase II reactions also generally increase water solubility, but there are exceptions. At one time, all Phase II reactions were considered to be pharmacologically deactivating, but this is now known not to be the case. For example, procainamide is acetylated by a Phase II mechanism to form N-acetylprocainamide, which has similar pharmacologic activity to the parent drug.

For a Phase II conjugation reaction to occur, a compound must be chemically active or chemically prepared by one or more of the many reactions described in detail below. If a compound is already chemically active (chemical activity is unrelated to pharmacologic activity), then the compound may undergo a Phase II reaction directly because the Phase I step is unnecessary. Two examples of compounds that take this route are sulfanilamide and procainamide.

Some examples of major Phase 1 oxidation reactions are described in detail in Schemes 9.2–9.9.

Phase I oxidation reactions thus include aliphatic and aromatic hydroxylation, N- and O-dealkylation, oxidative deamination, sulfoxidation, and N-hydroxylation. All of these reactions take place in the microsomal cell fraction, the endoplasmic reticulum. Alcohol dehydrogenase, on the other hand, occurs in the soluble cell fraction. Ethyl alcohol and benzyl alcohol are two typical substrates for this reaction.

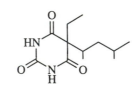 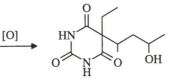

Pentobarbital   OH-Pentobarbital

SCHEME 9.2   Aliphatic hydroxylation.

 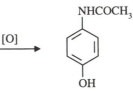

Acetanilide   p-OH-Acetanilide

SCHEME 9.3   Aromatic hydroxylation.

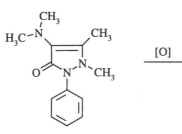
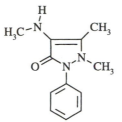

Aminopyrine   Methyl-4-aminopyrine

N-Dealkylation.

SCHEME 9.4

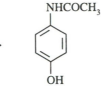

Phenacetin   *p*-OH-Acetanilide

O-Dealkylation.

SCHEME 9.5

Amphetamine   Phenylacetone

Oxidative Deamination.

SCHEME 9.6

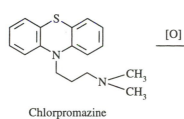
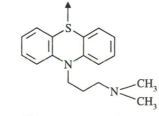

Chlorpromazine   Chlorpromazine sulfoxide

Sulfoxidation.

SCHEME 9.7

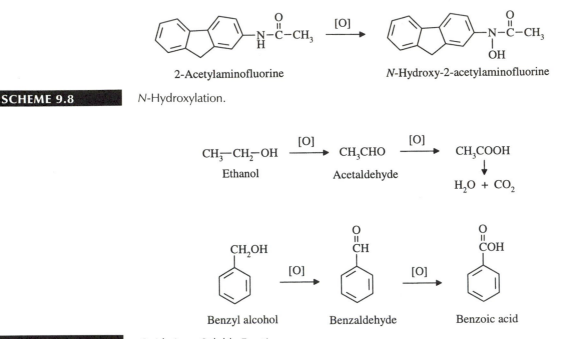

2-Acetylaminofluorine

N-Hydroxy-2-acetylaminofluorine

SCHEME 9.8   N-Hydroxylation.

$CH_3-CH_2-OH \xrightarrow{[O]} CH_3CHO \xrightarrow{[O]} CH_3COOH$

Ethanol   Acetaldehyde

$\downarrow$

$H_2O + CO_2$

Benzyl alcohol   Benzaldehyde   Benzoic acid

SCHEME 9.9   Oxidations–Soluble Fraction.

The ease with which a drug undergoes oxidative reactions, or any other metabolic reaction, may profoundly influence the biological half-life of the drug in the body. The two hypoglycemic agents, chlorpropamide and tolbutamide, have similar pharmacological activity. These agents also have similar chemical structures, except that tolbutamide has an additional methylene group in the side chain and also a methyl group *para* on the benzene ring, whereas chlorpropamide has a chlorine group *para* on the ring (Chart 9.1). The methyl group on tolbutamide is readily metabolized sequentially through the alcohol and the aldehyde to the carboxylic acid, which is then conjugated. The chlorine group on chlorpropamide, on the other hand, is resistant to metabolism. Thus, tolbutamide has a biological half-life in the body of approximately 4 h compared to 10–12 h for chlorpropamide, and tolbutamide has to be dosed more frequently in clinical practice.

There are only two major Phase I reduction reactions, nitro reduction and azo reduction, as shown in Schemes 9.10 and 9.11. Mammalian livers carry out these reactions far less effectively than the intestinal microflora. Therefore, Phase I reduction reactions tend to occur more commonly when drugs are administered orally.

The last major Phase I mechanism is hydrolysis. Esterases are ubiquitous enzymes that are active in plasma and other tissues as well as in liver. Esterases are capable of hydrolyzing a large number of esters to alcohols and free carboxylic acids. The example in Scheme 9.12 is interesting because rapid in vivo hydrolysis of procaine, a compound originally used for cardiac arrhythmias, led to the development of procainamide. The amide is more resistant to hydrolysis than the ester

Tolbutamide

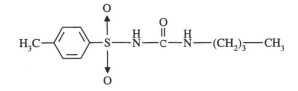

Chlorpropamide

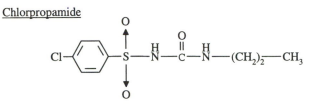

Structures of the hypoglycemic agents tolbutamide and chlorpropamide.

CHART 9.1

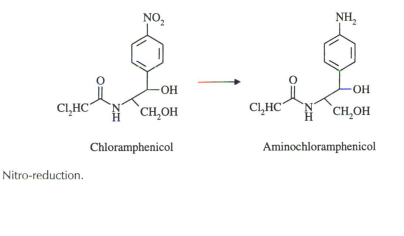

Chloramphenicol   Aminochloramphenicol

Nitro-reduction.

SCHEME 9.10

Prontosil   Triaminobenzene   Sulfanilamide

Azo-reduction.

SCHEME 9.11

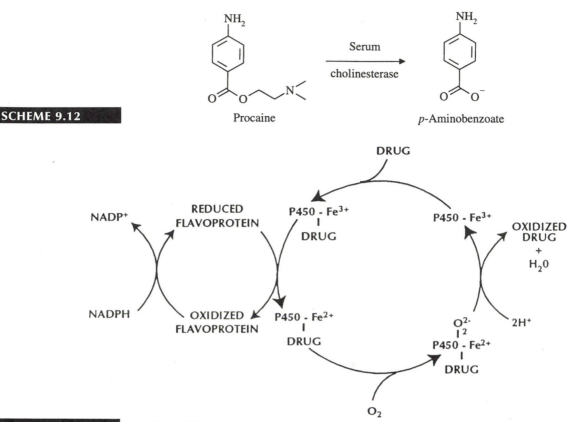

SCHEME 9.12

SCHEME 9.13   Outline of the electron transport chain involving cytochrome P-450 and the oxidation of drugs by the hepatic microsomal system. (Adapted with permission from references 2 and 3.)

and is therefore a far more useful drug. As already noted, procainamide is metabolized predominantly by acetylation rather than by hydrolysis.

Mechanisms of Oxidative Drug Metabolism.   A key element in the oxidative process is the hemoprotein cytochrome. This was so named for the development of an absorption peak at 450 nm when the hemoprotein reacts with carbon monoxide. Cytochrome P-450 plays a vital role in an electron-transport system in which electrons are lost by the drug during oxidation and passed along an electron-transport system to the final receptor, NADPH (nicotinamide adenine dinucleotide phosphate). In this process, "active oxygen", which is instrumental in the oxidative process, is generated. The role of cytochrome P-450 in oxidative metabolism, as currently understood, is shown in Scheme 9.13. NADPH acts conversely as an electron donor in a reverse process for reductive metabolic reactions (2, 3).

The Cytochromes P-450.   During the last decade there has been an explosive increase in the understanding and characterization of cytochrome P-450 enzymes. What was first thought to be a single enzyme has been shown to be a huge variety of enzymes, with different substrate specificities within and across species. The number of enzymes that are being identified, cloned, and expressed continues to increase. Several reviews have recently been published on the role and selectivity of cytochrome P-450 enzymes (4–6). The following brief treatment summarizes the current understanding of P-450 enzymes with particular relevance to humans.

Cytochrome P-450 (P-450) is now known to comprise a large number of isozymes, each representing a separate gene product. In the rat, more than 40 P-450s have been identified, while in humans the number is at least 25–30. The metabolism of drugs in humans is dominated by six major P-450 isozymes. These are listed in Table 9.1 together with some typical substrates (4). These isozymes are relatively abundant except for CYP1A1, which is found only in individuals exposed to tobacco smoke, polycyclic aromatic hydrocarbons, and related enzyme inducers.

Some compounds are substrates for several different isozymes whereas others are acted upon by only a single isozyme. The quinolone antibiotic enoxacin is acted upon almost exclusively by the isozyme CYP3A4.

Some isozymes exhibit genetic polymorphism in their levels of expression. For example, CYP2D6 is functionally deficient in approximately 7% of Caucasians, giving rise to impaired metabolism of debrisoquine and other drugs metabolized by this isozyme. Such individuals are at considerable risk from a variety of adverse drug reactions. CYP2C18/19 isozymes also exhibit genetic polymorphism, and mephenytoin metabolism is used as a marker for this isozyme in humans. Although CYP1A2 is not truly polymorphic in nature, it does exhibit a wide range of activity, varying 50–60-fold in humans. An excellent marker substrate for this isozyme is provided by caffeine, either through analysis of caffeine urinary metabolites (7) or the caffeine breath test. The latter specifically measures caffeine N-demethylation (8).

Knowledge of which of the P-450s is dominant in the metabolism of a particular drug is important in that it permits prediction of interactions with other drugs. If two drugs are acted upon by the same isozyme, then they are likely to compete for this isozyme in the body, and this competition may affect their rates of metabolism. Metabolic pathways may also be affected if a major isozyme is inhibited, thereby permitting a normally minor metabolic pathway to dominate.

Six Major Isozymes of Cytochrome P-450 Responsible for Drug Metabolism in Humans   **TABLE 9.1**

Isozyme	Typical Substrates
CYP1A2	Phenacetin, theophylline
CYP2C8/9/10	Hexobarbital, phenytoin, tolbutamide, warfarin
CYP2C18/19	Diazepam, mephenytoin
CYP2D6	Bufuralol, debrisoquin, encainide, imipramine, metoprolol, propranolol
CYP2E1	Acetaminophen, chlorzoxazone
CYP3A4	Alfentanol, cimetidine, cyclosporine, enoxacin, erythromycin, lidocaine, tacrine

Source: Adapted with permission from reference 4.

Similarly, if a drug induces an isozyme (and more on that later), then this may cause the metabolism of another drug that is acted upon by the same isozyme to be accelerated. This could also accelerate the metabolism of the inducing drug. Increased awareness of the nature and extent of P-450 activities in drug metabolism, accompanied by rapid advances in associated technology, has had a dramatic affect on the utilization of drug metabolism information during drug discovery and development.

Whereas in the past, elucidation of metabolic pathways of a new drug candidate was a slow and tortuous process, drug companies now use banks of pure expressed isozymes, typically CYP1A1, 1A2, 2A6, 2B6, 2C9, 2C19, 2E1, and 3A4, and also induced or noninduced fractions from human and nonhuman livers, to obtain complete metabolic drug profiles in vitro. These can be obtained before any in vivo studies are conducted to provide valuable predictive information regarding species differences or similarities in metabolism, and the specific isozymes involved. This information is also important to validate toxicity species relative to humans and to predict whether interactions might be expected with concomitantly administered drugs. Apart from the scientific value of this new information and technology, the practical and economic advantages of obtaining detailed metabolism information early in a drug development process are considerable and will be elaborated on in Chapter 21.

Phase II Mechanisms

Examples of Phase II conjugation reactions are given in Schemes 9.14–9.19. Unlike Phase I reactions, in which a large variety of different reactions may occur, Phase II reactions are relatively few, the principal ones leading to formation of the following compounds:

- Glucuronides of phenols, alcohols, carboxylic acids, and aromatic amines.

- Acetates of amines and sulfonamides.

- Sulfates, mainly of phenols.

- Glycine conjugates of aromatic carboxylic acids.

- Mercapturic acid conjugates of aromatic hydrocarbons, halogenated hydrocarbons, and halogenated nitrobenzenes.

- N-, S-, and O-methylation of phenols, amines, and sulfhydryl compounds.

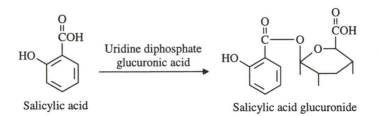

Salicylic acid   Salicylic acid glucuronide

SCHEME 9.14   Glucuronides, which occur with phenols, alcohols, carboxylic acids, and aromatic amines.

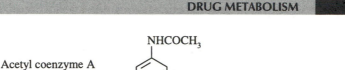

Sulfanilamide N^1,N^4-diacetylsulfanilamide

Acylation, which occurs mainly with amines and sulfonamides.

SCHEME 9.15

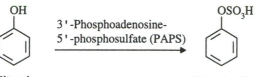

Phenol   Phenyl sulfate

Sulfation, which occurs mainly with phenols.

SCHEME 9.16

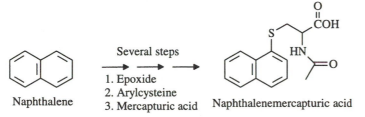

p-Aminosalicylic acid   *p*-Aminosalicyluric acid

Glycine conjugation, which occurs with aromatic carboxylic acids.

SCHEME 9.17

Naphthalene   Several steps   1. Epoxide   2. Arylcysteine   3. Mercapturic acid   Naphthalenemercapturic acid

Mercapturic acids, which occur with aromatic hydrocarbons, halogenated hydrocarbons, and halogenated aromatic hydrocarbons.

SCHEME 9.18

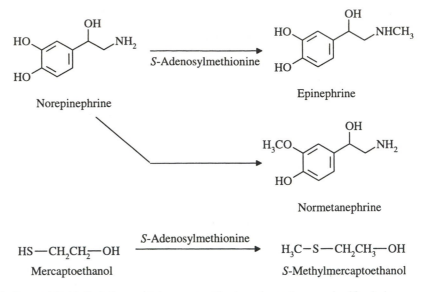

SCHEME 9.19   N-, S-, and O-Methylation, which occurs with phenols, amines, and sulfhydryl groups.

The conjugation steps, each specific for particular functional groups and under the control of a specific enzyme system, represent the final stage in the preparation of compounds for removal from the body. Phase II reactions generally reduce or completely inhibit pharmacologic activity and also increase aqueous solubility. However, there are several exceptions to this rule. Procainamide is metabolized by a Phase II reaction to form the pharmacologically active acetylprocainamide. Similarly, aminofluorene is acetylated to the carcinogenic metabolite N-acetylaminofluorene (2-AAF). Generally, however, Phase II reactions collectively represent a sophisticated and efficient process for terminating the activity of many drugs and removing them from the body. The formation of glycine conjugates tends to be more saturable than most other conjugation reactions because formation depends on the availability of free glycine in the body.

These mechanisms represent the most common drug metabolism reactions. Other systems may be specific for particular compounds, such as the action of xanthine oxidase in purine metabolism and the formation of nucleotides of some antineoplastic agents. Additional material on these subjects is available (9–12).

Species, Sex, and Age Differences

Species

When considering drug metabolism, particularly during development of new drugs, determining how constant the metabolic processes are within and between species is important. Within a species, reactions are generally qualitatively similar, and only quantitative differences that might be expected with normal biological variation are observed. However, exceptions do occur, as described in the section "The Cytochromes P-450" for drugs metabolized by the CYP1A2 isozyme and in cases of polymorphic metabolism. Vesell (13) examined metabolism rates and plasma half-lives of several drugs, including antipyrine, bishydroxycoumarin, and phenylbutazone in fraternal and identical twins. Differences between fraternal twins were between 6 and 22 times greater than between pairs of identical twins. This difference represents a good example of genetic rather than environmental causes of different metabolic rates. Probably the best example of genetically based variation in drug metabolism is the rate of acetylation of isoniazid and some other compounds. This metabolic step is distinctly bimodal between "fast" and "slow" acetylators. It is interesting that Caucasian and Black populations appear to have a predominance of slow acetylators, whereas Asians, Eskimos, and Native Americans appear to have a predominance of fast acetylators. Tests have been described to identify fast and slow acetylators (14, 15), and such tests may facilitate accurate dosing of isoniazid for the treatment of tuberculosis.

Differences in drug metabolism among species may have a significant impact on the validity of preclinical safety and tolerance studies. Although some drugs are handled similarly among species, others show marked species specificity. For example, a mitochondrial enzyme in the guinea pig and the rat converts cyclohexanecarboxylic acid to benzoic acid, whereas this aromatizing activity is absent from most other species, including humans. Humans and apes form glutamine conjugates of some acids, whereas other species do not. Birds and reptiles use ornithine in place of glutamine and glycine for conjugation of carboxylic acids. Dogs generally do not acetylate as efficiently as other species, with the notable exception of the anticonvulsant agents gabapentin and isobutyl γ-aminobutyric acid (GABA), which are both extensively acetylated by the dog but not by rodents, monkeys, or humans. Many other qualitative metabolic differences among species will no doubt be reported as the speed and efficiency of metabolism screening continue to improve.

Quantitative metabolic differences among species are common. Many drugs have far longer biological half-lives in humans than in other species. The plasma elimination half-life of phenylbutazone is 2–3 days in humans but only 6 h in dogs and guinea pigs, and only 3 h in rabbits. On the other hand, acetylation of aromatic amines is generally faster in humans than in dogs. The primary route of ephedrine metabolism in dogs and guinea pigs is N-demethylation to norephedrine, whereas in rats and humans norephedrine is a minor metabolite. Humans metabolize epinephrine primarily to 3-methoxy-4-hydroxymandelic acid, along with conjugated metanephrine. Cats, on the other hand, convert ephedrine primarily to 3-methoxy-4-hydroxypropylene glycol. Rats form both of these metabolites.

As knowledge of the pattern of P-450 isoforms in different species becomes more complete, it will be easier to understand and to predict species differences in drug metabolism. Some examples of different isoforms used to metabolize substrates in various species are given in Table 9.2 (16). The data in the table indicate both similarities and differences in isozyme involvement across species. These few examples give some idea of the dimensions and implications of species differences in drug metabolism. They emphasize the inherent danger of extrapolating metabolism, pharmacokinetics, and toxicity data from experimental animals to humans unless the relative behavior of drugs in the different species is established in some detail.

Sex

Some sex differences in drug metabolism, both quantitative and qualitative, have been observed in experimental animals, but such differences are usually negligible in humans, and of little or no clinical importance. Typically, plasma levels of the antidementia drug tacrine are somewhat higher in human females than in males from equivalent doses, possibly because of slightly lower expression of CYP3A4 in females. However, the difference in plasma levels appears to have minimal effect on drug activity and tolerance. Notwithstanding the general similarity in drug metabolism between human males and females, most regulatory agencies now insist that females be included in all clinical studies involving new drugs or drug formulations. The reasons for this are partly scientific and partly political.

Age

Age may play an important role in the rate of drug metabolism, but only in the very young and very old. A long plateau exists from early childhood well into old age, during which time the efficiency of drug metabolism is fairly constant, given good health. In the very young, the newborn, and even in the fetus, lack or immaturity of drug-metabolizing enzymes may constitute a problem. This is particularly important for the unborn child because drugs frequently cross the placenta efficiently, so that whatever the mother ingests, the fetus will probably be exposed to also. Typically, ingestion of alcohol and nicotine by the mother has been shown to have severe effects in the fetus. Other drugs may give rise to adverse effects in the fetus because of impaired metabolism or other mechanisms. These drugs include narcotic and non-narcotic analgesics, benzodiazepines, local anesthetics, neuromuscular blocking agents, anticholinergics, anti-

TABLE 9.2	**Metabolism of Some Substances by P-450 Isoforms in Different Species**			
	Principal Isoform			
Substrate, reaction	*Rat*	*Mouse*	*Rabbit*	*Human*
Ethoxyresorufin, *O*-dealkylation	1A	1A	1A	1A
Phenacetin, *O*-dealkylation	1A	1A	1A	1A
Coumarin, 7-hydroxylation	NK[a]	2A	2A	2A
Benzphetamine, *N*-demethylation	2B	2B	2B	3A?
S-Mephenytoin, 4-hydroxylation	3A	2C	2C	2C
Hexobarbital, 1'-hydroxylation	2C	2C	2C	2C
Tolbutamide, hydroxylation	2B	NK	3A	2C
Debrisoquin, 4-hydroxylation	2D	NK	2D	2D
Aniline, *p*-hydroxylation	2E	2E	2E	2E
Nifedipine, oxidation	2C	—	3A	3A
Lidocaine, *N*-deethylation	2C	—	3A	3A

[a]NK, not known, that is, the isozyme responsible has not been identified, but it is not the isoform catalyzing the reaction in other species

Source: Adapted with permission from reference 16.

thyroid drugs, hypoglycemic agents, anticoagulants, and some antimicrobial agents (*17, 18*).

Drug metabolism, like any other enzyme-mediated process, requires that an enzyme bind in some way to a substrate. Binding is generally exquisitely sensitive to substrate structure and can therefore be selective for enantiomeric or chiral molecules, giving rise to chiral recognition (*19, 20*). Chiral recognition, or selective metabolism of one steric form over another, occurs whether a drug exists as an enantiomer (one asymmetric center) or as a diastereomer (more than one asymmetric center).

Steric Factors

There are many examples of substrate specificity, and some of these have been reviewed (*20*). Typically, the (*R*) enantiomer of disopyramide is metabolized mainly by *N*-dealkylation, whereas the (*S*)-enantiomer is metabolized by aromatic oxidation (*21*). The enantiomeric antidepressant mianserin is metabolized in a similar stereoselective pattern, aromatic oxidation occurring with the (*S*)-enantiomer and *N*-demethylation with the (*R*)-enantiomer (*22*). Stereoselectivity in Phase I or Phase II metabolism reactions have similarly been demonstrated under a variety of conditions for such drugs as propafenone, verapamil, nicotine, warfarin, fenofibrate, and terbutaline.

The possibility of stereospecific metabolism, coupled with the additional possibility of stereospecific pharmacologic and toxicologic effects, has given rise to a natural preference among pharmaceutical companies, and also the regulatory agencies with whom they interact, to develop achiral molecules whenever possible. It is less complicated, less expensive, and faster.

Product stereoselectivity occurs when a chiral center is introduced into a molecule as a result of metabolism. Typically, phenytoin is hydroxylated predominantly to the (*S*)-enantiomer, while the 9-ketocannabinoid nabilone is metabolically reduced to the (*S, S*)-enantiomer faster than to the (*R, R*)-enantiomer.

Metabolic reactions can also give rise to chiral inversion. The nonsteroidal anti-inflammatory agent ibuprofen, like most other drugs of this class, is marketed as a racemate although the activity resides mainly in the (+)-(*S*)-enantiomer. In the body, chiral inversion occurs from the (−)-(*R*)- to the (+)-(*S*)-enantiomer, both for metabolites and unchanged drug. The related agents fenoprofen and benoxaprofen undergo similar chiral inversion in humans. Flunoxaprofen and ketoprofen show no such inversion in humans, but do in rats, another interesting example of species differences in metabolism.

Enzyme Induction

The rate at which an experimental animal or human metabolizes a drug is not always constant. Several drugs can increase the activity of drug-metabolizing enzymes. These drugs include pentobarbital, phenobarbital, the carcinogenic hydrocarbon benzo[*a*]pyrene, steroid hormones, ethyl alcohol, the anticonvulsant carbamazepine, and nicotine. These drugs may also stimulate their own metabolism; this process is called *autoinduction*. For example, chronic dosing of the anticonvulsant methsuximide can lead to induction of oxidative demethylation to the active metabolite desmethylmethsuximide.

If a drug is cleared from the body in unchanged form, prior exposure to the drug or to enzyme inducers is unimportant. However, if a drug is metabolized it is important to avoid, or at least be cognizant of, the possibility of prior drug exposure when establishing baseline metabolic or pharmacokinetic patterns, or dose–response relationships.

Enzyme induction occurs primarily after repeated administration of an inducing agent and is associated with increased liver weight and protein content. Liver concentrations of induced cytochrome P-450, NADPH–cytochrome C reductase, and cytochrome b_5 also increase.

The ability of a drug to act as an enzyme inducer is generally determined during early preclinical studies by standard in vitro or in vivo tests. A drug that acts as an inducer in animals generally acts as an inducer in humans. Enzyme induction is usually rapidly apparent from pharmacokinetic observations, again during preclinical studies. The ability of a drug to induce and/or autoinduce is generally concentration dependent, so that effects observed in animals, often at high doses, may be greatly reduced in humans. In particular, autoinduction is often observed in preclinical toxicology studies, where doses are often many times greater than clinical doses, and it may not occur at all in a clinical situation. Each drug needs to be considered on a case-by-case basis. Enzyme induction is often associated with possible drug carcinogenic potential, but this relationship is still a matter of debate.

Enzyme Inhibition

Drug-metabolizing enzymes can also be inhibited, and these effects tend to be more acute and more immediate than induction. Inhibition appears to be the result of direct competition between compounds for drug-metabolizing enzymes. The compound β-diethylaminoethyldiphenylpropyl acetate (SKF-525-A, Chart 9.2) is a potent enzyme inhibitor. The compound has little pharmacologic effect of its own, but it has the remarkable property of competitively inhibiting the metabolism of a large number of drugs, thereby increasing their biological half-lives and duration of pharmacologic effect. The inhibitory effect of SKF-525-A is not limited to oxidation because azo reduction and nitro reduction, which require NADPH but not active oxygen, are also inhibited by this compound. SKF-525-A is thus a powerful pharmacologic tool for the study of drug metabolism, but it is not used clinically. Other compounds that inhibit drug-metabolizing enzymes include allopurinol, chloramphenicol, warfarin, and some monoamine oxidase inhibitors.

An interesting example of the use of drug inhibition in clinical practice is that of disulfiram for the treatment of alcoholism. This compound (Chart 9.2) also has no useful apparent pharmacologic activity alone but is capable of blocking alcohol oxidation at the aldehyde step. This blocking causes accumulation of acetaldehyde in the body after ethyl alcohol ingestion, which leads to a violently unpleasant syndrome including severe nausea and vomiting. Disulfiram is used under medical supervision to deter alcohol consumption by association.

Summary

Drug metabolism is a primary mechanism for removal of drugs from the body.

SKF-525-A

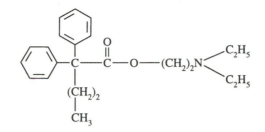

Disulfiram

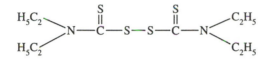

Structures of the metabolism enzyme inhibitors SKF-525-A and disulfiram.

CHART 9.2

░ Drugs may undergo Phase I or Phase II metabolism reactions depending on their structure.

░ The major site for drug metabolism is the microsomal fraction of the hepatocyte.

░ The cytochromes P-450 are a large family of isozymes with species and substrate specifically for drug metabolism.

░ Drug metabolism may vary among species and may be affected by age. Sex differences in drug metabolism in humans are minor and are generally considered not to be clinically important.

░ Drug metabolism may be influenced by steric factors both in terms of substrate and product specificity.

░ Metabolism can be induced or inhibited by various agents.

1. Williams, R. T. *Detoxication Mechanisms*; Wiley: New York, 1959.
2. Creasey, W. A. *Drug Disposition in Humans*; Oxford University: New York, 1979.
3. Hildebrandt, A.; Estabrook, R. W. *Arch. Biochem. Biophys.* **1977**, *143*, 66–69.
4. Guengerich, F. P. *Toxicol. Lett.* **1994**, *70*, 133–138.
5. Gonzalez, F. J.; Gelboin, H. V. *Drug. Metab. Rev.* **1994**, *26*, 165–183.
6. Caldwell, J. *Pharm. Forum* **1995**, *4*, 5–9.

References

7. Butler, M. A.; Lang, N. P.; Young, J. F.; Caporaso, N. E.; Vineis, P.; Hayes, R. B.; Teitel, C. H.; Massangill, P.; Lawsen, M. F.; Kadlubar, F. F. *Pharmacogenetics* **1992**, *2*, 116–127.

8. Kalow, W.; Tang, B-K. *Clin. Pharmacol. Ther.* **1991**, *49*, 44–48.

9. Goldstein, A.; Aranow, L.; Kalman, S. M. *Principles of Drug Action*; Harper Row: New York, 1969.

10. *Fundamentals of Drug Metabolism and Drug Disposition*; LaDu, B. N.; Mandell, H. G.; Way, E. L., Eds.; Williams and Wilkins: Baltimore, MD, 1971.

11. *Drug Metabolism in Man*; Gorrod, J. W.; LaDu, B. N., Eds.; Taylor Francis: London, 1978.

12. Jenner, P.; Testa, B. *Concepts in Drug Metabolism*; Marcel Dekker: New York, 1980.

13. Vesell, E. S. *Fed. Proc.* **1972**, *31*, 253–269.

14. Kalow, W.; Tang, B-K. *Clin. Pharmacol. Ther.* **1991**, *49*, 44–48.

15. Kalow, W.; Tang, B-K. *Clin. Pharmacol. Ther.* **1993**, *53*, 503–514.

16. Pelkonen, O.; Breimer, D. D. In *Pharmacokinetics of Drugs*; Welling P. G.; Balant, L. P., Eds.; Springer-Verlag: Heidelberg, Germany, 1994; pp 289–332.

17. Ward, R. M.; Singh, S.; Mirkin, B. L. T. In *Principles and Practices of Clinical Pharmacology*; Avery, G. S., Ed.; ADIS: New York, 1980; pp 76–96.

18. Weiss, C. F.; Glazko, A. J.; Weston, J. K. *N. Engl. J. Med.* **1960**, *262*, 787–794.

19. Testa, B. *Principles of Organic Stereochemistry*; Dekker: New York, 1979.

20. Testa, B.; Mayer, J. M. *Prog. Drug. Res.* **1988**, *32*, 249–303.

21. LeCorre, P.; Ratanasavanh, D.; Gibassier, D.; Barthel, A. M.; Sado, P.; LeVerge, R.; Guillouzo, A. *Colloq. INSERM.* **1988**, *164*, 321–324.

22. Riley, R. J.; Lambert, C.; Kitteringham, N. R.; Park, B. K. *Br. J. Clin. Pharmacol.* **1989**, *27*, 823–830.

Effect of Liver Disease on Drug Metabolism and Pharmacokinetics

10

Disease states affecting different organs may have a variety of effects on drug disposition and on pharmacologic action and pharmacodynamics. The two organs that are recognized as having the greatest influence on drug disposition, either directly or indirectly, are the liver and the kidney. Although most clinical studies carried out early in drug development are conducted in healthy subjects, it is recognized that renal or hepatic impairment may exert a considerable influence on both the disposition and the pharmacodynamic consequences of an administered drug (1).

The influence of renal insufficiency on drug disposition is well characterized in terms of markers of renal function, such as creatinine clearance, serum creatinine, and inulin clearance. In general, the impact of renal impairment on drug pharmacokinetics can be accurately predicted on the basis of the relative contributions of metabolism and renal excretion to overall drug clearance from the body. With few exceptions, the influence of renal failure on drug pharmacokinetics and pharmacodynamics is predictable and manageable in the clinical environment. The effect of renal failure on drug pharmacokinetics will be discussed in Chapter 12.

Unfortunately, such is not the case with liver disease. Despite a large number of studies, and attempts to establish guidelines for drug administration in patients with liver disease, it is still not possible to predict with any accuracy the effect of liver disease on the pharmacokinetics or pharmacodynamics of a particular drug. This is due not only to the overall structure and unique position of the liver in the cardiovascular system, but also to the heterogeneous nature of liver diseases and the different effects that they may have on the metabolism and distribution of particular drugs.

The purpose of this chapter is to review various approaches that have been used to better understand how liver impairment may affect drug disposition, to describe the effects of liver impairment on some classes of drugs, and to suggest guidelines for drug use in patients with liver failure.

Function and Structure of the Liver

As described in Chapter 9, the liver receives blood from both the splanchnic portal circulation and the hepatic arteries. It is this dual blood supply, one carrying absorbed substances from the gastrointestinal (GI) tract and the other arising from the systemic circulation, that places the liver in a unique position as an eliminating organ. By receiving portal blood, the liver sees substances before they enter the systemic circulation, and it can metabolize drugs with variable efficiency, influencing drug systemic availability. Blood reaching the liver via the hepatic arteries, on the other hand, carries substances that have already entered and been diluted into the systemic circulation so that, for those substances that reach the liver by this route, the liver acts purely as an eliminating organ.

As described in Chapter 9, drug metabolism reactions are conveniently differentiated into Phase I reactions, which involve oxidation, reduction, or hydrolysis, and Phase II reactions, which involve synthetic conjugation reactions. Phase I reactions are controlled predominantly by the cytochrome P-450 enzyme systems in the smooth endoplasmic reticulum located predominantly in the pericentral region of the liver. Reduction and hydrolysis reactions occur mainly in the cell cytoplasm. Phase II conjugation reactions occur at various sites in the liver. Glucuronidation, which is mediated by uridine diphosphate (UDP)-glucuronyl transferase enzymes, occurs in the rough endoplasmic reticulum (2). These enzymes occur throughout the liver, with highest concentrations in the periportal region. Other reactions, including sulfation, acetylation, and conjugation with glycine, glutamine, or glutathione, are generally thought to occur in the cytosolic fraction. The location of these various enzymes within the cell, and generally throughout the liver, plays an important role in the degree to which various types of liver disease affect their activity and, hence, drug metabolism and pharmacokinetics.

Types and Severity of Liver Disease

The function of the liver may be impaired by various conditions, such as nutritional status (3) and environmental factors (4), and by many disease states, such as acquired immunodeficiency syndrome (5), cirrhosis, acute hepatitis, biliary statis, hepatic neoplasm, and drug-induced liver disease. The results of liver function impairment may range from acute effects (6) to chronic changes associated with aging (7). However, the liver has marked excess capacity in most of its enzyme systems. For example, bilirubin elimination must fall below 10% of normal before jaundice develops. Similarly, in the normal liver, urea formation occurs at 60% capacity, while glucose maintenance requires only 20% of liver function, and albumin and clotting factors are synthesized by only a small percentage of liver cells at any one time (8). Excess liver capacity, together with the possible influence of a variety of enzyme-inducing substances such as alcohol and agricultural and industrial chemicals, makes it difficult to determine the extent of liver injury, which may manifest in terms of both liver function and various vascular changes.

Changes in Hepatic Function

Changes in hepatocyte function have been reported in liver disease, and such changes have been associated with what has been called the sick cell hypothesis (9). Hepatic enzyme levels are affected by liver cirrhosis but changes are not uniform among different enzyme classes. Liver disease is associated with changes in the hepatic sinusoids, including changes in the endothelial lining and in the deposition of mucopolysaccharides, collagen, and fibrin. These changes have been associated with an increased blood–hepatocyte barrier leading to hepatocellular necrosis (*10*). Other changes associated with hepatocyte function include development of diffuse fibrosis leading to resistance to blood flow.

Changes in Hepatic Vasculature

An alternative to the sick cell hypothesis for liver failure is the intact hepatocyte hypothesis (*11*). This hypothesis, similar to the intact nephron hypothesis associated with renal function, assumes that the diseased liver consists of regions of nonfunctional cells and other regions that are not involved at all in the disease process. As liver disease progresses, with associated fibrotic infiltration, there is altered vascular distribution with loss of tributaries of the portal vein, leading to diversion of blood flow within the liver. As a consequence of the resulting hepatic resistance, the portal venous pressure increases and extrahepatic collateral veins between portal and systemic circulations dilate to divert up to 60% of blood flow in the splanchnic and mesenteric veins away from the liver (*11*).

Although both the sick cell and intact hepatocyte theories are interesting alternatives for characterizing liver failure and its impact on drug metabolism and pharmacokinetics, neither one accurately predicts the impact of liver function on individual drugs. Such is the complex nature of liver failure and its relationship to drug disposition.

Changes in Renal Function

Renal blood flow may be reduced in cirrhosis. Glomerular function is variable in early cirrhosis but is usually decreased when ascites develops. Renal sodium retention is another important factor in the development of ascites. As liver disease progresses, these factors can lead to renal failure. Reduced renal clearance in patients with impaired liver function has been reported for several drugs, including furosemide, bumetanide, cimetidine, and ranitidine. However, renal elimination of drugs is not influenced by liver disease unless renal function itself is also impaired (*12*).

Ascites

Increased fluid retention by the kidneys and increased production of lymph both lead to the formation of ascites (*8*). If lymph production exceeds the capacity of lymph channels, lymph may enter the peritoneal cavity. This, together with sodium-related water retention by the kidneys, leads to the condition of peritoneal ascites. From a pharmacokinetic viewpoint, the condition of ascites may change drug disposition by increasing the overall drug distribution volume. This may be exacerbated by simultaneous reduction in drug binding to plasma proteins.

Effects on Pharmacodynamics and Pharmacokinetics

Pharmacodynamic Factors

Pharmacodynamic changes may occur in liver disease, and these may or may not be related directly to drug or metabolite dispositional changes. Because of the complexity of factors affecting drug disposition in liver failure, changes in pharmacodynamic response may occur at what appear to be "normal" therapeutic drug levels. Increased cerebral sensitivity to several compounds, including morphine, chlorpromazine, and the diazepines, has been demonstrated in patients with liver disease (*2*).

The three major mechanisms postulated to cause altered pharmacodynamic response in liver disease, particularly those associated with the central nervous system, are altered drug access to the site of action due to reduced binding to plasma proteins (*12*), altered blood–brain barrier permeability, and altered receptor response (*13*). Although increased response to benzodiazepines has been attributed to increased γ-aminobutyric acid (GABA) and benzodiazepine binding sites in hepatic encephalopathy (*14*), other studies have not confirmed these changes (*15*).

Pharmacokinetic Factors

Pharmacokinetic parameters most likely to be affected by impaired liver function are distribution, clearance, and elimination half-life. All of these may change in liver impairment, and observed effects on drug disposition may or may not be predictable. For drugs that are eliminated only by metabolism, equation 10.1 gives a quantitative relationship between hepatic clearance, Cl_H, intrinsic clearance of unbound or protein-free drug Cl'_{int}, hepatic blood flow, Q_H, and protein binding, β.

$$Cl_H = Q_H \frac{\beta Cl'_{int}}{Q_H + \beta Cl'_{int}} \qquad (10.1)$$

Equation 10.1 may be used to differentiate between those drugs that are highly or poorly extracted by the liver. From the equation, the hepatic clearance of highly extracted drugs ($Q_H << Cl'_{int}$) approaches hepatic blood flow, Q_H. The metabolism of such drugs is said to be hepatic blood flow limited, and hepatic clearance is relatively insensitive to changes in protein binding. On the other hand, the hepatic clearance of poorly extracted drugs ($Q_H >> Cl'_{int}$) may be approximated by β Cl'_{int}. For these drugs, not only is hepatic clearance relatively inefficient but it is also dependent on any changes in protein binding that may result from impaired liver function. The metabolism of such drugs is said to be enzyme capacity limited. Traditionally, those drugs whose hepatic extraction is greater than 70% are defined as flow limited, whereas those whose hepatic extraction is less than 30% are considered to be enzyme capacity limited. Many drugs have intermediate values and do not clearly fit into either of these two categories.

Enzyme-capacity-limited drugs can be further defined as those which are protein binding sensitive and those which are protein binding insensitive. Hepatic clearance for capacity-limited drugs that are extensively (>85%) bound to plasma proteins is likely to be significantly affected by changes in protein binding. The clearance of drugs that are only moderately bound, on the other hand, is likely to be relatively insensitive to changes in protein binding. As with the flow-rate- and capacity-limited classifications, many drugs have intermediate protein binding values and are likely to exhibit moderate sensitivity to changes in protein binding. Some drugs that fall into flow-limited, capacity-limited binding-sensitive, and capacity-limited binding-insensitive categories are given in Table 10.1 (*8*).

Although it is intellectually satisfying to classify drugs as in Table 10.1, the variable nature of liver disease, and also possible extrahepatic involvement, makes prediction from such classifications to individual drugs and patient situations difficult. For example, hepatic blood flow is likely to progressively decrease in cirrhosis, but it is likely to be unchanged or it may even increase in viral hepatitis. Similarly, hepatic cell mass is generally not affected in moderate cirrhosis and decreased in severe cirrhosis, and it may be unaffected, increased, or decreased in hepatitis. Intrinsic hepatocyte function, on the other hand, uniformly decreases in severe cirrhosis and in viral and alcoholic hepatitis (*16*).

Distribution Volume

The distribution volume relates the amount of drug in the body to the concentration of drug in plasma or serum. Most distribution volumes described in the literature are "apparent" in that they are calculated from concentrations of total drug (free plus protein bound) rather than drug that is unbound and free to equilibrate between plasma or serum and other body tissues and fluids. The fraction of drug that can penetrate into tissues is sensitive both to plasma protein binding and tissue binding, and also to body composition, all of which are likely to be affected by liver disease.

Elimination Half-life

The elimination half-life, $t_{1/2}$, of a drug is related to the distribution volume, V, and plasma clearance, Cl_p, by Equation 10.2.

$$t_{1/2} = \frac{V_d \cdot \ln 2}{Cl_p}$$

(10.2)

Equation 10.2 reflects the complex relationship between liver function and drug pharmacokinetics. Reduced intrinsic hepatic clearance is likely to directly prolong the half-life of a highly metabolized drug. On the other hand, particularly for capacity-limited protein-binding-sensitive drugs, reduced protein binding in liver disease would tend to increase hepatic extraction efficiency and thereby have the

Examples of Flow-Limited, Capacity-Limited Binding-Sensitive, and Capacity-Limited Binding-Insensitive Drugs

TABLE 10.1

Flow-Limited	Capacity-Limited	
	Binding-Sensitive	*Binding-Insensitive*
Acebutolol	Carbamazepine	Adriamycin
Chlormethiazole	Cefoperazone	Aminopyrine
Ketoconazole	Chloramphenicol	Amylobarbital
Labetolol	Chlordiazepoxide	Antipyrine
Lidocaine	Diazepam	Caffeine
Meperidine	Digitoxin	Cyclophosphamide
Metoprolol	Diphenylhydantoin	Hetobarbital
Morphine	Fenprofen	Isoniazid
Pentazocine	Ibuprofen	Phenacetin
Pethidine	Nafcillin	Phenobarbital
Propranolol	Prednisone	Pindolol
Propoxyphene	Tolbutamide	Theophylline
Tricyclic antidepressants	Valproic acid	
Verapamil	Warfarin	

Source: Adapted with permission from reference 8.

opposite effect, reducing the half-life. Reduced binding may also increase the drug distribution volume and hence the amount of drug that is free to distribute into tissues. This, in turn, will tend to increase the drug elimination half-life. These different factors are important in interpreting blood level data in liver disease (16).

Protein Binding

The majority of acidic drugs bind specifically to plasma albumin, whereas basic drugs tend to bind to both albumin and α_1-acid glycoproteins. Although synthesis of albumin is commonly decreased in liver disease, indirectly giving rise to possibly increased intrinsic hepatic extraction and also increased drug distribution volume, many acute conditions affecting the liver give rise to increased synthesis of acute-phase α_1-acid glycoproteins, which would have the opposite effect. In some liver diseases, such as cirrhosis, levels of α_1-acid glycoprotein may be reduced, giving rise to increased fraction of free drug (17).

Presystemic Clearance

As the liver receives blood from the regions of the upper and middle GI tract via the portal system, any drug absorbed from this region is liable to undergo first-pass hepatic clearance. Drugs that undergo extensive first-pass metabolism often exhibit low and variable bioavailability.

In the case of a high-extraction drug, only a small fraction of absorbed compound will likely reach the systemic circulation. On the other hand, for a drug with low intrinsic clearance a relatively large fraction will reach the systemic circulation. Regardless of whether a drug has high or low clearance, any reduction in intrinsic clearance will decrease the first-pass effect and give rise to increased systemic availability. The presence of extrahepatic and possibly also intrahepatic shunts in liver disease will also diminish the first-pass effect, resulting in increased oral systemic bioavailability. Increased systemic bioavailability of orally administered drugs in cirrhotic patients has been demonstrated for several compounds, including meperidine, pentazocine, propranolol, and chlormethiazole.

Markers of Liver Disease

No single diagnostic test adequately characterizes liver function and its relationship to drug pharmacokinetics. Some diagnostic liver function tests involve monitoring bile pigments, serum bile acids, bilirubin, albumin, ferritin, lipids, ammonia, and the enzymes alkaline phosphatase (ALP), aspartate aminotransferase (AST), alanine aminotransferase (ALT), lactate dehydrogenase (LDH), and γ-glutamyltransferase (GGT) (18). Other tests use contrast cholangiography, radiology, and nuclear medicine.

Although all of these tests provide information regarding cellular function or vascular status, they are not generally useful for predicting how liver failure may affect the metabolism or pharmacokinetics of particular drugs. Some studies have demonstrated significant correlations between liver function tests and drug metabolism, whereas others have found no relationships.

Good correlations have been obtained among elimination rate of amylobarbital, procainamide, and phenylbutazone and serum albumin levels. Similarly, procainamide and antipyrine elimination half-lives vary directly with prothrombin activity, and negative correlations have been obtained between the clearance of several compounds, including antipyrine and clindamycin, and serum bilirubin, and between clindamycin clearance and SGOT (serum glutamic–oxaloacetic transaminase) values. On the other hand, no correlations were obtained among clearance of diazepam, chlordiazepoxide, and meperidine and conventional liver function tests. Thus, correlations among conventional liver diagnostic tests and drug metabolism and pharmacokinetics are generally unreliable.

The clearances of compounds such as antipyrine, aminopyrine, phenacetin, and caffeine have been examined as indices of drug metabolism capacity. With antipyrine, the rate of metabolism is monitored from disappearance of drug from plasma as a result of multiple metabolic pathways. With aminopyrine, phenacetin, and caffeine, on the other hand, procedures are available to determine metabolic efficiency by measuring exhaled $^{13}CO_2$ or $^{14}CO_2$ in the breath resulting from metabolic demethylation of administered radioactively labeled compound (19, 20).

When adjusting drug doses in patients with liver impairment, it is necessary to know the extent to which hepatic metabolism contributes to drug elimination and also the possible pharmacodynamic consequences of impaired liver function.

Examples of Effects of Liver Disease on the Pharmacokinetics of Some Drug Classes

Cardiovascular Agents

Digitoxin is cleared from the body mainly by hepatic metabolism, whereas digoxin is cleared principally unchanged in the urine. Consistent with these facts, digitoxin elimination is impaired in liver failure but digoxin elimination is unchanged. The clearance of lidocaine is reduced in chronic liver disease, and clearance of d-propranolol is reduced in cirrhosis. Both of these drugs are highly extracted by the liver so that reduced clearances may be associated with reduction in hepatic blood flow in hepatic disease, although reduced intrinsic clearance may also contribute. The free fraction of circulating d-propranolol increases in liver cirrhosis, and this may compensate, at least in part, for decreased intrinsic clearance but may also give rise to increased central nervous system activity in cirrhotic patients. The oral bioavailability of d-propranolol is increased approximately 40% in cirrhotic patients compared to normal controls. Because of their decreased clearance, doses of lidocaine and d-propranolol should be reduced in patients with liver impairment.

Disopyramide clearance is decreased in cirrhosis, leading to a recommendation for dose reduction, and the elimination half-life of procainamide is increased. Quinidine elimination half-life is also increased and protein binding is decreased in cirrhosis, leading to a recommendation that the drug dose should be lowered by 70% with drug level monitoring in severe liver disease.

Elimination of encainide, lorcainide, and flecainide, all extensively metabolized by the liver, is reduced in liver impairment. Although doses of lorcainide and flecainide should be reduced in liver impairment (21, 22), dose adjustment of encainide is probably not necessary because circulating concentrations of the active metabolite are correspondingly lower. Clearance of mexiletine and the β-receptor antagonists propranolol, metoprolol, and labetol is decreased in cirrhosis (23). Pindolol and bisoprolol are metabolized in the liver and the kidneys, so that dose adjustment in patients with liver impairment is probably unnecessary. Atenolol and sotalol are metabolized predominantly in the kidneys, so that dose adjustment for these agents is probably also unnecessary in liver impairment.

However, atenolol has been reported to cause renal insufficiency in patients with chronic liver disease (*24*).

The calcium channel blocking agents appear to be uniformly affected by liver impairment. Clearance of verapamil, nifedipine, diltiazem, nicardipine, nisoldipine, nitrendipine, isradipine, amlodipine, and felodipine is reduced in liver impairment, and dose reduction of these drugs is warranted (*25*). The angiotensin-converting enzyme inhibitors, on the other hand, are relatively unaffected by impaired liver function. Only small pharmacokinetic changes have been reported for captopril, enalopril, and lisinopril. In cases of ascites, captopril has been claimed to impair sodium excretion, thus contraindicating this agent in patients with cirrhosis and ascites (*26*).

Drugs Acting on the Central Nervous System

Among the benzodiazepines, the presence of acute viral hepatitis or cirrhosis reduces drug clearance of diazepam and chlordiazepoxide by approximately 50%. These changes may be accompanied by changes in distribution volume, but they are nonetheless associated with increased elimination half-lives (*27*). The influence of liver impairment on the clearance and elimination half-life of lorazepam is variable, whereas the clearance and half-life of oxazepam is unaffected by either cirrhosis or acute viral hepatitis. Oxazepam may be the benzodiazepine of choice in patients with hepatic impairment (*28*). Hepatic impairment may substantially impair the capacity of the liver to eliminate barbiturates and narcotic analgesics. In patients with compensated or decompensated cirrhosis or with acute viral hepatitis, the clearance of hexobarbital is reduced by 50%, and pentobarbital half-life is increased moderately in patients with cirrhosis and hepatic neoplasm but is not affected in patients with acute viral hepatitis.

Narcotic analgesics may cause excessive sedation in cirrhosis and may also induce or exacerbate encephalopathy. Oral bioavailability of pethidine and pentazocine is increased in cirrhosis, and systemic clearance of pethidine is decreased by 50% in acute or chronic hepatic disease (*29*). Both oral and parenteral doses of these agents should be reduced in liver impairment. Although morphine pharmacokinetics are unchanged in mild liver disease because of extensive extrahepatic metabolism, excessive sedation nonetheless occurs in cirrhosis. In a study in cirrhotic patients, systemic clearance of morphine was unchanged relative to controls, probably because of the extrahepatic conjugation or protected intrahepatic conjugation (*30*). Clearance of phenytoin and valproate is decreased in cirrhosis and hepatitis.

Antimicrobial Agents

Many antibiotics are water soluble and are excreted for the most part unchanged in urine; hence, they are not susceptible to changes in liver function impairment unless there are secondary changes in renal function. The penicillins, ampicillin and carbenicillin, require no dose adjustment in liver disease unless there is associated renal impairment. On the other hand, the broad spectrum penicillins and mezlocillin require dose reduction in cirrhosis (*31*).

Many of the first- and second-generation cephalosporins are excreted mainly unchanged in urine and, except in cases of renal involvement, can be dosed regardless of hepatic function. Among the third-generation cephalosporins, ceftriaxone, cefotaxime, and cefoperazone, elimination involves liver function, and clearance of these agents may be reduced in patients with liver disease.

The macrolide antibiotics erythromycin and roxithromycin do not require dose adjustment in compensated liver disease, but changes in binding of erythromycin to serum α_1-acid glycoprotein may influence its pharmacokinetics in liver impairment (32). Vancoymcin clearance is reduced and its elimination half-life prolonged in liver disease. A 50% dose reduction is recommended (33). The pharmacokinetics of the antituberculosis agents ethambutol, streptomycin, and cycloserine is unaffected by liver disease in the absence of renal involvement. Isoniazid acetylation is affected only in cases of severe liver impairment.

Other Drugs

The impact of liver disease on drugs in other therapeutic areas is similarly variable and unpredictable. Liver disease diminishes the clearance or prolongs elimination half-lives of antipyrine, paracetamol, theophylline, and heparin (28). Among the anti-inflammatory agents, liver disease has little effect on conversion of prednisone to the active form prednisolone. However, increased side effects have been reported from both prednisone and prednisolone in liver failure, and dose reduction is recommended (34). Cyclosporin elimination is delayed in moderate liver impairment, but its elimination is variable in primary biliary cirrhosis. Frequent monitoring of cyclosporin blood levels is recommended in patients with liver disease. Tolbutamide metabolism is not altered in acute viral hepatitis. It has been suggested that acute hepatic disease may in fact accelerate elimination of tolbutamide because of reduced protein binding (35). However, in another report the elimination half-life of tolbutamide was doubled in patients with liver cirrhosis (36).

Conclusions

Relationships between impaired liver function and the pharmacokinetic and pharmacodynamic activity of therapeutic agents are complex. The complexity results from many contributing factors, including the unique position of the liver in the cardiovascular system, the different types and severity of liver disease, the degree of involvement of hepatic metabolism and biliary excretion in drug elimination, the degree of first-pass hepatic clearance of orally administered compounds, the degree of extrahepatic involvement, metabolic pathways, and pharmacodynamic contributions. It is perhaps not surprising that it is still extremely difficult to predict the effect of liver impairment on drug disposition in a patient or patient population. Too little is known regarding the relative contributions of altered hepatic blood flow, reduced intrinsic cellular metabolism activity, and the relative contributions of reduced clearance, altered binding to plasma proteins, and altered distribution volume to accurately predict pharmacokinetic and pharmacodynamic effects together with appropriate dosage regimen changes.

Current knowledge nonetheless permits some generalizations. If a drug is cleared from the body by metabolism, then its clearance is likely to be reduced in liver disease and the effect will increase with decreasing liver function. Oral availability of drugs that are subject to high hepatic clearance is likely to be increased in patients with liver disease. Similarly, binding of highly protein bound drugs will tend to decrease, which may compensate for reduced intrinsic clearance. However, reduced binding may increase drug distribution volume, which will have the opposite effect. The presence of ascites, and possibly also encephalopathy, in liver failure further complicates prediction of pharmacokinetic changes and pharmacokinetic–pharmacodynamic relationships. The use of exogenous substances as markers has been only moderately successful in predicting the effect of liver impairment on drug metabolism and pharmacokinetics. The high current interest in breath tests using radiolabeled markers may improve pharmacokinetic prediction in the future.

Summary

Liver disease may vary in terms of type and severity.

Liver disease may affect intrinsic hepatic function and also hepatic and extrahepatic vasculature.

Reduced renal function and ascites are often secondary to impaired liver function.

Liver disease may affect both pharmacokinetic and pharmacodynamic factors.

Correlations between markers of liver disease and pharmacokinetic effects are variable, and often poor.

Representative data from therapeutic drug classes illustrate the complex and unpredictable relationship between liver function and drug pharmacokinetics.

References

1. Welling, P. G.; Pool, W. F. In *Drug Induced Hepatotoxicity: A Handbook of Experimental Pharmacology*; Cameron, R. G.; Feuer, G.; de la Iglesia, F., Eds.; Springer-Verlag: Berlin, 1995.
2. Secor, J. W.; Schenker, S. *Adv. Intern. Med.* **1987**, *32*, 379–406.
3. Hathcock, J. N. *Drug–Nutrient Interactions* **1985**, *4*, 217–234.
4. O'Mahony, M. S.; Woodhouse, K. W. *Pharmac. Ther.* **1994**, *61*, 279–287.
5. Lee, B. L.; Wong, D.; Benowitz, N. L.; Sullam, P. M. *Clin. Pharmacol. Ther.* **1993**, *53*, 529–535.
6. Lewis, J. H.; Zimmerman, H. J. *Med. Clin. North Am.* **1989**, *73*, 775–792.
7. Mooney, H.; Roberts, R.; Cooksley, W. G. E.; Halliday, J. W.; Powell, L. W. *Clin. Gastroenterol.* **1985**, *14*, 757–771.
8. Arns, P. A.; Wedlund, P. J.; Branch, R. A. In *The Pharmacologic Approach to the Critically Ill Patient*; Chernow, B., Ed.; Williams and Wilkins: Baltimore, MD, 1988; pp 85–111.
9. McLean, A. J.; Morgan, D. J. *Clin. Pharmacokinet.* **1991**, *21*, 42–69.
10. Sotaniemi, E. A.; Niemelä, O.; Risteli, L.; Stenbäck, F.; Pelkonen, R. O.; Lahtela, J. T.; Risteli, J. *Clin. Pharmacol. Ther.* **1986**, *40*, 46–55.
11. Branch, R. A.; Shand, D. G. *Clin. Pharmacokinet.* **1976**, *1*, 264–279.
12. Rowland, M.; Tozer, T. N. *Clinical Pharmacokinetics: Concepts and Applications*; Lea Febiger: Philadelphia, PA, 1985; pp 189–190.
13. McConnell, J. B.; Curry, S. H., Davies, M.; Williams, R. *Clin. Sci.* **1982**, *63*, 75–80.
14. Zeneroli, M. L. *J. Hepatol.* **1985**, *1*, 301–312.

15. Butterworth, R. F.; Lovoie, J.; Giguere, J. F.; Pomier-Layrargues, G. *Hepatolog.* **1988**, *8*, 1084–1088.

16. Blaschke, T. F. *Clin. Pharmacokinet.* **1977**, *2*, 32–44.

17. Barre, J.; Houin, G.; Brunner, F.; Bree, F.; Tillement, J-P. *Int. J. Clin. Pharm. Res.* **1983**, *3*, 215–226.

18. Chopra, S.; Griffin, P. H. *Am. J. Med.* **1985**, *79*, 221–230.

19. Vesell, E. S. *Ann. N. Y. Acad. Sci.* **1984**, *428*, 293–307.

20. Goldberg, D. M.; Brown, D. *Clin. Biochem.* **1987**, *20*, 127–148.

21. Klotz, U.; Fischer, C.; Müller-Seydlitz, P.; Schultz, J.; Müller, W. A. *Clin. Pharmacol. Ther.* **1979**, *26*, 221–227.

22. McQuinn, R. L.; Pentikäinen, P. J.; Chang, S. F.; Conrad, G. J. *Clin. Pharmacol. Ther.* **1988**, *44*, 566–572.

23. Watson, R. G. P.; Bastain, W.; Larkin, K. A.; Hayes, J. R.; McAinsh, J. A.; Shanks, R. G. *Br. J. Clin. Pharmacol.* **1987**, *24*, 527–535.

24. Kirch, W.; Rose, I.; Demers, H. G.; Leopold, G.; Pabst, J.; Ohnhaus, E. E. *Clin. Pharmacokinet.* **1987**, *13*, 110–117.

25. Siersema, P. D.; Wilson, J. H. P. *Eur. J. Int. Med.* **1992**, *3*, 197–212.

26. Daskalopoulos, G.; Pinzani, M.; Murray, N.; Hirschberg, R.; Zipser, R. D. *J. Hepatol.* **1987**, *4*, 330–336.

27. Roberts, R. K.; Wilkinson, G. R.; Branch, R. A.; Schenker, S. *Gastroenterology,* **1978**, *75*, 479–485.

28. Williams, R. L.; Mamelok, R. D. *Clin. Pharmacokinet.* **1980**, *5*, 528–547.

29. Pond, S. M.; Tong, T.; Benowitz, N. L.; Jacob, P. *Aust. N. Z. J. Med.* **1980**, *10*, 515–519.

30. Patwardhan, R.; Johnson, R.; Sheehan, J.; Desmond, P.; Wilkinson, G.; Hoyumpa, A.; Branch, R.; Schenker, S. *Gastroenterol* **1981**, *8*, 1344.

31. Bunke, C. M.; Aronoff, G. R.; Brier, M. E.; Sloan, R. S.; Luft, F. C. *Clin. Pharmacol. Ther.* **1983**, *33*, 73–76.

32. Barre, J.; Mallat, A.; Rosenbaum, J.; Deforges, L.; Houin, G.; Dhumeaux, D.; Tillement, J-P. *Br. J. Clin. Pharmacol.* **1987**, *23*, 753–757.

33. Brown, N.; Ho, D. H. W.; Fong, K. L.; Bogerd, L.; Maksymiuk, A.; Bolivar, R.; Fainstein, V.; Bodey, G. P. *Antimicrob. Agents Chemother.* **1983**, *23*, 603–609.

34. Renner, E.; Horber, F. F.; Jost, G.; Frey, B. M.; Frey, F. J. *Gastroenterol* **1986**, *90*, 819–828.

35. Williams, R. L.; Blaschke, T. F.; Meffin, P. J.; Melmon, K. L.; Rowland, M. *Clin. Pharmacol. Ther.* **1977**, *21*, 301–309.

36. Ueda, H.; Sakurai, T.; Ota, M.; Nakajima, A.; Kamii, K.; Maezawa, H. *Diabetes* **1963**, *12*, 414–419.

Renal Excretion

hapters 9 and 10 described how drugs and other foreign substances can be chemically altered in the body, usually to less active and more water soluble derivatives. Whether a drug undergoes metabolism or not, either unchanged drug or metabolites must eventually be removed from the body. Although several nonrenal routes may be involved, such as the lungs, saliva, sweat, bile, and direct excretion into the gastrointestinal (GI) tract, the major excretory pathway for most substances is via the kidneys. Even when a compound is extensively metabolized in the liver or elsewhere, the metabolites are usually voided via the kidneys.

Just as the liver is uniquely situated in the cardiovascular system to play a dual role contributing to both presystemic and systemic clearance, so are the kidneys ideally situated and constructed as the major organs of drug and metabolite elimination.

The human kidneys are two bean-shaped organs situated at the back of the abdominal cavity, each embedded in a mass of fatty tissue, the adipose capsule, and further protected by a sheath of fibrous tissue, the renal fascia. Each human kidney is about 11 cm long, 5.0–7.5 cm wide, 2.5 cm thick, and weighs between 120 and 160 g. The kidney comprises an outer cortex and an inner medulla. The major functional unit of the kidney is the nephron, which consists of a renal tubule and its blood supply. The nephron, of which there are a million or more in each human kidney, is shown diagrammatically in Figure 11.1 (1).

The nephron begins as a closed, invaginated layer of epithelium, the renal capsule or Bowman's capsule, named after Sir William Bowman, the English anatomist and ophthalmologist. The inner layer of the globe-like expansion adjoins a capillary tuft, or glomerulus. The glomerulus consists of capillary loops that are completely contained within Bowman's capsule except at the point where afferent

Structure and Function of the Kidney

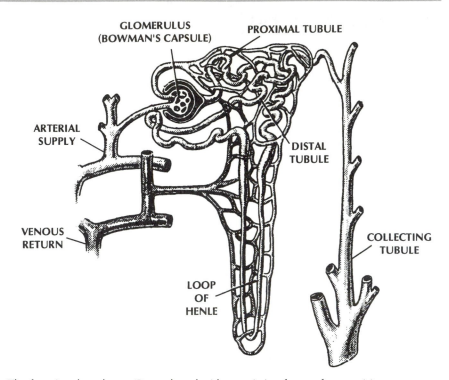

FIGURE 11.1   The functional nephron. (Reproduced with permission from reference 1.)

and efferent arterioles enter and leave the capillary tuft. After a neck, or constriction, below the capsule the tubule forms the proximal convoluted tubule followed by the loop of Henle and distal convoluted tubule, which opens into the collecting ducts. These in turn converge into the calyces of the kidney through which urine is collected to pass into the ureter and hence to the bladder. Bowman's capsule and the proximal and distal convoluted tubules are located in the cortex of the kidney, whereas the ascending and descending limbs of the loop of Henle are located in the outer or inner medulla. The initial collecting tubules are located in the renal cortex but pass from the cortex through the medulla as they converge into the renal calices (2).

Renal blood flow rate is about 1.2 L/min or 1730 L/day. This represents 25% of total cardiac output. Blood enters the kidneys via the renal arteries from the abdominal aorta. Before or immediately after entering the kidney at the hilum, each artery divides into several branches, which enter the renal parenchyma separately. When these arteries reach the boundary between cortex and medulla, they divide laterally and form the arch, or arcuate arteries. From these arches, the interlobular arteries enter the cortex, giving off at intervals minute afferent arterioles, which branch out as capillaries of a glomerulus. These capillaries reunite to form efferent arterioles. The efferent arterioles break up in a close meshwork, or plexus of capillaries, which are in close proximity to both the convoluted tubule in the cortex and the loop of Henle in the medulla. These capillaries unite to form inter-lobular veins and medullary veins, which pour their contents into the arcuate veins lying between the cortex and the medulla. The arcuate veins converge to form the interlobular veins. These merge into the renal vein, which leaves the kidney and opens into the inferior vena cava (2).

The 1.2 L of blood that perfuses the kidneys each minute contains about 650 mL of plasma water. Of this plasma water, about 130 mL is filtered at the glomerulus. This filtration results in about 170–190 L of plasma water being filtered each day. Of this volume, approximately 1.5 L of fluid is excreted as urine each day, the remainder being absorbed back into the circulation. Thus, the kidneys receive a large proportion of the circulation and have developed efficient methods to filter and secrete substances and to excrete them in a concentrated urine. Each part of the nephron is so constructed and located within the kidney to play a unique role regulating the elimination of substances from the body and maintaining homeostasis.

Bowman's capsule is concerned with filtration of plasma water and its contents into the tubule. This filtration is passive and relatively nonspecific because it is not influenced by fat solubility, ionization, or, for the most part, molecular size. An upper limit to molecular size exists because proteins, particularly soluble plasma proteins, are not filtered at the glomerulus. Therefore, any drug that is protein-bound cannot be filtered directly at the glomerulus.

The proximal tubule is concerned principally with reabsorption of sodium chloride and water. However, from a pharmacokinetic viewpoint, the most important function of the proximal tubule is active secretion of various drugs from the plasma via the capillaries adjacent to the nephron into the kidney tubule. This second mechanism of drug clearance from the body is more drug-specific than filtration. Secretion is an active, energy-consuming process, and different mechanisms exist for secreting acids (or anions) and bases (or cations), including quaternary ammonium compounds. Although drugs that are bound to plasma proteins cannot be actively secreted, the process of secretion is so rapid that bound drug may dissociate from protein and be secreted during a single pass through the kidney. Secretion may be so efficient that virtually all of the drug in blood may be removed during a single pass through the kidney regardless of whether drug is bound to plasma proteins or even located in red blood cells. Apart from active drug secretion, reabsorption may also occur in the proximal tubule. In fact, reabsorption occurs all along the nephron. In some cases reabsorption may be an active process, such as the reabsorption of glucose and probenecid. But for a vast majority of drugs and other foreign substances, reabsorption is passive and, like most other passive membrane-transport processes, favors un-ionized, fat-soluble compounds.

Henle's loop is a fine structure that passes from the renal cortex into the medulla and then returns to the cortex. This loop is concerned with sodium chloride and water reabsorption and is important in producing a concentrated, hypertonic urine, but it does not appear to play a major role in drug excretion.

In the distal tubules, acidification of urine (pH ~ 6) and further sodium chloride and water reabsorption occur. However, the most important activity in the distal tubules with respect to the rate and efficiency of drug excretion is reabsorption. The collecting ducts are concerned with water reabsorption and act under the control of the pituitary hormone vasopressin. Collecting ducts do not appear to play a major role in drug elimination (3, 4).

From a pharmacokinetic viewpoint, each of the three major sites in the nephron is associated with particular functions related to drug excretion:

1. Passive filtration at the glomerulus, which is restricted to drug that is unbound to plasma proteins.

2. Active secretion at the proximal kidney tubule, which is essentially independent of protein binding but dependent upon molecular structure.

3. Passive reabsorption throughout the nephron, which is restricted, for the most part, to un-ionized, fat-soluble compounds.

Clearance

Renal clearance can be determined by means of specific marker compounds. For example, creatinine, a substance that is produced in the body from muscle metabolism, and also inulin, a synthetic polymer hydrocarbon, are both filtered at the glomerulus, but are neither secreted nor reabsorbed in the tubules. Approximately 130 mL of plasma water is filtered by the kidneys, and is therefore efficiently cleared of creatinine and inulin (regardless of concentration), during each minute. Thus, the *plasma clearance*, or the volume of plasma that is cleared of creatinine and inulin and of any other substance that is eliminated by this mechanism per unit time, is equal to the glomerular filtration rate (GFR), which is approximately 130 mL/min.

Other compounds, for example, *p*-aminohippurate (PAH) and some penicillins and cephalosporins, are not only filtered at the glomerulus, but are also actively secreted at the proximal convoluted tubule. Because this process is efficient, all of the drug may be removed from the blood or plasma during a single pass through the kidney, and the clearance is equal to the renal plasma flow, which is 650 mL/min. Thus, PAH is typically used to measure effective renal plasma flow (ERPF) or effective renal blood flow (ERBF).

Glucose is filtered at the glomerulus and is not secreted in the proximal tubule, but is normally completely reabsorbed at the distal tubule. Glucose clearance under normal circumstances is essentially zero. However, reabsorption of glucose is an active process, and there is a threshold to the reabsorption rate that may be exceeded in cases of glucose overload. For example, glucose may be passed in urine in the condition of diabetes. Passive reabsorption, on the other hand, is not saturable, and the requirement of fat solubility for reabsorption to occur is the essence of the relationship between drug metabolism (producing a more water-soluble derivative) and renal excretion in removing substances from the body.

The semipermeable nature of the membranes of kidney tubules can be used to treat such conditions as drug overdose. For example, acidification of urine by administering ammonium chloride will favor reabsorption of acidic drugs but will inhibit reabsorption of basic drugs that will be more ionized at the acidic pH. On the other hand, basification of urine by administering sodium bicarbonate will favor reabsorption of basic drugs but will inhibit reabsorption of acidic drugs that will be more ionized at the basic pH. Therefore, a useful approach in cases of drug overdose is to make the urine acidic for basic drugs and to make the urine alkaline for acidic drugs to facilitate their removal from the body. This approach may be used in conjunction with diuresis to increase renal excretion of compounds.

Although protein-bound drugs or metabolites may be actively secreted in the proximal tubules, these drugs will not filter at the glomerulus. If this is the case, why is it that protein-bound drugs that are not secreted do not stay in the blood for a very long time? The answer is that such drugs may be voided quite rapidly in urine because of the continuous nature of kidney perfusion by circulating blood. During a 1-min period in a healthy kidney, 130 mL of plasma water, out of a total renal plasma flow of 650 mL, will be filtered. Because this process is simple filtration, the concentration of unbound drug in the unfiltered plasma water does not change, and the bound–unbound drug equilibrium is not perturbed. No further drug will dissociate from protein. However, of the 130 mL of plasma water that is filtered each minute, approximately 128 mL is reabsorbed from the loop of Henle, distal tubule, and collecting ducts. If the water is reabsorbed and the filtered drug, or any portion of filtered drug, remains in the tubule because it cannot be reabsorbed, then the effective concentration of free drug in plasma will be reduced, and protein-bound drug will dissociate to retain the bound–unbound equilibrium. Thus, although the percentage of binding remains constant, as is normally the case, the concentration of total drug will continually drop with each pass through the kidney until all drug is eventually removed.

Most compounds that are eliminated from the body by the kidneys undergo glomerular filtration. Active secretion by the kidney is less common. Some acids and bases that undergo active renal tubular secretion are given in Table 11.1.

Although large proteins such as albumin (MW 69,000) do not filter at the glomerulus, small peptides do. Once these peptides enter the kidney tubule, they may be reabsorbed by means of a peptide transporter that resides in the proximal tubular cells. This transporter is specific for peptides consisting of 2–4 amino acids. It can transport peptides against a concentration gradient, and it is coupled to a proton gradient (5–7). Current evidence indicates that coupling occurs between peptides and histidyl and thiol groups on the transporter.

Renal Clearance and Plasma Clearance

The terms *renal clearance* and *plasma clearance* need to be differentiated. These terms are sometimes used interchangeably, but they describe two entirely different phenomena. The clearances considered so far have been renal clearances. Renal clearance describes the volume of plasma cleared of drug per unit time as a result of urinary excretion. Renal clearance is determined from the relationship between the amount of compound voided in urine during a given time and the mean concentration of compound in plasma during the time. Renal clearance can be calculated by using equation 11.1.

$$Cl_r = \frac{UV'}{C} \qquad (11.1)$$

where Cl_r is renal clearance in milliliters of plasma per minute, U is the concentration of drug in urine, V' is urine flow rate, and C is the mean concentration of drug in plasma during the urine collection interval. The symbol V' is used to differentiate urine volume from the drug distribution volume, V.

The numerator in equation 11.1 represents the quantity of drug eliminated in urine in units of mass per unit time. Dividing the numerator by the concentration, C, yields the renal clearance in units of volume per unit time. The volume referred to in renal clearance is volume of plasma. Ideally, the value of C is held constant during the entire urine collection interval, and this constant value may be achieved by using an endogenous substance such as creatinine, or by continuous infusion of an exogenous marker substance such as inulin.

Clearance may also be calculated during a period of changing plasma levels, for example, following bolus intravenous injection of a drug. This method of calculating clearance is frequently used out of sheer necessity, but it may yield unreliable values because of bladder holdup, changing tissue distribution during the collection interval, and various other factors. Clearances should be measured several times through the use of sequential urine collection intervals, and the results should be averaged to provide a best estimate.

To determine plasma clearance, a model must be proposed that differentiates between drug voided in urine and drug eliminated by other routes. A typical situation is one in which drug is removed from the body both by urinary excretion of unchanged drug and by hepatic metabolism. If both phenomena are nonsaturable and therefore first-order in nature, then first-order rate constants k_e and k_m can be assigned to urinary excretion of unchanged drug and drug metabolism, respectively. If the arithmetic sum of k_e and k_m equals k_{el}, which is the rate constant for loss of drug from plasma by all elimination routes, then the model for drug loss

---

TABLE 11.1   **Drugs That Undergo Active Renal Tubular Secretion**

Acids	Bases
Acetazolamide	Dihydromorphine
Cephalosporins	Dopamine
Chlorpropamide	Hydrazine
Chlorothiazide	Mepiperphenidol
Ethacrynic acid	Methyguanidine
Furosemide	Morphine
Hydrochlorothiazide	Neostigmine
Indomethacin	Piperidine
Methotrexate	Procaine
Oxalic acid	Quinidine
Penicillin	Quinine
Probenecid	Serotonin
Salicylic acid	Thiamine
Spironolactone	
Sulfonamides	

can be shown as in Scheme 11.1. By using this model, the rate of drug loss from the body due to both urinary excretion and metabolism can be predicted by using equation 11.2, and the rate of drug loss due to urinary excretion alone can be predicted from equation 11.3.

$$-\left(\frac{dA}{dt_{total}}\right) = k_{el}A = k_{el}CV \tag{11.2}$$

$$-\left(\frac{dA}{dt_{renal}}\right) = k_e A = k_e CV \tag{11.3}$$

Comparing equations 11.1 and 11.3 and rearranging yield equation 11.4.

$$k_e CV = UV' \tag{11.4}$$

Dividing both sides of equation 11.4 by the plasma concentration term, C, yields equation 11.5.

$$k_e V = \frac{UV'}{C} \tag{11.5}$$

This equation provides an expression for the renal clearance, UV'/C, in terms of the first-order urinary excretion constant, k_e, and the drug distribution volume in the body, V. However, the term $k_e V$ describes only the fraction of plasma clearance due to renal excretion. If the drug is not metabolized and is eliminated entirely as unchanged drug, then $k_m = 0$ because no metabolism occurs, $k_{el} = k_e$, and equation 11.5 can be rewritten as equation 11.6.

$$k_{el}V = \frac{UV'}{C} = Cl_p \tag{11.6}$$

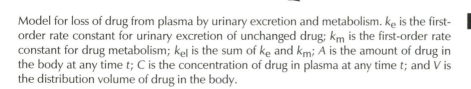

Model for loss of drug from plasma by urinary excretion and metabolism. k_e is the first-order rate constant for urinary excretion of unchanged drug; k_m is the first-order rate constant for drug metabolism; k_{el} is the sum of k_e and k_m; A is the amount of drug in the body at any time t; C is the concentration of drug in plasma at any time t; and V is the distribution volume of drug in the body.

SCHEME 11.1

However, if metabolism or any other nonrenal elimination pathway occurs, then $k_{el} > k_e$, because of the contribution of k_m, and the true plasma clearance, $k_{el} V$, will be greater than the renal clearance, UV'/C.

Thus renal clearance, which is due to the renal elimination of unchanged drug, is distinguished from plasma clearance, which is due to all elimination processes. Metabolism clearance, frequently called hepatic clearance, can similarly be described using the product $k_m V$. Clearance by other organs can also be described simply by substituting the appropriate rate constant. However, the total plasma clearance, $k_{el} V$, will always be equal to or greater than the renal clearance or any other organ clearance.

Another factor often complicates accurate interpretation of clearance values. Both equations 11.1 and 11.2 contain the term C, which represents the concentration of drug in plasma. Because drugs bind to plasma proteins to varying degrees, the question arises as to whether the value of C should represent the concentration of total drug or the concentration of free drug. In other words, should the clearance value be uncorrected, or should the clearance value be corrected for plasma-protein binding?

This question is difficult to answer because clearance depends on the precise mechanism for drug elimination, and the mechanism is frequently unknown. For example, if a drug has a renal clearance of 130 mL/min, this value could arise because the drug is removed from the plasma solely by glomerular filtration, in which case protein-bound drug plays no part in the elimination process, or because the drug is actively secreted in the proximal tubules but is subsequently partially reabsorbed at the distal tubules, in which case protein-bound drug does play a role in the elimination process. If the actual mechanism of elimination can be defined, then a reasonable rule of thumb is that if protein-bound drug does not participate in the elimination process, then the corrected clearance should be used. In other words, the concentration of unbound drug goes into the denominator of equation 11.1 and into the right side of equation 11.2. On the other hand, if protein-bound drug does participate in the elimination process, then the uncorrected clearance should be used, and the concentration of total drug should be used in the equations.

Relationship Among Clearance, Drug Elimination Rate, and Half-Life

By definition, the biological or *plasma half-life* of a drug is the time taken for the drug concentration in plasma to be reduced by one-half. In all first-order processes, the half-life is a constant and is related to the first-order rate constant, k_{el}, as in equation 11.7.

$$t_{1/2} = \frac{\ln 2}{k_{el}} = \frac{0.693}{k_{el}} \tag{11.7}$$

Thus, if a drug has an elimination half-life of 1 h, then the drug concentration in plasma will be reduced by one-half each hour. If, at time 0, the concentration is 100 μg/mL, then at 1 h the concentration will be reduced to 50 μg/mL, at 2 h to 25 μg/mL, and at 3 h to 12.5 μg/mL.

A relationship between drug clearance and half-life can be established from equation 11.7. By combining equations 11.6 and 11.7, equation 11.8 is obtained, which rearranges to equation 11.9.

$$Cl_p = \frac{0.693V}{t_{\frac{1}{2}}} \tag{11.8}$$

$$t_{\frac{1}{2}} = \frac{0.693V}{Cl_p} \tag{11.9}$$

This equation shows that the half-life of a drug is directly related to the drug distribution volume, but inversely related to drug clearance. The effect that volume has on drug elimination rate or half-life for a given clearance is demonstrated in Figure 11.2.

In this figure, Drug A distributes into a volume of 3 L and Drug B distributes into a 10-fold greater volume of 30 L. Although the clearance of the two drugs is identical, 10% of the total body load of Drug A is cleared during 1 min, compared to only 1% of Drug B. Therefore, the elimination rate of Drug A will be 10 times greater than that of Drug B, and the elimination half-life of Drug A will be 10 times shorter than that of Drug B, despite identical clearance values. This relationship can be summarized by noting that clearance describes an intrinsic ability of an organ or organ system to remove drug from the body. Clearance is completely independent of drug distribution volume. On the other hand, the half-life, or elimination rate constant, is a complex function of both clearance and distribution volume of a drug or its metabolites. Although equation 11.6 might be interpreted as implying a relationship between volume and clearance, this interpretation is incorrect. Any change in volume would result in a change in the elimination rate constant, not the clearance. These interesting relationships are summarized in Table 11.2. For a given clearance, the half-life of a drug increases with increasing distribution volume, and for a given distribution volume, the half-life decreases with increasing clearance. The shortest drug half-life is obtained with a drug that has a high clearance and a small distribution volume. The longest half-life is obtained with a drug that has a low clearance and a large distribution volume.

Influence of Distribution Volume on Drug Half-Life   **TABLE 11.2**

Cl_p	Drug Half-Life in Plasma (h)		
	V = 5 L	V = 25 L	V = 100 L
50 mL/min (reabsorption)	1.2	5.8	23.1
130 mg/min (filtration)	0.4	2.2	8.9
650 mL/min (secretion)	0.09	0.44	1.8

Note: Cl_p denotes plasma clearance.

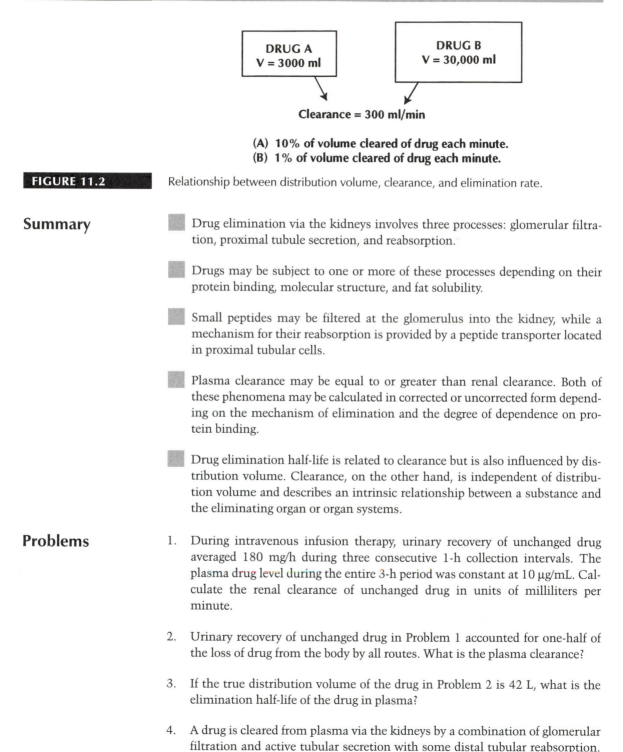

DRUG A	DRUG B
V = 3000 ml	V = 30,000 ml

Clearance = 300 ml/min

(A) 10% of volume cleared of drug each minute.
(B) 1% of volume cleared of drug each minute.

FIGURE 11.2   Relationship between distribution volume, clearance, and elimination rate.

Summary

Drug elimination via the kidneys involves three processes: glomerular filtration, proximal tubule secretion, and reabsorption.

Drugs may be subject to one or more of these processes depending on their protein binding, molecular structure, and fat solubility.

Small peptides may be filtered at the glomerulus into the kidney, while a mechanism for their reabsorption is provided by a peptide transporter located in proximal tubular cells.

Plasma clearance may be equal to or greater than renal clearance. Both of these phenomena may be calculated in corrected or uncorrected form depending on the mechanism of elimination and the degree of dependence on protein binding.

Drug elimination half-life is related to clearance but is also influenced by distribution volume. Clearance, on the other hand, is independent of distribution volume and describes an intrinsic relationship between a substance and the eliminating organ or organ systems.

Problems

1. During intravenous infusion therapy, urinary recovery of unchanged drug averaged 180 mg/h during three consecutive 1-h collection intervals. The plasma drug level during the entire 3-h period was constant at 10 μg/mL. Calculate the renal clearance of unchanged drug in units of milliliters per minute.

2. Urinary recovery of unchanged drug in Problem 1 accounted for one-half of the loss of drug from the body by all routes. What is the plasma clearance?

3. If the true distribution volume of the drug in Problem 2 is 42 L, what is the elimination half-life of the drug in plasma?

4. A drug is cleared from plasma via the kidneys by a combination of glomerular filtration and active tubular secretion with some distal tubular reabsorption.

If the drug is highly bound to plasma proteins, should the plasma clearance be corrected for protein binding?

References

1. Smith, H. W. *The Kidney, Structure and Function in Health and Disease;* Oxford University Press: New York, 1951; p 8.
2. Miller, M. A.; Drakontides, A. B.; Leavell, L. C. *Anatomy and Physiology,* 17th ed.; MacMillan: New York, 1967; pp 540–551.
3. Weiner, I. M. *Ann. Rev. Pharmacol.* **1967**, *7*, 39–56.
4. Cafruny, E. J. *Ann. Rev. Pharmacol.* **1968**, *8*, 131–150.
5. Tirrupathy, C.; Ganapathy, V.; Leibach, F. H. *Biochem. Biophys. Acta* **1991**, *1069*, 14–20.
6. Brandsch, M.; Brandsch, C.; Hopfer, V.; Ganapathy, V.; Leibach, F. H. *FASEB J.* **1995**, *9*, A1773.
7. Liu, W.; Liang, R.; Ramamoorthy, S.; Fei, Y. J.; Ganapathy, M. E.; Hediger, M. A.; Ganapathy, V.; Leibach, F. H. *FASEB J.* **1995**, *9*, A310.

Drug Elimination in Renal Impairment

12

ll of the relationships described in Chapter 11 are based on normally functioning kidneys. Most compounds, whether drug or metabolite, are eventually cleared from the body via the kidneys. If kidney function becomes impaired, the kidneys lose the ability to eliminate substances to varying degrees. When considering the possible impact of impaired renal function on drug or metabolite clearance, and the necessity of dose adjustment in this condition, several questions must be answered: (1) to what extent are the drug and/or metabolites cleared by the kidneys, (2) are the metabolites pharmacologically active, and (3) how much excess accumulation of drug, metabolite, or both can be tolerated before the dose has to be modified to avoid toxic side effects? Much has been written on this subject (1–3). Space permits description of only some basic concepts associated with drug disposition and dosage in renal failure, and some of the problems associated with this area of research.

Renal failure can result from a variety of pathological conditions. If impairment of renal function is rapid in onset and of relatively short duration, then the renal failure is described as acute. The primary cause of this condition may be acute congestive heart failure, shock, acute tubular necrosis, or hypercalcemia. The condition is generally reversible, although complete restoration of renal function may take from 6 to 12 months.

Chronic renal failure is distinguished from the acute condition in that it is almost always caused by intrinsic renal disease and is characterized by slow, progressive development. Unlike the acute condition, chronic renal impairment is generally irreversible. The degree of loss of kidney functional capacity in the chronic condition is best described in terms of the *intact nephron hypothesis*, which considers that the diseased kidney comprises normal nephrons and nephrons that are essentially nonfunctional because of the pathological condition. Progressive impairment of renal function is reflected in an increased fraction of nonfunctional nephrons. This concept is analogous to, and somewhat better documented than, the intact hepatocyte hypothesis discussed in Chapter 10. The intact nephron hypothesis implies that the efficiency with which a compound is handled by the kidneys is related to overall kidney function regardless of the mechanism(s) by which it is handled by the kidney. Considerable evidence has shown that, regardless of whether renal clearance of a drug involves passive filtration, active secretion, active or passive reabsorption, or a combination of these, the impact of kidney failure on drug clearance can be accurately predicted from a simple measure of renal function. This is the creatinine clearance, which is described in this chapter.

The prolonged and progressive nature of chronic renal failure is of particular concern in some elderly patients who may require a variety of medications. The inability of these patients to adequately excrete drugs and drug metabolites, and the influence of their uremic condition on the function of other physiological systems, requires careful drug dosage adjustment to achieve adequate therapeutic blood and urine concentrations without toxicity. Some common causes of kidney failure are given in Table 12.1 (4).

Methods of Measuring Renal Function

Renal function can be measured in several ways. The most common method and one which, as described above, provides almost universal prediction of drug or metabolite clearance in cases of impaired kidney function is *creatinine clearance*. Creatinine is formed from muscle metabolism in the body and circulates in the plasma of individuals with normal renal function at a concentration of approximately 1 mg/dL. Creatinine is cleared via the kidneys by filtration to yield a normal creatinine clearance of about 130 mL/min. This value depends partially on body size, degree of activity, muscle mass, and age. Various nomographs have been described to account for these.

As kidney function declines, for whatever reason, glomerular filtration rate, and hence creatinine clearance, will also decline. If the kidneys are working with

| TABLE 12.1 | **Some Common Causes of Kidney Failure** |

Cause	Description
Pyelonephritis	Inflammation and deterioration of the pyelonephrons due to infection, antigens, or other idiopathic causes.
Hypertension	Chronic overleading of the kidney with fluid and electrolytes may lead to kidney insufficiency.
Diabetes mellitus	Disturbance of sugar metabolism and acid-base balance may lead to or predispose a patient to degenerative renal disease.
Nephrotoxic drugs or metals	Certain drugs taken chronically may cause irreversible kidney damage, e.g., the aminoglycosides, phenacetin, and heavy metals such as mercury and lead.
Hypovolemia	Any condition that causes a reduction in renal blood flow will eventually lead to renal ischemica and damage.

Source: Reproduced with permission from reference 4.

only 50% efficiency, the creatinine clearance will drop to 50 to 60 mL/min, depending on age and other factors. According to the intact nephron hypothesis, other kidney functions will also decline, including tubular secretion. The decline in kidney function leads to the reasonable assumption that, provided a compound is cleared via the kidneys, its clearance will be affected to a similar extent as creatinine clearance.

Creatinine clearance, Cl_{cr}, like any other renal clearance value, is generally calculated from the relationship described in equation 12.1.

$$Cl_r = \frac{\text{Rate of urinary excretion of creatinine}}{\text{Serum concentration of creatinine}} \qquad (12.1)$$

Numerical values are substituted into equation 12.2 to show a normal creatinine clearance of 130 mL/min.

$$Cl_{cr} = \frac{1.3 \text{ mg/min}}{0.01 \text{ mg/mL}} = 130 \text{ mL/min} \qquad (12.2)$$

In many cases it is not possible, for practical reasons, to obtain quantitative urine collections in order to determine creatinine clearance by this method. To address this problem, several mathematical approaches have been described to estimate creatinine clearance based on serum creatinine levels and some additional patient information (5). Although these methods contain certain approximations and assumptions, they nonetheless provide reasonable estimates of creatinine clearance that can be used for drug dosage adjustment.

Probably the most common approach, and one that has stood the test of time, is that of Cockcroft and Gault (6). With this method, creatinine clearance, Cl_{cr}, is obtained for males from equation 12.3.

$$Cl_{cr} = \frac{(140 - \text{Age})(BW)}{72 S_{cr}} \qquad (12.3)$$

where Age is the patient's age in years, BW is body weight in kilograms, and S_{cr} is the patient's measured serum creatinine in milligrams per deciliter. The value 140 is derived from a regression of 24-hour creatinine excretion per kilogram of body weight in males, and the value 72 is average weight for males in kilograms (6). For females, the value obtained from equation 12.3 is multiplied by 0.85.

Thus, for a 70-kg, 50-year-old male with a serum creatinine of 1.5 mg/dL, the estimated creatinine clearance is obtained from equation 12.4.

$$Cl_{cr} = \frac{(140 - 50)(70)}{72 \times 1.5} = 58 \text{ mL/min} \qquad (12.4)$$

Similarly, for a 50-kg, 30-year-old female with a serum creatinine of 1.3 mg/dL, the estimated creatinine clearance is obtained from equation 12.5.

$$Cl_{cr} = 0.85 \frac{(140 - 30)(50)}{72 \times 1.3} = 50 \text{ mL/min} \tag{12.5}$$

Use of Creatinine Clearance to Predict the Effect of Renal Impairment on Drug Elimination

The use of creatinine clearance as a measure of kidney function to predict the impact of kidney failure on drug elimination rate is based on a simple, linear relationship between creatinine clearance and the rate of drug elimination. Note that as drug half-life increases with decreasing creatinine clearance, the extent of increase depending on the proportion of drug cleared by the kidneys, the relationship between drug half-life and creatinine clearance cannot be used to accurately estimate drug disposition in renal impairment. The relationship between these parameters is asymptotic, as shown in Figure 12.1. Thus in cases of severe renal impairment, where dose adjustment is most critical, changes in creatinine clearance give rise to large changes in drug elimination half-life, often with wide fluctuations, making use of this relationship impractical.

Drug elimination rate, on the other hand, is linearly related to creatinine clearance, based on equation 12.6.

$$k = k_{nr} + X \cdot Cl_{cr} \tag{12.6}$$

where k is the drug elimination rate constant in reciprocal hours, k_{nr} is the drug elimination rate constant with zero kidney function in reciprocal hours, Cl_{cr} is creatinine clearance in milliliters per minute, and X is a constant. This method is simple and of general application. Suppose that a drug is cleared entirely via the kidneys. When creatinine clearance falls to zero in complete renal impairment, k

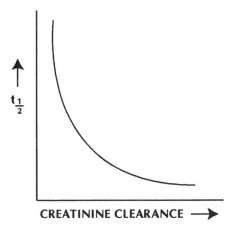

CREATININE CLEARANCE ⟶

FIGURE 12.1   Relationship between creatinine clearance and half-life for a drug that is cleared via the kidneys.

will become zero because k_{nr} is zero. Consider, on the other hand, a drug that is cleared in equal proportions by the kidneys and by the liver (metabolism or biliary excretion). Then in complete renal shutdown, the value of k will reduce to k_{nr}, which is one-half the value of k in a normal individual, and the elimination rate in complete renal failure is one-half the normal value. Thus, if the elimination rate of a drug, the amount of drug cleared by the kidneys, and the amount of drug cleared by other processes under normal conditions are known, then the extent to which drug elimination will be impaired in declining renal function can be predicted with reasonable accuracy.

Some drugs for which this information is available are given in Table 12.2, and the relationship between creatinine clearance and elimination rate for each of the drug groups is illustrated in Figure 12.2 (1). The changes in drug half-life are shown on the right side of the figure. From Table 12.2 and Figure 12.2, drugs in Group A, such as minocycline, rifampicin, lidocaine, and digitoxin, are cleared to a very small extent by the kidneys in normal individuals. Elimination of these drugs is therefore not influenced by renal function, and dosage adjustment in renal impairment is not necessary. Drugs in Group L, on the other hand, are handled almost exclusively by the kidneys. Elimination of these drugs is markedly impaired in declining renal function, and dosage adjustment in renal impairment is likely to be mandatory.

Many options and opinions exist as to the optimal way to adjust drug dosage in renally compromised patients, but all options have the common objective of reducing the drug dosage to maintain therapeutic drug levels and avoid undue accumulation. In essence, drug input has to be reduced to the same extent that drug elimination is reduced by renal function impairment.

The following two alternatives form the basis of most methods of dosage adjustment. The first is to reduce the drug dose size but to maintain the same dosage interval as in normal renal function. The degree of dose reduction is given by equation 12.7.

Methods of Dosage Adjustment

$$\text{Uremic dose} = \text{normal dose}\,\frac{k_u}{k_n} \qquad (12.7)$$

where k_u and k_n are, respectively, drug elimination rate constants in uremic and normal individuals. The dose is reduced according to the ratio of the observed k in the uremic individual to that in a normal person. If k is halved, then the dose must also be reduced by one-half.

The second alternative is given by equation 12.8.

$$\text{Uremic dosage interval} = \text{normal dosage interval}\,\frac{t_{1/2u}}{t_{1/2n}} \qquad (12.8)$$

From this equation the dosing interval must be increased according to the increase in elimination half-life in the uremic patient while giving the same quantity of drug in each dose to that given to a normal patient.

TABLE 12.2

Elimination Rate Constants for Various Drugs

Group	Drug	k_n (h^{-1})	k_{nr} (h^{-1})	k_{nr}/k_n (%)
A	Minocycline	0.04	0.04	100.0
	Rifampicin	0.25	0.25	100.0
	Lidocaine	0.39	0.36	92.3
	Digitoxin	0.114	0.10	87.7
B	Doxycycline	0.037	0.031	83.3
	Chlortetracycline	0.12	0.095	79.2
C	Clindamycin	0.16	0.12	75.0
	Chloramphenicol	0.26	0.19	73.1
	Propranolol	0.22	0.16	72.8
	Erythromycin	0.39	0.28	71.8
D	Trimethoprim	0.054	0.031	57.4
	Isoniazid (fast)	0.53	0.30	56.6
	Isoniazid (slow)	0.23	0.13	56.5
E	Dicloxacillin	1.20	0.60	50.0
	Sulfadiazine	0.069	0.032	46.4
	Sulfamethoxazole	0.084	0.037	44.0
F	Nafcillin	1.26	0.54	42.8
	Chlorpropamide	0.020	0.008	40.0
	Lincomycin	0.15	0.06	40.0
G	Colistimethate	0.154	0.054	35.1
	Oxacillin	1.73	0.008	33.6
	Digoxin	0.021	0.06	33.3
H	Tetracycline	0.120	0.033	27.5
	Cloxacillin	1.21	0.31	25.6
	Oxytetracycline	0.075	0.014	18.7
I	Amoxicillin	0.70	0.10	14.3
	Methicillin	1.40	0.19	13.6
J	Ticarcillin	0.58	0.066	11.4
	Penicillin G	1.24	0.13	10.5
K	Cefazolin	0.32	0.02	6.2
	Cephaloridine	0.51	0.03	5.9
	Cephalothin	1.20	0.06	5.0
	Gentamicin	0.30	0.015	5.0
L	Flucytosine	0.18	0.007	3.9
	Kanamycin	0.28	0.01	3.6
	Vancomycin	0.12	0.004	3.3
	Tobramycin	0.32	0.010	3.1
	Cephalexin	1.54	0.032	2.1

Note: k_n is the observed elimination rate constant under conditions of normal renal function, and k_{nr} is the observed elimination rate constant in the absence of or under conditions of severely impaired renal function.

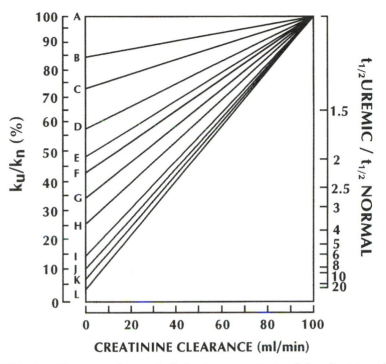

FIGURE 12.2

Changes in the percentage of normal elimination rate constant (left ordinate), and the consequent geometric increase in elimination half-life (right ordinate), as a function of creatinine clearance. The letters A–L refer to the drug groups listed in Table 12.2 (Reproduced with permission from reference 1).

Application of these two approaches for gentamicin in a patient with a creatinine clearance of 10 mL/min is shown in Figure 12.3. Although both methods yield the same mean serum level, the shapes of the serum drug profiles are quite different. It is worth noting that, regardless of the method of dosage adjustment, achieving an identical drug blood profile in a uremic patient to that in a normal individual is impossible.

The relative merits of the two approaches, and in fact the decision whether to adjust drug dosage at all in renal impairment, depend on the type of drug. For example, most penicillins and cephalosporins have wide therapeutic indices, are relatively nontoxic, and generally do not require dose adjustment except in cases of severe impairment. The aminoglycosides, on the other hand, are potentially highly toxic, and dosage has to be carefully titrated to renal function in order to avoid toxicity.

The previous discussion has provided only a brief glimpse at some pharmacokinetic aspects of the clinical problem of dosage adjustment in renal failure. Other aspects that confound the task of providing useful, nontoxic therapy for specific therapeutic classes to uremic individuals have not been discussed. Similarly, the problems of altered distribution and protein binding in renal failure are not covered here but are described in detail elsewhere (1, 4).

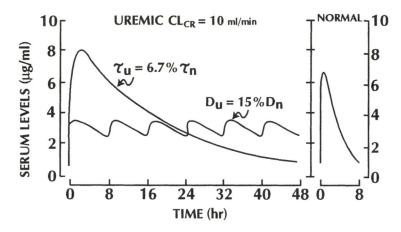

| **FIGURE 12.3** | Predicted steady-state serum levels of gentamicin in a 70-kg normal subject and in a patient with a creatinine clearance of 10 mL/min. The dose D_n, dosing interval τ, and elimination rate constant k_n in normal renal function are 1.7 mg/kg, 8 h, and 0.347 h^{-1}, respectively. V_n is 14 L and is assumed to be unchanged in renal failure. Adjusted doses and dosage intervals for the uremic patient are obtained from Figure 12.2. |

Maintenance of Patients with End-Stage Kidney Disease

Hemodialysis

Intermittent hemodialysis is one method used to replace the function of the kidneys in patients with severely impaired renal function. During a dialysis period, any drugs or metabolites that are in the body are liable to be filtered out of the bloodstream in the same way as endogenous metabolic by-products and other biological substances.

Removal of drugs and metabolites from blood plasma occurs during dialysis through passive diffusion across the filter membrane into the dialysate. Diffusion properties are based on drug characteristics such as molecular size, protein binding, and distribution volume, on the characteristics of the dialyzing filter, such as pore size and surface area, and on blood and dialysate flow rate (7). Drugs with a molecular size greater than 500 daltons are not removed well by dialysis. Similarly, drugs that are highly bound to plasma proteins do not dialyze well as only free drug is removed. Drugs with large distribution volumes dialyze poorly despite the fact that their molecular size may be in the ideal range to cross the dialysis membrane. This is because only a small percentage of the total body drug load is contained within the plasma volume and the dialysis system "sees" only this small percentage of drug. Typically, digoxin dialyzes poorly although its molecular size is ideal for dialyis. With a total apparent distribution volume of 500 L, less than 1% of the body digoxin load is in the "dialyzable" compartment at any one time, resulting in inefficient removal. On the other hand, the aminoglycoside antibiotics have distribution volumes of less than 1 L/kg and are rapidly removed by dialysis. Some typical drugs regarded as dialyzable and nondialyzable are given in Table 12.3 (8).

If a dialysis patient is receiving medication, then, depending on its dialyzability, circulating drug is likely to exhibit quite different pharmacokinetic profiles during the periods when the patient is on and off dialysis. If the elimination rate of

Some Drugs Regarded as Dialyzable and Nondialyzable

TABLE 12.3

Dialyzable	Nondialyzable
Acyclovir	Amphotericin B
Carbenicillin	Clindamycin
Cefadroxil	Cloxacillin
Gentamicin	Digoxin
Kanamycin	Erythromycin
Metronidazole	Glutethimide
Primidone	Miconazole
Ticarcillin	Propranolol
Tobramycin	Quinidine
Vidarabine	Trimethoprim

Source: Adapted with permission from reference 8.

a compound is markedly reduced in renal impairment, and the compound is efficiently dialyzable, then blood drug levels will be markedly reduced during dialysis and a supplemental dose may be required following the dialysis period to regain the desired drug levels during the interdialysis period. This situation is shown in Figure 12.4. Some theoretical and practical approaches to this problem have been described (*1, 8*).

Continuous ambulatory peritoneal dialysis (CAPD) was introduced as an alternative to hemodialysis for maintenance of patients with end-stage renal disease (*9*). Patients on CAPD may receive medication by conventional routes or, in cases of peritonitis, by direct introduction of drug into the peritoneal cavity together with the dialysate infusion. The concept of adding medication to the dialysate has raised some interesting questions regarding the feasibility of adding drugs to peritoneal dialysate for systemic activity as well as peritoneal sterilization and, conversely, the feasibility of using conventional doses (oral, intravenous) to obtain effective peritoneal levels to treat peritonitis often associated with CAPD patients. Two studies have examined these different aspects of drug disposition. The first study showed that effective systemic concentrations of vancomycin can be achieved by administering a suitable loading dose followed by lower maintenance doses directly into the peritoneal cavity together with the peritoneal dialysate (*10*). In the second study, however, orally administered cephradine did not result in adequate drug concentrations in the peritoneum to treat all organisms commonly associated with peritonitis (*11*). Although studies continue in this area, these studies suggest that direct introduction of drug into the peritoneal cavity of CAPD patients may be useful as a method of achieving both local and systemic drug activity. On the other hand, systemic administration may be inadequate to treat CAPD-related peritonitis.

Continuous Ambulatory Peritoneal Dialysis

Drug elimination may be markedly affected by renal function impairment, depending on the proportion of drug that is cleared by the kidneys.

Summary

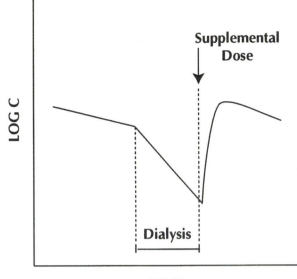

FIGURE 12.4   Interdialysis and intradialysis profiles for a dialyzable drug showing a supplemental dose administered immediately postdialysis.

The degree of renal function impairment can be estimated by the use of markers. The most common of these is clearance of endogenous creatinine.

Drug dosage adjustment may be necessary to avoid toxic overdose in cases of renal impairment. The degree of dose adjustment can be calculated on the basis of the fraction of drug cleared by the kidneys and degree of reduction in creatinine clearance.

Drugs and/or metabolites may be removed during dialysis, depending on their dialyzability. For a highly dialyzable drug, supplemental drug dosage may be required postdialysis.

Problems

1. A 40-year-old, 50-kg male patient has a serum creatinine of 1.6 mg/dL. Calculate his estimated creatinine clearance.

2. A 50-year-old, 45-kg female patient has a serum creatinine of 2.4 mg/dL. Calculate her estimated creatinine clearance.

3. Gentamicin has an elimination half-life of 2 h ($k = 0.35$ h^{-1}) in normal renal function (creatinine clearance = 130 mL/min). Calculate the estimated half-life when creatinine clearance is reduced to (i) 40 mL/min and (ii) 10 mL/min. Gentamicin is cleared from the body exclusively via the kidneys.

4. If the dosage of gentamicin in normal renal function is 50 mg every 8 h, what would the revised dose, or dosage interval, be in Problem 3 (ii)?

References

1. Welling, P. G.; Craig, W. A. In *Effects of Disease States on Drug Pharmacokinetic;* Benet, L. Z., Ed.; American Pharmaceutical: Washington, DC, 1976; pp 155–187.
2. Fillastre, J-P.; Singlas, E. *Clin. Pharmacokinet.* **1991,** *20,* 293–310.
3. Singlas, E.; Fillastre, J-P. *Clin. Pharmacokinet.* **1991,** *20,* 389–410.
4. Shargel, L.; Yu, A. B. C. *Applied Biopharmaceutics and Pharmacokinetics;* Appleton-Century-Crofts: New York, 1980; pp 102-115, 187–203.
5. Talbert, R. L. *J. Clin. Pharmacol.* **1994,** *34,* 99–110.
6. Cockcroft, D. W.; Gault, M. H. *Nephron* **1976,** *16,* 31–47.
7. Matzke, G. R.; Millikin, S. P. In *Applied Pharmacokinetics Principles of Therapeutic Drug Monitoring,* 3rd ed.; Evans, W. E.; Schentag, J. J.; Jusko, W. J., Eds.; Applied Therapeutics: Vancouver, WA, 1992; pp 8.1–8.49.
8. Lee, C. S.; Marbury, T. C. *Clin. Pharmacokinet.* **1984,** *9,* 42–46.
9. Popovich, R. P.; Moncrief, J. W.; Decherd, J. B.; Bomar, J. B.; Pyle, W. K. *Am. Soc. Artif. Intern. Organs J.* **1976,** *5,* 64.
10. Rogge, M. C.; Johnson, C. A.; Zimmerman, S. W.; Welling, P. G. *Antimicrob. Agents Chemother.* **1985,** *27,* 578–582.
11. Johnson, C. A.; Welling, P. G.; Zimmerman, S. W. *Nephron* **1984,** *38,* 57–61.

The Mathematics
of Pharmacokinetics

The One-Compartment Open Model with Intravenous Dosage

13

n order to understand the mathematical approaches used throughout this book, a basic knowledge of calculus is needed. Initially some kinetic expressions will be derived. However, with some exceptions, mathematical derivation will be kept to a minimum. Helpful integrating procedures, such as the Laplace transform, must be used to solve rate equations for complex pharmacokinetic expressions. However, the intent of this book is not to teach mathematics but to provide a basic understanding of pharmacokinetics and its uses. Therefore, only minor emphasis will be placed on derivations, and major emphasis will be placed on the meaning and application of pharmacokinetic principles.

Drug input, elimination, and transfer between pharmacokinetic compartments will be assumed to be first-order and linear. This assumption is consistent with the modeling approach. In later chapters, departures from this general approach will be described, but the principal arguments will be developed assuming first-order, nonsaturable, and either reversible kinetics (e.g., between spatial compartments) or irreversible kinetics (e.g., between chemical compartments, and also absorption and elimination).

To reiterate a comment in Chapter 1, the pharmacokinetic compartment can be used to describe both spatial and chemical states. For example, if a drug appears to distribute in a heterogeneous manner in the body so that overall drug distribution can be described in terms of two distinct body volumes, then the concentration of drug in these volumes and its distribution between them are described in terms of two spatial compartments. On the other hand, if a drug forms a metabolite, particularly if the metabolite is active, which makes it of interest, then the metabolite is considered to be a separate chemical compartment regardless of whether the metabolite occupies the same or different body fluids and tissues as the parent drug. Spatial and chemical compartments can coexist in the same kinetic model. For any drug that is metabolized, coexistence is necessarily the case.

Consider the simplest model of all, the one-compartment open model. Despite its associated simplifications and assumptions, this model is the most common for describing drug profiles in blood, plasma, serum, or urine after oral or intramuscular doses. Following intravenous bolus doses, an additional drug distribution phase is often more readily discernible. This situation will be discussed in more detail later. In the simple one-compartment model, however, the drug is assumed to rapidly distribute into a homogeneous fluid volume in the body regardless of the route of administration (1, 2).

Pharmacokinetic rate constants are based on transfer of amounts of drugs. Rate constants are subsequently applied to concentration changes by dividing the expressions by the appropriate distribution volumes. Also, on a microscopic basis, most pharmacokinetic rate constants describe a multiplicity of events. For example, an absorption rate constant is possibly influenced by dissolution, stomach emptying, splanchnic blood flow, and a variety of other factors. However, despite the gross simplifications involved, observed rate constants describe the overall rate-limiting process, be it absorption, distribution, metabolism, or excretion. How much more mechanistic information can be obtained from such rate constants depends on the drug and the enthusiasm and ingenuity of the investigator.

The One-Compartment Open Model with Bolus Intravenous Injection

This model, which has been summarized by Gibaldi and Perrier (3), is depicted in Scheme 13.1. Because of the generally heterogeneous nature of the body, and the impact of this on drug distribution, this model is relatively rare. However, examples in the literature include plasma concentrations of prednisolone following bolus intravenous administration to a kidney transplant patient (4), and of tritium following intravenous administration of tritiated Hirulog 1 (BG 8967), a synthetic thrombin inhibitor (5, 6).

The box, or compartment, represents the drug distribution volume, and other values and rate constants are defined in the caption. The value k_{el} is equal to the sum of all elimination rate constants, including those for drug eliminated via sweat, bile, lungs, etc. However, in this example only two routes of elimination are assumed, urinary excretion and metabolism. The curved arrow leading into the compartment represents instantaneous introduction of drug.

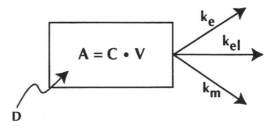

SCHEME 13.1   One-compartment open model with bolus intravenous injection: D is the dose, A is the amount of drug in the body, C is the concentration of drug in body fluids, and V is the drug distribution volume.

Using this model, equation 13.1 can be written in the following form.

$$\frac{dA}{dt} = -(k_e + k_m)A = -k_{el}A \tag{13.1}$$

where A is the amount of drug in the body, t is time, k_e is the rate constant for urinary excretion, and k_m is the rate constant for metabolism. Equation 13.1 describes the rate of loss of drug from the body. This equation is rearranged to

$$\frac{dA}{A} = -k_{el}dt \tag{13.2}$$

Equation 13.2, when integrated between the limits of zero and finite time, with the value of A varying from A_0, the initial amount of drug in the body, to some value less than A_0, becomes

$$\ln A - \ln A_0 = -k_{el}t \tag{13.3}$$

The natural logarithms appear in this expression because the integral of the reciprocal of any single value X is equal to the natural logarithm of X. Rearrangement of equation 13.3 yields

$$\ln\left(\frac{A}{A_0}\right) = -k_{el}t \tag{13.4}$$

If both sides of equation 13.4 are made a power of e, as in equation 13.5, equation 13.6 is obtained.

$$e^{\ln\left(\frac{A}{A_0}\right)} = e^{-k_{el}t} \tag{13.5}$$

$$\frac{A}{A_0} = e^{-k_{el}t} \quad \text{or} \quad A = A_0 e^{-k_{el}t} \tag{13.6}$$

Equation 13.5 converts to equation 13.6 because e to the power of the natural logarithm of X is equal to X ($e^{\ln X} = X$). This is analogous to logarithms to the base 10. To use a numerical example, the logarithm to the base 10 of 100 is equal to 2, and 10^2 is 100. Thus, 10 raised to the power of the logarithm of 100 is equal to 100, or 10 raised to the power of the logarithm of X is equal to X.

Equation 13.6 can be converted into concentration terms by dividing both sides of the expression by the distribution volume, V, as in equation 13.7, to yield equation 13.8.

$$\frac{A}{V} = \left(\frac{A_0}{V}\right)e^{-k_{el}t} \tag{13.7}$$

$$C = C_0 e^{-k_{el}t} \tag{13.8}$$

where C is the concentration of drug in the body and C_0 is the initial concentration of drug at zero time. Equation 13.3 can similarly be converted to concentration form as in

$$\ln C = \ln C_0 - k_{el}t \quad \text{or} \quad \log C = \log C_0 - \frac{k_{el}t}{2.3} \tag{13.9}$$

Conversion from natural logarithms to logarithms to the base 10 in equation 13.9 is obtained from the simple relationship that $\ln X = 2.3 \log X$.

What information can be obtained about a drug by using some of these expressions? From equations 13.8 and 13.9, a plot of the logarithm of drug concentration against time will be linear. Logarithms to the base 10 will be used in this book because logarithmic graph paper is printed that way, and it is thus more convenient.

In Figure 13.1, the slope of the line, which will be linear if the data fit the model, gives the elimination rate constant k_{el}, and the extrapolated intercept at time zero gives C_0. Actually, the intercept is the logarithm of C_0, but as the actual concentration values are plotted on semilogarithmic graph paper, the paper converts actual values into logarithmic values. Actual concentration values can therefore be read directly from the plots.

The elimination half-life of the drug can also be obtained from the relationship in equation 13.10.

$$t_{1/2} = \frac{\ln 2}{k_{el}} = \frac{0.693}{k_{el}} \tag{13.10}$$

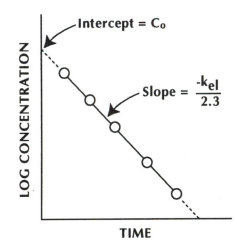

FIGURE 13.1   Plot of logarithm of drug concentration vs. time following intravenous bolus injection.

Equation 13.10 is valid for any first-order rate constant. However, instead of finding the elimination rate constant and then calculating the half-life, obtaining these values in reverse order is usually more convenient when analyzing data graphically. For example, the elimination half-life can be obtained by selecting any time interval during which the value of C is reduced by one-half. Whichever values of C are used, the time interval for C to be reduced by one-half will be the same. The value of k_{el} is then obtained from equation 13.10.

If the administered dose D is divided by the extrapolated value C_0, and if the reasonable assumption is made that all of the injected dose was absorbed, then the drug distribution volume is obtained from

$$V = \frac{D}{C_0} \qquad (13.11)$$

A word of caution is appropriate here. During this and subsequent exercises, the simplifying assumption is made that drugs are not bound, or are bound to only a negligible extent, to plasma and tissue proteins or other macromolecules. This assumption saves considerable time and keeps the mathematics relatively simple. However, if binding does occur, then appropriate adjustments may be made to such parameters as distribution volume, as described in Chapter 8.

The drug elimination half-life, overall elimination rate constant k_{el}, and its distribution volume have now been calculated from the data in Figure 13.1. Multiplying the distribution volume, V, by the elimination rate constant, k_{el}, as in equation 13.12, yields the plasma clearance, Cl_p.

$$Cl_p = Vk_{el} \qquad (13.12)$$

Knowing also the renal clearance and differentiating it from other clearance processes would be useful information. This information cannot be obtained from plasma data alone because the information in Figure 13.1 indicates only how rapidly drug is leaving the body. The figure provides no information regarding the route of elimination. However, if all the drug that is excreted in unchanged form in the urine, A_u^∞, were collected, then the renal clearance can be obtained from

$$\frac{Cl_r}{Cl_p} = \frac{k_e V}{k_{el} V} = \frac{k_e}{k_{el}} = \frac{A_u^\infty}{D} \qquad (13.13)$$

where Cl_p is the plasma clearance, Cl_r is the renal clearance, and A_u^∞ is the total amount of drug excreted in urine.

The renal clearance is thus related to plasma clearance in direct proportion to the ratio of total urinary recovery of unchanged drug to the administered dose. As discussed previously, renal clearance may be equal to or less than plasma clearance, but never greater. That is, k_e can never be greater than k_{el}. Once k_e is obtained, k_m can be calculated simply by subtracting k_e from k_{el}, as in equation 13.14.

$$k_m = k_{el} - k_e \qquad (13.14)$$

Another useful pharmacokinetic parameter that can be obtained from intravenous data, or from any other data for that part, is the area under the plasma level curve, AUC.

The total area under the plasma curve, that is, the area from zero to infinite time, is obtained mathematically by integrating the terms in equation 13.8 between zero and infinite time. This integration, after appropriate cancellations, yields

$$
\begin{aligned}
\mathrm{AUC}^{0 \to \infty} = \int_0^\infty C = C_0 \int_0^\infty e^{-k_{el}} dt \\
= -\frac{C_0}{k_{el}} \left(e^{-k_{el}\infty} - e^{-k_{el}0} \right) \\
= -\frac{C_0}{k_{el}} (0 - 1) = \frac{C_0}{k_{el}}
\end{aligned}
\tag{13.15}
$$

Because C_0 can be expressed as D/V, equation 13.15 can be written as equation 13.16.

$$
\mathrm{AUC}^{0 \to \infty} = \frac{D}{V k_{el}} = \frac{D}{Cl_p}
\tag{13.16}
$$

This expression shows that the area under the plasma curve is equal to the dose divided by plasma clearance. Perhaps more importantly, plasma clearance can be obtained by dividing the dose by the AUC. However, the area must be the total area. If a truncated area is used, and this is frequently all that can be determined by direct observation of the data, overestimation of the plasma clearance will result.

Renal clearance can also be obtained with this approach, provided urinary recovery of unchanged drug is known. Renal clearance is readily obtained from

$$
Cl_r = \frac{A_u^t}{\mathrm{AUC}^{0 \to t}}
\tag{13.17}
$$

Equation 13.17 is analogous to a rearranged form of equation 13.16. Thus, renal clearance is calculated by dividing the quantity of drug recovered in urine up to a certain time by the area under the plasma curve up to the same time. (This time can be infinity but need not be.) The calculation for renal clearance has the advantage over calculation for plasma clearance in that truncated areas and partial urine collections can be used. If the values in equation 13.17 are extrapolated to infinity, then equation 13.18 results.

$$
Cl_r = \frac{A_u^\infty}{\mathrm{AUC}^{0 \to \infty}}
\tag{13.18}
$$

Equation 13.8 shows that renal clearance and plasma clearance differ only in terms of the difference between the administered dose and urinary recovery of unchanged drug, A_u^∞.

Equations 13.15 and 13.16 describe the area under the plasma curve following bolus intravenous injection. In many cases, however, area values are measured directly from the data, for example, in model-independent kinetics, and several methods are available to do this. These include the trapezoidal rule and the log trapezoidal rule. The simple trapezoidal rule is described here because it is most commonly used. The trapezoidal rule is quick and accurate. The accuracy of the method is directly related to the number of data points used in its calculation.

The Trapezoidal Rule

A trapezoid is a four-sided figure with two sides parallel and two sides nonparallel. When the length of one of the sides is reduced to zero, the trapezoid becomes a triangle. If plasma data are plotted on regular graph paper, the area under the plasma profile can be divided into a series of trapezoids, and the areas of the individual trapezoids can be calculated and summed.

The data in Table 13.1 constitute a typical drug profile that might be obtained following bolus intravenous injection of a drug that has a biological half-life of 1 h. If these data are plotted on regular graph paper, and if the data points are joined by straight lines, a series of trapezoids is obtained, terminating with a triangle for the 8–12 h interval. Calculating the area for each segment of the curve and cumulatively adding each successive segment yield the trapezoidal area shown in the third column of the table. In this example, the sampling time has been extended until no detectable drug remains in the plasma. Unfortunately, this situation does not usually occur in practice. In most cases, the plasma sampling time is not extended for a sufficiently long period to allow plasma drug levels to decline to zero, so that the area calculated by the trapezoidal rule is the area from time zero to some time t when drug levels are still present. Thus, a truncated area is obtained, as in Figure 13.2. The 4-h plasma sample still contains drug, so the total area under the plasma level curve cannot be calculated.

The truncated area is useful for many types of calculations, but the complete area under the curve is more useful. For example, the area from time zero to infinity is required to calculate plasma clearance and total absorption and to construct Wagner–Nelson absorption plots, which will be discussed shortly. So it is important to be able to extend the truncated area to infinite time.

A Typical Drug Profile Following Bolus IV Injection   **TABLE 13.1**

Time (h)	Concentration ($\mu g/mL$)	Cumulative AUC ($\mu g \cdot h/mL$)
0	25.0	
0.25	21.0	5.75
0.50	17.6	10.58
1.0	12.5	18.11
2.0	6.25	27.49
3.0	3.13	32.18
4.0	1.56	34.53
6.0	0.40	36.49
8.0	0.10	36.99
12.0	0.0	37.19

The simplest and most common method to achieve this is by end correction. This involves calculating the area from the last sampling time, C_t, to infinite time and adding this value to the truncated area. To calculate the terminal area from time t to infinity, this portion of the drug profile is treated as if it were a separate profile where $C_0 = C_t$. Then, just as the area to time infinity for the total curve can be calculated using equation 13.15, so the terminal area can be calculated by dividing C_t by k_{el}, as in

$$\text{AUC}^{t \to \infty} = \frac{C_t}{k_{el}} \tag{13.19}$$

Adding this calculated area to the truncated area obtained by the trapezoidal rule will yield the area from zero to infinite time. End correction has been used here to correct or complete an area following intravenous dosing. However, end correction can be used equally well after any route of administration provided sufficient drug concentrations are obtained to identify and characterize the elimination rate constant, k_{el}.

Referring to the data in Table 13.1, suppose that the sampling time had not been extended beyond 4 h so that the measured trapezoidal area also extended only to 4 h. How would the end correction be calculated to obtain the total area from zero to infinite time?

The first step is to plot the data on semilogarithmic graph paper to obtain the linear descending portion of the drug concentration in plasma profile, and thereby calculate the drug half-life and elimination rate constant. Then by substituting appropriate values into equation 13.19, equation 13.20 is obtained.

$$\text{AUC}^{4 \to \infty} = \frac{C_t}{k_{el}} = \frac{1.56}{0.693} = 2.25\,\mu g \cdot h/mL \tag{13.20}$$

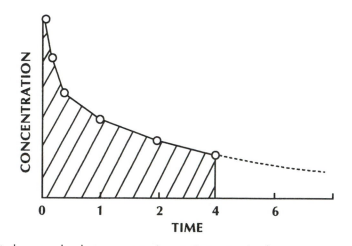

FIGURE 13.2   Truncated area under drug concentration vs. time curve in plasma.

In this case, C_t from Table 13.1 is 1.56 µg/mL, k_{el} is 0.693 h^{-1}, and C_t divided by 0.693 equals 2.25 µg · h/mL. Adding this value to the area from 0 to 4 h, which is 34.53, yields an area to infinite time of 36.8 µg · h/mL, as in equation 13.21.

$$\text{AUC}^{0\rightarrow 4} + \text{AUC}^{4\rightarrow\infty} = \text{AUC}^{0\rightarrow\infty} = 34.53 + 2.25$$
$$= 36.8 \mu g \cdot h/mL \qquad (13.21)$$

The area obtained from equation 13.21 is inconsistent with the total area of 37.19 µg · h/mL in Table 13.1. Clearly, both values cannot be correct. The inconsistency between the two values lies in the 12-h data point in the table. The 12-h sample contained no detectable drug, but that does not indicate when the levels actually reached zero. It could have been reached at 9, 10, 10.5, 11, or 12 h, or any other time between 8 and 12 h postdose. The 12-h sample simply indicated that the drug level was 0 at that time; therefore, the triangle used to measure the small terminal area was the largest possible value. The method of end correction shows that the value was an overestimate. This overestimation may appear to be unimportant because the last portion of the AUC represents only a small fraction of the total area. However, plasma samples that are obtained during the latter period of plasma profiles are frequently separated by longer time intervals than earlier sampling times. Therefore, the terminal area may constitute a large percentage of the total area. In these cases, calculating the terminal portion of the area as accurately as possible is important.

Before leaving this topic it is important to review briefly one aspect of area values pertinent to later discussions of bioavailability. Equation 13.16 showed that the area under the plasma curve is equal to the dose divided by the plasma clearance. Rearrangement of equation 13.16 leads to

$$\text{AUC}^{0\rightarrow\infty}(k_{el}) = \frac{D}{V} \qquad (13.22)$$

Equation 13.22 shows that the administered dose of drug, expressed as a concentration term D/V, is equal to the area under the curve multiplied by the elimination rate constant, k_{el}. This relationship is important because D/V is an expression of bioavailability, or absorption efficiency. In a later chapter, similar expressions will be described for situations in which the fraction of drug absorbed is unknown, such as after oral dosing. If the fraction is unknown, the expression D/V becomes FD/V, where F is the fraction of dose absorbed. Comparison of FD/V values obtained from two different oral dosages of a drug gives a measure of relative bioavailability, whereas comparison of FD/V from an oral dose to D/V from an intravenous dose of the same drug gives the absolute bioavailability from the oral dose. Whatever is being compared, the D/V or FD/V value is equal to the product of the area under the plasma curve and the elimination rate constant, not just the area under the curve.

There are two reasons for this equality. The first reason is that it is mathematically correct. The second reason is that it is intuitively correct. Whatever the quantity of drug absorbed, the area under the drug concentration versus time curve will be inversely proportional to the elimination rate constant. If the rate constant is increased, the area will decrease, and vice versa. Because the elimination rate constant has nothing to do with drug absorption or drug bioavailability, the contribution of that rate constant to the magnitude of the area should be removed by multiplying the area by the elimination rate constant.

The values of the elimination rate constant following oral and intravenous doses or following different formulations of the same drug should not differ, but they often do. Even when comparing two oral dosage forms, small differences in k_{el} could alter the area value to a sufficient extent to affect bioequivalence calculations. Normalizing for k_{el} removes this potential error.

Urinary Excretion Kinetics

In many instances, urinary excretion data may provide useful information in addition to blood or plasma data. For example, the degree of metabolism, the bioavailability of a drug following oral doses, renal clearance, and a variety of other useful parameters can be calculated from urinary excretion of drug or its metabolites. Plasma concentrations of some drugs are low and difficult to measure. In these cases, urinary excretion data may provide the only means of examining drug pharmacokinetics. For the diuretic agents chlorothiazide and hydrochlorothiazide, urinary excretion data have been shown to be as good as, or better than, plasma data for bioavailability and bioequivalence determinations (7, 8).

Loss of drug from the body following intravenous bolus injection was described by the expression $A = A_0 e^{-k_{el}t}$ (equation 13.6), where A_0 is the amount of drug initially in the body. In that expression, the only elimination rate constant reflecting drug loss from the body was the overall rate constant, k_{el}. The rate of appearance of unchanged drug in the urine, dA_u/dt, can be obtained from the product of the quantity of drug in the body and the rate constant for urinary excretion, k_e, as in equation 13.23.

$$\frac{dA_u}{dt} = k_e A \tag{13.23}$$

The value A is not a constant in this expression but is a constantly changing value; therefore, equation 13.23 can be rewritten as

$$\frac{dA_u}{dt} = k_e A_0 e^{-k_{el}t} \tag{13.24}$$

Integration of equation 13.24 yields

$$A_u = \frac{k_e A_0}{k_{el}} \left(1 - e^{-k_{el}t}\right) \tag{13.25}$$

This equation describes the cumulative excretion of unchanged drug in urine with respect to time. Consider the limits of this expression when $t = 0$ and when $t = \infty$. When $t = 0$, that is, immediately after the intravenous dose is given, $e^{-k_{el}t}$ is equal to unity because any number to the power of zero equals unity, and the parenthetical term in equation 13.25 becomes 0, as does A_u. If t is then allowed to increase to an infinitely large value, that is, a sufficiently long time for the urinary excretion of unchanged drug to be completed, then the expression $e^{-k_{el}t}$ becomes $e^{-k_{el}\infty}$, which equals zero, and A_u then becomes equal to $A_0 k_e/k_{el}$, as in

$$A_u^\infty = \frac{A_0 k_e}{k_{el}} \tag{13.26}$$

Equation 13.26 can then be substituted into equation 13.25 to obtain a new expression:

$$A_u = A_u^\infty \left(1 - e^{-k_{el}t}\right) \tag{13.27}$$

Equations 13.25 and 13.27 thus describe the continuous cumulative excretion of unchanged drug in urine, as in Figure 13.3. Note two interesting characteristics about these equations. First, total urinary recovery of unchanged drug is equal to the dose, A_0, multiplied by the quotient k_e/k_{el}. Clearly, if k_e is equal to k_{el}, that is, the drug is removed from the body solely by urinary excretion, then $k_e/k_{el} = 1$ and $A_u = A_0$. Thus, the quantity of drug recovered in urine is controlled by the dose and also by the ratio k_e/k_{el}.

The second characteristic is that, whereas the rate constant for urinary excretion is k_e, the time course of urinary recovery, as described by equations 13.25 and 13.27, is the overall rate constant, k_{el}. The value of k_e plays its part in the constant term in equation 13.28 and thus determines the proportion of drug that is recovered unchanged in urine.

A considerable amount of information can be obtained from urinary excretion data. If sufficient collection points are taken, types of information can be obtained that are similar to those from plasma data, with the exception of the drug distribution volume. This parameter cannot be estimated from urinary excretion data alone because the calculations are restricted to the quantity of drug voided in urine.

One useful method of analyzing urine data is the construction of sigma-minus plots (9). This method, as the name implies, is based on the relationship between the total quantity of drug eliminated in urine minus the quantity that has been

Construction of Sigma-Minus Plots

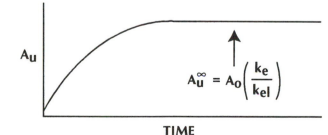

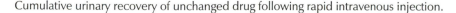

Cumulative urinary recovery of unchanged drug following rapid intravenous injection.   **FIGURE 13.3**

eliminated up to a certain time. Another way of describing this plot is the quantity of drug remaining to be eliminated in urine at any time.

The equation used to construct this plot is derived directly from equation 13.27. Rearranging this equation leads to

$$A_u^{\infty} - A_u = A_u^{\infty} e^{-k_{el}t} \tag{13.28}$$

Taking logarithms of both sides of the equation yields

$$\log(A_u^{\infty} - A_u) = \log A_u^{\infty} - \frac{k_{el}t}{2.3} \tag{13.29}$$

Thus, a plot of the logarithm of the quantity of drug remaining to be eliminated in urine, $A_u^{\infty} - A_u$, versus time yields a straight line with a slope of $-k_{el}/2.3$ and an intercept of $\log A_u^{\infty}$. This type of plot from urinary excretion data yields the overall elimination rate constant, k_{el}, that is, the rate constant for total loss of drug from the body by all routes. It does not yield the urinary excretion rate constant, k_e. This is always the case regardless of the relative proportions of drug cleared by the urine and by other routes. The value of k_e can of course be obtained by multiplying k_{el} by A_u^{∞}/A_0 or A_u^{∞}/D, as in

$$k_e = k_{el} \frac{A_u^{\infty}}{A_0} = k_{el} \frac{A_u^{\infty}}{D} \tag{13.30}$$

The metabolism rate constant, k_m, is then obtained, if metabolism occurs, by subtracting k_e from k_{el} as in equation 13.31.

$$k_m = k_{el} - k_e \tag{13.31}$$

Some drugs that are frequently given by intravenous bolus injection, and their elimination half-lives, are given in Table 13.2.

Before leaving this topic, a word of caution is appropriate. Although urinary excretion data can provide useful information, in particular cumulative drug or metabolite excretion patterns or both, obtaining accurate estimates of kinetic parameters from urinary collection data is limited by the practical limits of sampling frequency and bladder holdup. One way to avoid this type of problem is to catheterize the bladder to obtain continuous urinary recovery. Bladder catheterization is not a significant problem in some experimental animal studies, but it is less convenient in human studies. An alternative approach to achieve adequate urine flow, and hence shorter and more accurate collection intervals, is to water-load the subject. This procedure presents the disadvantage that induced diuresis may change elimination kinetics, particularly for those compounds whose elimination may be urine flow rate-dependent.

This situation is more complex than the intravenous bolus case because it includes a time-dependent drug input or absorption component (*10*). An assumption intrinsic to this model is that the rate of drug absorption is zero-order, that is, the rate is a constant and is independent of drug concentration at the absorption site. Zero-order absorption commonly occurs with intravenous infusions; therefore, understanding the kinetics underlying these types of infusions is important. Infusions may be over a prolonged period, for example, an intravenous drip or infusion pump, where drug levels need to be sustained over prolonged time periods, or infusions may be relatively short, perhaps for 10–30 min. Short infusions are frequently used for drugs such as the aminoglycoside and cephalosporin antibiotics to avoid toxicity due to the bolus effect and to reduce pain or embolism, or both, at the site of administration.

Zero-order absorption kinetics has recently become more common because of the recent proliferation of enteral and parenteral sustained-release products, many of which are intended to release drug at a zero-order rate. Even conventional dosage forms of some compounds have been claimed to yield zero-order absorption. So this model is important for several different types of drug dosage forms and administration routes.

The model is depicted in Scheme 13.2. All of the symbols are identical to the intravenous bolus case except for the additional zero-order rate constant, k_0.

From the model, the rate equation 13.32 can be written to describe the rate of change in the amount of drug in the body with respect to time.

$$\frac{dA}{dt} = k_0 - k_{el}A \tag{13.32}$$

In this equation, k_0 is a constant input rate that will continue as long as drug is available to be absorbed, or until the infusion is stopped. The infusion rate may be written also as D/T, where D is the dose and T is the total absorption or infusion time. Although the input rate is constant, the elimination rate is variable and

Zero-Order Drug Input and First-Order Elimination

Some Drugs That Are Commonly Given by Intravenous Injection and Their Elimination Half-Lives

TABLE 13.2

Drug	$t_{1/2}$
Aminophylline	6 h
Cefazolin	2 h
Cimetidine	1.5–2 h
Diazepam	1–2 days
Digoxin	2 days
Fluorouracil	10 min
Furosemide	20–30 min
Gentamicin	3–4 h
Lidocaine	1.5–2 h
Phenytoin	~20 h

Note: The half-life of phenytoin is dose-dependent because of saturable elimination.

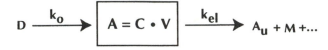

SCHEME 13.2 One-compartment open model with zero-order absorption and first-order elimination: k_0 is the zero-order rate constant for drug administration.

obeys first-order kinetics, as in the bolus intravenous case. The elimination rate is dependent on the product of the rate constant, k_{el}, and the amount of drug in the body, A.

During the initial period of zero-order input, the amount of drug in the body, A, will be small. Thus the product, $k_{el}A$, will also be small, the rate of drug input will exceed the rate of drug output, and the quantity of drug in the body will increase. As the value of A increases, the product $k_{el}A$ will also increase, so the overall rate of drug elimination will approach and eventually become equal to the rate of input. A steady state is then achieved in which the rate of absorption equals the rate of elimination.

The infusion may be stopped either before or after the amount of drug in the body has reached steady state. During the resulting postabsorption phase, drug levels will decline at a first-order rate as in the intravenous bolus case. The two possible situations are shown in Figure 13.4.

Integration of equation 13.32 yields equation 13.33, which in concentration terms becomes equation 13.34.

$$A = \frac{k_0}{k_{el}}\left(1 - e^{-k_{el}t}\right) \tag{13.33}$$

$$C = \left(\frac{k_0}{V} \cdot k_{el}\right)\left(1 - e^{-k_{el}t}\right) \tag{13.34}$$

Although two rate constants are involved in the overall drug profile, only one time-dependent function, $e^{-k_{el}t}$, is involved. If t=0, $e^{-k_{el}t}$ becomes unity and $C = 0$. As the time after the start of infusion increases, the value $e^{-k_{el}t}$ becomes progressively smaller, the value $(1 - e^{-k_{el}t})$ increases, and the accumulation curve in Figure 13.3 is obtained. If the infusion is continued for a sufficiently long period so that $e^{-k_{el}t}$ approaches or becomes zero, then the parenthetical term becomes unity, and $C = k_0/Vk_{el}$ as in

$$C_{ss} = \frac{k_0}{Vk_{el}} \tag{13.35}$$

Because the steady-state concentration is described, C is now expressed as C_{ss}.

Thus, at steady state, as in Figure 13.4a, the concentration of drug in the distribution volume is equal to the infusion rate, k_0, divided by the plasma clearance,

Vk_{el}. Because the constant k_0 has units of mass per unit time, and because plasma clearance is commonly expressed in terms of volume per unit time, C_{ss} has units of mass per volume, or concentration.

The relationship in equation 13.35 provides considerable information. For example, knowledge of plasma clearance and drug infusion rate permits calculation of the steady-state drug concentration in plasma. Alternatively, if k_0 and C_{ss} are known, then plasma clearance can be calculated. Similarly, if k_0, C_{ss}, and the elimination $t_{1/2}$ are known, then the distribution volume can be calculated.

If both sides of equation 13.35 are multiplied by the distribution volume, V, equation 13.36 is obtained.

$$A_{ss} = \frac{k_0}{k_{el}} \qquad (13.36)$$

This equation describes the amount of drug in the body at steady state, A_{ss}, in terms of the absorption and elimination rate constants. Thus, the total body drug load can be determined by dividing the zero-order infusion rate constant by the first-order elimination rate constant. A_{ss} can therefore be determined without knowing C_{ss}.

As previously noted from equations 13.34 and 13.35, steady-state drug levels are dependent on both the infusion and elimination rate constants. Faster infusion yields higher blood levels; faster elimination yields lower blood levels. However, from equation 13.34, the time dependency of the accumulation process is clearly dependent only on the elimination rate constant, k_{el}. No matter how fast a drug is infused, the time to reach steady state is governed exclusively by the elimination rate constant. How long it will take for a drug level to reach steady state can be determined from equation 13.34. Because this equation is exponential, steady state will theoretically take a very long time to achieve. However, because pharmacokineticists have to consider practicalities, 95% of steady state may be considered a reasonable approximation given the normal variability of biological data. How long does a drug need to be infused before drug levels in the blood reach 95% of

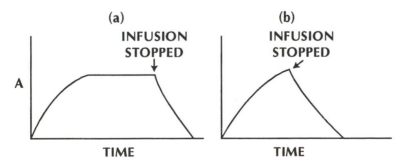

(a) INFUSION STOPPED

(b) INFUSION STOPPED

A

TIME

TIME

Time course of the quantity of Drug A in the body during and following zero-order infusion. In (a), drug levels had reached steady state before the infusion was stopped, and in (b), the levels had not reached steady state.

FIGURE 13.4

steady-state values? To calculate this, equation 13.34 can be rewritten in the form of equation 13.37.

$$95 = 100(1 - e^{-k_{el}t}) \tag{13.37}$$

From this equation, and from the relationship $0.693/k_{el} = t_{1/2}$, t becomes equal to $4.3t_{1/2}$. Thus, whatever the infusion rate, 4.3 (or approximately 4.5) drug elimination half-lives will be needed to reach 95% of steady-state values. Consider some examples. If a drug has a biological half-life of 1 h, then the drug would need to be infused for about 4.5 h before steady-state levels are approached. If the infusion rate were doubled, steady-state levels would also be doubled, but the new levels would not be reached any sooner. If on the other hand, the drug half-life were 24 h, 4.5 days would be needed to approach steady state.

Regardless of the infusion rate, the time required to reach steady state, or the time period during which circulating drug levels are below or have not yet approached those levels required for therapeutic efficacy, cannot be reduced provided k_{el} remains constant. This situation occurs not only with drugs that have long biological half-lives, but also with drugs whose elimination half-lives are normally short but may be prolonged in disease conditions such as renal failure.

An appreciation of the need to understand the important relationships in equations 13.34 and 13.37 can be obtained by considering the cardiac glycoside digoxin. This drug frequently has to be administered by infusion. If a patient received an intravenous infusion of digoxin, and a digoxin blood concentration was determined 2 h after the start of infusion, could the drug concentration at that time be assumed to be at steady state? Digoxin has a biological half-life of approximately 2 days, so that the time required for blood levels to reach steady state during a continuous infusion is 9 days. After 2 h of infusion, the digoxin concentration can be calculated from equation 13.34 by setting k_{el} equal to 0.693/48, or 0.014 h^{-1}, t equal to 2 h, and k_0/Vk_{el} equal to 1, or 100%. Substituting these numerical values into equation 13.34, as in equation 13.38, yields a concentration value at 2 h that is only 2.8% of the steady-state value.

$$C - 100(1 - e^{-0.014 \times 2}) = 2.8\% \tag{13.38}$$

If it had been assumed that steady state had been achieved at 2 h into the infusion, and drug levels at that time were within the therapeutic range, then continuing the infusion beyond 2 h would undoubtedly have produced toxic drug concentrations.

In cases of prolonged drug accumulation, a bolus loading dose often has to be administered to achieve required therapeutic levels earlier. Under most circumstances, a convenient approach is to administer a loading dose that will instantaneously achieve the drug concentration, or total body load, that gives the desired therapeutic response at steady state. This situation is achieved by administering a bolus dose equal to k_0/k_{el}. Because k_0 has units of mass per unit time and because k_{el} has units of reciprocal time, the quotient has units of mass.

Consider the time course of events if a bolus dose of magnitude k_0/k_{el} is administered as a bolus injection and at the same time an infusion is started at a zero-order rate, k_0. The resulting time course of the quantity of drug in the body is given by

$$A = \frac{k_0}{k_{el}} e^{-k_{el}t} + \frac{k_0}{k_{el}}(1 - e^{-k_{el}t}) \qquad (13.39)$$

The two components to equation 13.39 are a decreasing component that results from the elimination of the bolus dose, and an increasing component that results from the infusion.

Equation 13.39 can be simplified to

$$A = \frac{k_0}{k_{el}} \qquad (13.40)$$

This equation indicates that the quantity of drug in the body is constant from the time that the bolus-loading dose is administered until the end of the infusion dose. This simple yet elegant relationship is due to control of both the decline in drug levels following the loading dose and the increase in drug levels from the infusion by the elimination rate constant, k_{el}. Therefore, the decreasing and increasing drug profiles are mirror images of each other, and the sum of the quantities of drug in the body from the loading and infusion doses will always equal k_0/k_{el}, as shown in Figure 13.5.

This is, of course, a simple example. For very potent drugs, dividing the loading dose into two smaller doses to approach steady state in a stepwise, cautious fashion is more appropriate. Two smaller doses will reduce initial drug levels, but will essentially achieve the same goal of rapidly attaining steady-state levels while minimizing toxic effects.

The situation for oral doses is more complicated because the bolus dose will probably be absorbed by a first-order process rather than be instantaneously available in the circulation. However, provided the absorption rate of this dose is fast compared to the elimination rate, the concepts embodied in equation 13.39 still apply.

What happens to drug levels when the infusion, or any zero-order drug absorption process, stops? Consider Figure 13.4 and equations 13.34 and 13.35.

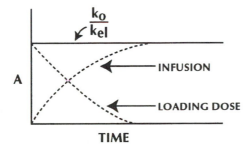

Quantity of drug in the body following a loading intravenous bolus dose, k_0/k_{el}, and a zero-order infusion initiated at the same time as the bolus injection.

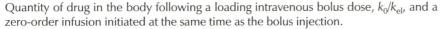

FIGURE 13.5

Assume the situation in Figure 13.4a, that is, steady-state drug levels are achieved during infusion. In this case, the declining drug levels obtained after infusion has stopped are described by equation 13.41.

$$C = \frac{k_0}{Vk_{\mathrm{el}}} e^{-k_{\mathrm{el}}t'} \tag{13.41}$$

where t' is the time elapsed since the end of the infusion. The drug profile at the end of the infusion is essentially identical to that after an intravenous bolus dose except that C_0 or D/V from the intravenous bolus dose case is replaced by k_0/Vk_{el}.

Assume now the situation in Figure 13.4b. In this case, steady state has not been achieved before zero-order infusion is stopped; therefore, at that time the drug level is given by equation 13.42.

$$C = \frac{k_0}{Vk_{\mathrm{el}}} (1 - e^{-k_{\mathrm{el}}T}) \tag{13.42}$$

where T is the total zero-order infusion time. The postabsorptive drug concentrations in this situation are described by

$$C = \frac{k_0}{Vk_{\mathrm{el}}} (1 - e^{-k_{\mathrm{el}}T}) e^{-k_{\mathrm{el}}t'} \tag{13.43}$$

where t' has the same meaning as in equation 13.41. This situation is again identical to that following an intravenous bolus dose except that C_0 from the intravenous case has been replaced by the more complex term in equation 13.43.

The situation described by equation 13.43 is common for orally administered dosage forms that are designed to release drug in zero-order fashion. The most suitable drug candidates for sustained release, as discussed earlier, have elimination half-lives between 2 and 8 h. For those drugs with the shortest half-life in this range, 4.5 half-lives, or 9 h, would be needed to approach steady-state levels in the body. On the other hand, drugs with a half-life of 8 h would take 36 h to approach steady-state values. As far as drug absorption is concerned, residence time in the gastrointestinal (GI) tract may be considered to be approximately 12 h (as mentioned earlier, there is much disagreement on this value). Therefore, only those drugs that have short half-lives of approximately 2 h will achieve steady-state blood levels from a single oral dose, and postabsorption blood profiles can be described by equation 13.41. Drugs with elimination half-lives longer than 2 h will not achieve steady-state levels after a single oral dose, and their postabsorptive blood profiles are likely to be described by equation 13.43. The problem of GI transit time, and its limiting influence on the available time for release of orally administered drugs, has led to several formulations designed specifically, with variable degrees of success, to remain in the GI tract for longer periods and so allow more time for drug release. Some drugs that are frequently given by zero-order intravenous infusion are listed in Table 13.3. Some recent publications have described the pharmacokinetics of nitroglycerin and its metabolites during intravenous infusion to conscious dogs (11), and pharmacokinetically assisted (12) and computer-assisted (13) continuous infusion of fentanyl to human patients.

Summary

Following bolus intravenous injection of a drug that obeys one-compartment model kinetics, plasma profiles can be used to estimate the elimination rate constant, half-life, apparent distribution volume, and plasma clearance.

Urinary excretion data can be utilized to calculate renal clearance.

Areas under drug profiles in plasma can be determined analytically, or measured by the trapezoidal rule or log trapezoidal rule. The area under the curve, normalized by elimination rate constant where appropriate, provides a measure of drug systemic availability.

For an intravenously administered drug, sigma-minus plots of urinary excretion data can be used to calculate rate constants for drug excretion and metabolism.

Zero-order drug absorption may occur, not only with intravenous infusions, but also with a variety of sustained-release dosage forms.

Drug concentrations in plasma during zero-order administration are controlled by the rates of administration and elimination. The time to reach steady-state drug levels is controlled only by the elimination rate constant.

Bolus loading dose for an infused drug can be calculated from the ratio of infusion and elimination rate constants.

Problems

1. Equal quantities of two different drugs are administered to a patient by rapid intravenous (IV) injection. Neither drug binds to proteins. The concentration of Drug A in plasma immediately after dosing is 10 μg/mL, and the concen-

Some Drugs That Are Commonly Given by Intravenous Infusion and Their Elimination Half-Lives

TABLE 13.3

Drug	$t_{1/2}$
Aminophylline	6 h
Amphotericin	1–2 days
Cyclophosphamide	3–6 h
Digoxin	2 days
Gentamicin	3–4 h
Heparin	1–2 h
Lidocaine	1.5–2 h
Nitroglycerin	~30 min
Verapamil	3–6 h

tration of Drug B is 20 µg/mL. Drug A has a biological half-life double that of Drug B. Which drug has the greater plasma clearance?

2. Following a 250-mg intravenous bolus dose of a drug, the resulting drug concentrations in plasma declined in an apparent monoexponential manner. Extrapolation of the log–linear elimination slope to time zero yielded a concentration of 25 µg/mL. At 2 h after dosing, the concentration had declined to 12.5 µg/mL. What is the plasma clearance of the drug?

3. If the drug in problem 2 is cleared from the body 25% via the kidneys and 75% via hepatic metabolism, what will be the ratio of its renal clearance to plasma clearance?

4. One hour after a bolus IV injection of 100 mg of a drug that has a biological half-life of 4 h, the concentration of unchanged drug in serum was 4.0 µg/mL. What is the apparent distribution volume of the drug?

5. A 200-mg dose of a drug is injected intravenously into a human subject. The drug rapidly distributes into an apparently homogeneous fluid volume. Ten hours after injection, the plasma concentration of drug is 3.28 µg/mL. At 16.92 hours after injection, the plasma concentration of drug is 1.64 µg/mL. Calculate (i) the elimination half-life, (ii) elimination rate constant, (iii) volume of distribution, and (iv) plasma clearance.

6. Following IV bolus injection, the following drug plasma concentrations were obtained:

Time (h)	Concentration (µg/mL)
0	100
0.5	77.9
1	60.7
2	36.8
3	22.3
4	13.5

Calculate (i) the trapezoidal area under the plasma curve from $t = 0$ to $t = 4$ h, and (ii) the area from zero to infinite time.

7. After an intravenous bolus injection of 250 mg of a drug, which distributes into a single homogeneous volume of 12 L in the body and is cleared completely as unchanged drug in urine, the following urinary excretion data were obtained:

Time After Dosing (h)	Cumulative Urinary Recovery of Drug (mg)
0.5	16.9
1.0	32.7
2.0	61.1
4.0	107.2

8.0	168.5
12.0	103.4
24.0	241.3
48.0	249.7
96.0	250.0

Calculate the following:

(i)   By sigma-minus plot, the drug elimination half-life.

(ii) Plasma clearance.

(iii) Renal clearance.

(iv) The concentration of drug in plasma immediately after and at 1 h after dosing.

(v)   The area under the drug plasma curve from zero to infinite time.

8. A drug is administered to a 70-kg male patient by means of a constant-rate IV infusion at 0.5 mg/min. The drug has a biological half-life of 4.5 h and appears to distribute into a space equivalent to that of blood volume (5 L). Calculate the following:

(i)   The steady-state plasma level achieved with the infusion.

(ii) The time taken to reach 95% and 99% of steady-state levels.

9. Another 70-kg male patient receives the same infusion as in Problem 8. However, this patient has also been taking another drug that has induced his drug-metabolizing enzymes. This drug causes the half-life of the infused drug to be reduced to 1.5 h. Recalculate the following:

(i)   The steady-state plasma levels achieved in this case.

(ii) The time taken to reach 95% and 99% of steady-state levels.

10. The steady-state levels obtained in Problem 9 are now too low for therapeutic efficacy. How would the constant-rate infusion have to be adjusted in order to achieve the steady-state levels of drug in Problem 8? Does altering the infusion rate in this way influence the time required to reach steady state?

11. From the data in Problem 10, calculate the drug plasma concentration at 1 h, 6 h, and 12 h after the infusion has stopped.

12. For the situation described in Problem 8, how much drug would have to be given as an initial bolus injection to achieve the required steady-state plasma level instantaneously?

References

1. Shargel, L.; Yu, A. B. C. *Applied Biopharmaceutics and Pharmacokinetics*; Appleton-Century-Crofts: New York, 1980; pp 28–37.

2. Wagner, I. G.; Northam, J. I. *J. Pharm. Sci.* **1967,** *56*, 529–531.

3. Gibaldi, M.; Perrier, D. *Pharmacokinetics*; Marcel Dekker: New York, 1982; pp 1–4.

4. Gambertoglio, J. G.; Amend, W. J. C., Jr.; Benet, L. Z. *J. Pharmacokinet. Biopharm.* **1980,** *8*, 1–52.

5. Fox, I.; Dawson, A.; Loynds, P.; Eisner, J.; Findlen, K.; Levin, E.; Hanson, D.; Mant, T.; Wagner, J.; Maragonore, J. *Thromb. Haemostasis* **1993,** *69*, 157–163.

6. Wagner, J. G. *Pharmacokinetics for the Pharmaceutical Scientist*; Technomic: Lancaster, PA, 1993; pp 1–4.

7. Osman, M. A.; Patel, R. B.; Irwin, D. S.; Craig, W. A.; Welling, P. G. *Biopharm. Drug. Dispos.* **1982,** *3*, 89–94.

8. Barbhaiya, R. H.; Craig, W. A.; Corrick-West, H. P.; Welling, P. G. *J. Pharm. Sci.* **1982,** *71*, 245–248.

9. Wagner, J. G. *Fundamentals of Clinical Pharmacokinetics*; Drug Intelligence: Hamilton, IL, 1975; pp 77–78.

10. Rodriguez, N.; Madsen, P. O.; Welling, P. G. *Antimicrob. Agents Chemother.* **1979,** *15*, 465–469.

11. Lee, F. W.; Salmonson, T.; Benet, L. Z. *J. Pharmacokinet. Biopharm.* **1993,** *21*, 533–550.

12. Glass, P. S. A.; Jacobs, J. R.; Smith, L. R.; Ginsburg, B.; Quill, T. J.; Bai, S. A.; Renes, J. G. *Anesthesiology* **1990,** *73*, 1082–1090.

13. Shafter, S. L.; Varvel, J. R.; Aziz, N.; Scott, J. C. *Anesthesiology* **1990,** *73*, 1091–1102.

The One-Compartment Open Model with First-Order Absorption and Elimination

14

hen a drug is taken orally, or by intramuscular or subcutaneous injection, the resulting drug profile in plasma can frequently be described by a pharmacokinetic model that incorporates first-order absorption and elimination. First-order absorption and elimination occur, or appear to occur, in the great majority of cases after oral dosing, regardless of whether the dose is given as a solution, suspension, capsule, tablet, or controlled-release product.

Interpretation of the absorption phase of such drug profiles in terms of a first-order rate constant is intuitive when drug is given as a solution. One might expect that the rate of absorption would be dependent upon the mucosal–serosal concentration gradient generated by the solution dose. But the absorption process for solid oral dosage forms is more complex and may be influenced by the dissolution rate of solid product, stomach emptying, membrane transport, and a variety of other factors. It is surprising that, in most instances, the first-order approach is a reasonable approximation to the overall absorption process and often gives a good description of the data. Other absorption models have been proposed for particular drugs and formulations. However, these models are frequently difficult to prove, and the first-order absorption model still appears to provide the most generally used description of the absorption phase.

The model is shown in Scheme 14.1. This model has two major differences from the models for intravenous administration. The first difference is the introduction of a first-order absorption constant, k_a, so that the rate of absorption becomes dose dependent. The second difference is the additional parameter, F, which describes the absorption efficiency, or the fraction of the dose, D, that is absorbed into the systemic circulation.

General Aspects of First-Order Absorption and Elimination

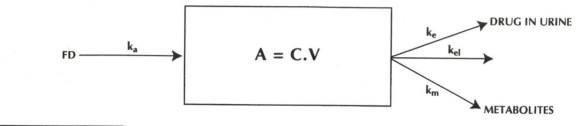

SCHEME 14.1 One-compartment open model with first-order absorption and elimination, where F is the fraction of the dose, D, absorbed from the dosage site into the systemic circulation, and k_a is the first-order rate constant for drug absorption.

After intravenous injection, the parameter F is not pertinent because the availability of administered drug is almost always 100%; therefore, F is equal to unity. However, after oral doses, and also after intramuscular doses in some cases, bioavailability is not always 100%. Complete absorption from oral doses tends to be the exception rather than the rule. Incomplete absorption might be expected because of limited dissolution, degradation or metabolism occurring in the gastrointestinal (GI) tract, incomplete membrane penetration, and also presystemic hepatic clearance. After intramuscular doses, more efficient absorption might be expected, but this is not always the case. Incomplete absorption from intramuscular doses may result from degradation of drug at the intramuscular site, drug precipitation, or slow release of a portion of the drug giving rise to low and perhaps undetectable drug levels during prolonged periods. Intramuscularly dosed phenobarbital has been shown to be only 80% bioavailable compared to oral doses in humans (1), and intramuscularly dosed promethazine has been shown to be approximately 70–80% bioavailable in dogs compared to intravenously dosed drug (2). All of the above factors may influence the magnitude and interpretation of the absorption rate constant, k_a.

Suppose that a drug is at the absorption site and is simultaneously being absorbed at a rate governed by an intrinsic absorption rate constant, k_{ab}, and being enzymatically degraded at the absorption site at a rate governed by a rate constant, k_d. The overall rate of drug loss from the absorption site is then governed by the sum of k_{ab} and k_d. Because k_a is used to describe the overall loss of drug from the absorption site, the amount of drug, X, remaining at the absorption site at any time is described by

$$X = FDe^{-(k_{ab}+k_d)t} = FDe^{-k_a t} \tag{14.1}$$

In Chapter 13 the apparent rate constant for appearance of intravenously dosed drug in the urine was shown to be equal to the overall elimination rate constant, k_{el}. Similarly, the apparent rate constant for appearance of orally or intramuscularly dosed drug into the circulation is equal to the overall rate constant for loss of drug from that absorption site by all processes. In other words, the rate constant that is obtained from the drug-concentration curve in plasma is not necessarily the intrinsic absorption rate constant but may be a constant related to overall loss of drug from the absorption site. An observed k_a may actually be the sum of k_{ab}, k_d, and any other rate constant that contributes to loss of drug from the absorption site. Note that the absorption half-life, $t_{1/2(abs)}$, is obtained from $t_{1/2(abs)} = 0.693/k_a$.

An interesting analogy can be drawn with ocular drug administration (3). When a drug solution is applied to the surface of the eye, for example from an eye dropper, more than 95% of the drug is washed from the eye surface by tear movement and is washed down the nasolachrymal duct. Thus, the overall rate of loss of drug from the absorption site at the surface of the eye is high. Absorption of drug into the eye will continue only as long as drug is available at the absorption site. Because overall loss of drug from the eye surface is approximately a first-order process, the apparent rate constant for drug penetration into the eye is very fast, and the absorption rate constant calculated from drug levels within the eye may overestimate the actual intrinsic absorption rate constant by a factor of 20 or more. This concept is worth remembering when considering drug absorption kinetics.

From Scheme 14.1, equation 14.2 can be written to describe the rate of change in the amount of drug, A, in the body.

$$\frac{dA}{dt} = k_a X - k_{el} A \qquad (14.2)$$

In this equation, X is the amount of drug remaining to be absorbed as described in equation 14.1. By substituting for X from equation 14.1 and then integrating, equation 14.3 is obtained.

$$A = FD \frac{k_a}{k_a - k_{el}} (e^{-k_{el}t} - e^{-k_a t}) \qquad (14.3)$$

This equation can then be converted to describe time-dependent drug concentrations by dividing both sides by the distribution volume, V, to obtain equation 14.4, which describes the drug profile shown in Figure 14.1.

$$C = \frac{FD}{V} \left(\frac{k_a}{k_a - k_{el}} \right) (e^{-k_{el}t} - e^{-k_a t}) \qquad (14.4)$$

Now that the kinetic parameters associated with the one-compartment model with first-order absorption and elimination have been identified, the next step is to understand how numerical values are assigned to these parameters from a drug-concentration profile. Understanding the variable relationship between k_a, the absorption rate constant, and k_{el}, the elimination rate constant, is important. Three different situations can occur:

1. k_a may be greater than k_{el}.

2. k_a may be less than k_{el}.

3. The two constants may have the same, or approximately the same, numerical value.

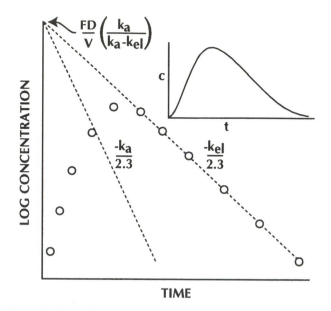

FIGURE 14.1 Plot of logarithm of plasma drug concentration vs. time for a drug that obeys one-compartment model kinetics with first-order absorption and elimination. The concentration is plotted on a linear scale in the inset.

In most instances, the first case occurs. The absorption rate constant (or the apparent absorption rate constant) of a drug is usually greater than the elimination rate constant. However, for drugs that have short biological half-lives of perhaps 1 h or less, and also for drugs that are absorbed slowly, the elimination rate constant may be greater than the absorption rate constant. Also, in many cases the values of the two constants may be so similar that they become analytically indistinguishable.

Graphical Estimation of Parameters

Case 1. In the most usual case, k_a is greater than k_{el}. Equation 14.4 shows that both of the exponential terms equal unity at time zero; C is therefore equal to 0. As time increases after dosing, the values of the exponential terms gradually diminish from unity toward zero, but the exponential term containing k_a (in this case, the larger rate constant) will approach zero at a faster rate than the term containing k_{el}. Therefore, after a certain time, the exponential term containing k_a becomes so small that equation 14.4 reduces to

$$C = \frac{FD}{V}\left(\frac{k_a}{k_a - k_{el}}\right)e^{-k_{el}t} \tag{14.5}$$

This equation, in logarithmic form, becomes

$$\log C = \log\left[\left(\frac{FD}{V}\right)\left(\frac{k_a}{k_a - k_{el}}\right)\right] - \left(\frac{k_{el}}{2.3}\right)t \tag{14.6}$$

Equation 14.6 describes a linear relationship between the logarithm of C and time, which has a slope of $-k_{el}/2.3$ and an intercept of $(FD/V)[k_a/(k_a - k_{el})]$. This linear postabsorptive phase is shown in Figure 14.1.

A suitable method to obtain k_a is the method of curve stripping, or residuals. Subtracting equation 14.4 from equation 14.5 yields the residual equation 14.7, which can be written in logarithmic form as equation 14.8.

$$R = \frac{FD}{V}\left(\frac{k_a}{k_a - k_{el}}\right)e^{-k_a t} \tag{14.7}$$

$$\log R = \log\left[\left(\frac{FD}{V}\right)\left(\frac{k_a}{k_a - k_{el}}\right)\right] - \left(\frac{k_a}{2.3}\right)t \tag{14.8}$$

Equation 14.8 is identical to equation 14.6 except that the slope is a function of k_a instead of k_{el}. Equation 14.8 was obtained by subtracting the equation describing the entire drug profile from the equation describing the extrapolated terminal elimination phase. Equation 14.8 can also be obtained graphically as shown in Figure 14.1 by subtracting the actual data points during the absorptive phase from the values on the extrapolated elimination slope. If these residuals are then plotted against time, an estimate of k_a is obtained. The intercepts of the terminal slope, from equation 14.6, and the residual slope, from equation 14.8, at time zero are identical. If they were not identical in practice, or at least nearly identical given usual biological data noise, then the model may be incorrect; nonlinear kinetics may be operative, or an absorption lag time may exist. Both of these situations will be discussed later.

From the curve-stripping procedures, estimates of k_a, k_{el}, and their associated half-lives have been obtained. The numerical value of the common intercept from both the terminal and residual slopes is also known. Therefore, the values of k_a and k_{el} can be substituted into the measured intercept value to solve for FD/V. The drug-concentration profile has thus been described mathematically in terms of the model using two rate constants and a concentration term, FD/V, which describes overall drug systemic availability.

To further resolve the function FD/V, solving for F, the fraction of dose absorbed, or V, the drug distribution volume in the body, is necessary. However, neither of these parameters can be solved for unless additional data, such as urinary excretion or intravenous plasma level data, are available. If the quantity of drug absorbed, F, is known by comparison with intravenous data, then V can be calculated. On the other hand, if V is known, then the value F can be calculated. However, in the absence of such data, the function FD/V cannot be further resolved. But even if unresolved, FD/V is a useful parameter, as will be demonstrated later in this chapter.

Case 2.   In the second kind of relationship between absorption and elimination rate constants, k_a is less than k_{el}. The equation that describes the drug-concentration profile in this case is identical to the first case because the form of the equation is independent of numerical values of the constants. However, the numerical values do influence how the equation is used to analyze the data.

Equation 14.4 is again appropriate for the analysis. However, in this case k_a is less than k_{el}, so both of the parenthetical terms in the equation become negative. Equation 14.4 is rearranged to equation 14.9, in which both parenthetical terms become positive.

$$C = \frac{FD}{V}\left(\frac{k_a}{k_{el} - k_a}\right)\left(e^{-k_a t} - e^{-k_{el} t}\right) \tag{14.9}$$

Because k_{el} is greater than k_a, the product $k_{el}t$ will increase with increasing time values at a faster rate than the product $k_a t$, so that the exponential term $e^{-k_{el}t}$ will be the first of the two exponential terms to approach zero and drop out of the equation. At that time, equation 14.9 reduces to equation 14.10, which can be written in logarithmic form as equation 14.11.

$$C = \frac{FD}{V}\left(\frac{k_a}{k_{el} - k_a}\right)e^{-k_a t} \tag{14.10}$$

$$\log C = \log\left[\left(\frac{FD}{V}\right)\left(\frac{k_a}{k_{el} - k_a}\right)\right] - \left(\frac{k_a}{2.3}\right)t \tag{14.11}$$

Thus, the terminal declining phase of the drug-level curve, which in the first example was controlled by the elimination rate constant, k_{el}, is now controlled by the absorption rate constant, k_a.

If residuals are taken between equations 14.10 and 14.9 in analogous fashion to that described previously between equations 14.4 and 14.5, the residual is described by equation 14.12, which in logarithmic form becomes equation 14.13.

$$R = \frac{FD}{V}\left(\frac{k_a}{k_{el} - k_a}\right)e^{-k_{el} t} \tag{14.12}$$

$$\log R = \log\left[\left(\frac{FD}{V}\right)\left(\frac{k_a}{k_{el} - k_a}\right)\right] - \left(\frac{k_{el}}{2.3}\right)t \tag{14.13}$$

Thus, the residual slope gives the elimination rate constant k_{el}. Also, the common intercept of the terminal and the residual lines at the y intercept at time zero differs from the analogous intercept when k_a was greater than k_{el}, because the values of k_a and k_{el} are transposed in the denominator.

This model, in which k_a is less than k_{el}, has been called the "flip-flop" model, for obvious reasons. Failure to recognize the possibility of this model may cause gross misinterpretation of data. This model is not rare, and it may apply to drugs that have short elimination half-lives, drugs that exhibit prolonged absorption, and combinations of these.

The situation just described leads to the question of which model is correct. How can the larger of the two rate constants, k_a or k_{el}, be determined? Sometimes assigning correct values is straightforward. For example, with drugs like digoxin or phenobarbital, which have elimination half-lives of 2 and 4 days, respectively, and even with drugs like tetracycline, which has a half-life of 8 to 10 h, it is intuitively obvious which of the rate constants is k_a and which is k_{el}. Because of the limited residence time the drug has in which to be absorbed from the GI tract, drug levels cannot possibly be prolonged over many hours or days in the situation where k_a is small relative to k_{el}, at least with GI absorption. Absorption would stop when drug was voided from the GI tract, the absorption phase would be truncated, and a much less prolonged profile than that actually observed would result. However, for drugs that have relatively short elimination half-lives of less than 4–6 h, correctly identifying the absorption and elimination rate constants may be difficult. The method of analyzing data with this model is summarized in Figure 14.2, which is analogous to Figure 14.1.

Case 3.   In the third kind of relationship between absorption and elimination rate constants, k_a is equal to k_{el}. Theoretically the two values will never be the same. However, the values may often be very similar and, with normal biological and individual variation, the values frequently become indistinguishable. The methods used for the first two situations cannot be used in this case. If k_a is set

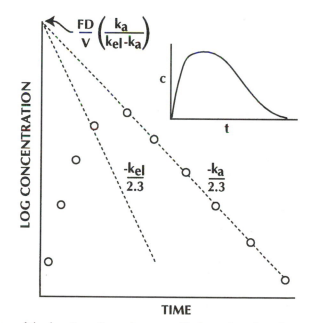

Interpretation of the log C vs. time plasma profile for a drug that obeys one-compartment model kinetics and for which the first-order absorption rate constant, k_a, is smaller than the first-order elimination rate constant, k_{el}. The concentration is plotted on a linear scale in the inset.

FIGURE 14.2

equal to k_{el} in equation 14.4 or in equation 14.9, then both equations become equal to zero. The correct rate equation for this model is equation 14.14, where k is the common value of the absorption and elimination rate constants.

$$\frac{dA}{dt} = kX - kA = k(X - A) \tag{14.14}$$

If X is defined as in equation 14.1, equation 14.14 becomes

$$\frac{dA}{dt} = k(FDe^{-kt} - A) \tag{14.15}$$

Integrating this expression and converting to concentration terms yield

$$C = \frac{FDkte^{-kt}}{V} \tag{14.16}$$

This equation is quite different from equations 14.9 and 14.14. The coefficient now contains the value of time and is therefore no longer a constant. Thus, although there is only one exponential term, which normally gives rise to a straight line when data are plotted against time on semilogarithmic graph paper, a straight line cannot be obtained during any phase of the drug level profile, be it absorption or elimination. Instead, a continuous convex curve is obtained. This situation frequently occurs. In practice, locating a linear elimination phase in a drug profile is frequently difficult. Maybe the linear elimination phase is not there! A true terminal linear phase may not exist, but probably the curve is completely curvilinear because of the similarity or identity of the absorption and elimination rate constants.

Other Parameters Associated with First-Order Absorption and Elimination

Further examination of the two cases where k_a does not equal k_{el} is necessary because these cases are the most common. Apart from the constants that have already been extracted from the data, such as k_a, k_{el}, their associated half-lives, and also FD/V, several other useful parameters can be derived. Three parameters that are frequently used, particularly in bioavailability or bioequivalence studies, are the maximum drug concentration in plasma, C_{max}, the time of the maximum drug concentration, T_{max}, and the area under the drug-concentration profile, $AUC^{0 \to t}$ or $AUC^{0 \to \infty}$. Together, C_{max}, T_{max}, and AUC provide an almost complete description of a drug-concentration profile.

The maximum drug concentration in the bloodstream, C_{max}, occurs at time T_{max}, which is the time when the drug profile is at its peak and the slope of the drug profile is changing sign, that is, the first derivative of equation 14.4 and equation 14.9 equals zero, as in equation 14.17.

$$\frac{FD}{V}\left(\frac{k_a}{k_a - k_{el}}\right)\left(-k_{el}e^{-k_{el}T_{max}} + k_a e^{-k_a T_{max}}\right) = 0 \tag{14.17}$$

The first part of this equation, $(FD/V)[k_a/(k_a - k_{el})]$, cannot equal zero because it comprises a constant. If this constant were zero, then the drug level would be zero for all values of t. Therefore, the exponential terms in parentheses must equal zero, as in

$$k_{el}e^{-k_{el}T_{max}} = k_a e^{-k_a T_{max}} \tag{14.18}$$

By simple rearrangement and taking logarithms, this expression can be written as

$$T_{max} = \left(\frac{1}{k_a - k_{el}}\right)\ln\frac{k_a}{k_{el}} \tag{14.19}$$

This equation shows that T_{max} is a function of the relative magnitude of the absorption and elimination rate constants k_a and k_{el}. The data in Table 14.1 show that the value of T_{max} is inversely related to both rate constants; T_{max} increases when the absorption rate is decreased, and also when the elimination rate is decreased, but not to quite the same extent. A fourfold change in the value of k_a causes a 2.5-fold change in the value of T_{max}. On the other hand, a tenfold change in k_{el} is required to have a similar effect on the value of T_{max}.

The value of C_{max} can be determined in a similar fashion to that for T_{max} and is given by

$$C_{max} = \frac{FD}{V}\left(\frac{k_a}{k_{el}}\right)^{\frac{k_{el}}{k_{el}-k_a}} \tag{14.20}$$

This expression is different in many ways from equation 14.19 for T_{max}. The primary difference is that the concentration term FD/V is included in addition to the rate constants. In other words, C_{max} is both drug concentration dependent and absorption efficiency dependent, whereas T_{max} is not. The data in Table 14.2 show that, whereas the value of T_{max} is inversely related to the absorption and elimination rate constants, the value of C_{max} is positively related to k_a but inversely related to k_{el}. This relationship makes good intuitive sense. A slower absorption rate constant produces a flatter drug profile and a lower value for C_{max}. A slower

Influence of k_a and k_{el} on T_{max}   **TABLE 14.1**

k_a (h⁻¹)	k_{el} (h⁻¹)	T_{max} (h)
2.0	0.1	1.6
1.0	0.1	2.6
0.5	0.1	4.0
1.0	0.5	1.4
1.0	0.1	2.6
1.0	0.05	3.2

Note: Data generated from equation 14.19.

elimination rate constant, on the other hand, permits blood levels to increase to a greater extent during the absorption phase, resulting in a higher C_{max} value.

Drug Absorption Plots

Searching for a satisfactory description of drug absorption as a function of time, Wagner and Nelson developed a method that applies only to the one-compartment model (4). For the two-compartment model, the alternative Loo–Riegelman method should be used (5). Because the Loo–Riegelman method is simply a modification of the one-compartment model approach, attention will focus here on the Wagner–Nelson method. To repeat the Wagner–Nelson derivation here would not be useful. The derivation is simple, based on mass balance principles, and culminates in equation 14.21.

$$\text{percent absorbed} = \frac{C_t + k_{el}\int_0^t C\,dt \times 100}{k_{el}\int_0^\infty C\,dt} \quad (14.21)$$

The numerator in this equation is the sum of the drug concentration at time t and the product of the drug elimination rate constant and the area under the drug-concentration curve from zero to time t. The denominator is the elimination rate constant multiplied by the area from zero to infinite time. The denominator is thus the limiting value for the numerator. That is, when C has dropped to zero, the drug-concentration profile is concluded, and all drug that is going to be absorbed will have been absorbed. Thus, after all drug is absorbed, equation 14.21 must equal unity, or 100%.

Although the details and implications of the Loo–Riegelman absorption model are beyond the scope of this text, it is interesting to note that the Loo–Riegelman equivalent to equation 14.21 is equation 14.22.

$$\text{percent absorbed} = \frac{C_{1(t)} + C_{2(t)} + k_{el}\int_0^t C_{1(t)}\,dt}{k_{el}\int_0^\infty C_{1(t)}\,dt} \quad (14.22)$$

In this equation, C_1 and C_2 are drug concentrations in the first and second compartments, respectively (see Chapter 17 and reference 5). Although the Loo–

TABLE 14.2	Influence of k_a and k_{el} on C_{max}		
	k_a (h^{-1})	k_{el} (h^{-1})	C_{max} $(\mu g/mL)$
	2.0	0.1	0.68
	1.0	0.1	0.46
	0.5	0.1	0.21
	1.0	0.5	0.01
	1.0	0.1	0.46
	1.0	0.05	0.67

Note: Data generated from equation 14.20; FD/V set at 100 $\mu g/mL$.

Riegelman method, and variations of it, may give improved fits to drug absorption data for drugs that obey two- or multicompartment kinetics, it often requires additional data, for example intravenous data, to set up the model parameters. The relative advantages of the two models have been discussed by Wagner (6) and by Gibaldi and Perrier (7).

The example in Table 14.3 demonstrates how the Wagner–Nelson method is used. As shown in the table, the Wagner–Nelson method consists of listing the drug concentration in plasma at each sampling time; calculating the cumulative area under the blood curve up to that time; multiplying each area by the elimination rate constant, k_{el}, which is obtained from the terminal log–linear portion of the curve; and adding the drug concentration in plasma at the particular sampling time. Thus, a series of gradually increasing values is obtained. In this example, an average maximum value of 9.407 is obtained. Each of the preceding values is then expressed as a percentage of the maximum. The cumulative percentage of the drug that is going to eventually be absorbed is represented at each time point. In this case, the drug is 69.7% absorbed at 3 h, 86% at 8 h, and 100% at 15 h. A plot of the percentage absorbed versus time provides a useful picture of how rapidly absorption occurs.

One shortcoming of the Wagner–Nelson method is that, because the numerator in equation 14.21 will always eventually equal the denominator, the cumulative absorption will always eventually equal unity, or 100% (8). Even if a drug is

Wagner–Nelson Method to Calculate Cumulative Drug Absorption — **TABLE 14.3**

t (h)	C ($\mu g/mL$)	$\int_0^t C \cdot dt^a$ ($\mu g/mL \cdot h$)	$C_t + k_{el} \int_0^t C \cdot dt^b$ ($\mu g/mL$)	$\dfrac{C_t + k_{el} \int_0^t C \cdot dt \times 100}{k_{el} \int_0^\infty C \cdot dt}$ (percent)
0	0	0	0	0
1	2.28	1.140	2.416	25.7
2	3.69	4.125	4.181	44.4
3	5.52	8.730	6.559	69.7
4	5.52	14.25	7.216	76.7
5	5.08	19.55	7.406	78.7
6	4.91	24.545	7.831	83.2
8	4.10	33.555	8.093	86.0
10	3.38	41.035	8.263	87.8
12	3.33	47.745	9.012	95.8
15	2.66	56.730	9.411	100.0
24	0.80	72.300	9.404	
28	0.49	74.88	9.401	Average = 9.407
32	0.31	76.48	9.411	

Note: Each value in the fourth column is expressed as a percentage of the asymptotic value, 9.407, to yield the percentage absorbed shown in the fifth column.

[a] Cumulative areas were estimated by trapezoidal rule from observed t and C values.

[b] Value of k_{el} estimated by method of least squares from ln C and t in the 15–32-h range was 0.119 h.

Source: Adapted with permission from reference 6.

only 10% bioavailable, the Wagner–Nelson plot will have an asymptote of 100%. This may be confusing when considering the relative bioavailability of different forms of a drug. For example, if one generic product was actually less efficiently absorbed than another generic product, the Wagner–Nelson plot would clearly indicate different absorption rates (if they were different) but could be interpreted to imply that each product was 100% absorbed.

A resolution to this problem is provided by a modification of the Wagner–Nelson method that is based on the cumulative relative fraction absorbed (CRFA) (9). The difference between the CRFA and the Wagner–Nelson methods is illustrated by equation 14.23, where A is the standard formulation and B is the test formulation.

$$\text{CRFA} = \frac{C_{t(B)} + k_{el(B)}\int_0^t C_{(B)}dt \times 100}{k_{el(A)}\int_0^\infty C_{(A)}dt} \tag{14.23}$$

In this equation, the numerator contains the values for the test product B, and the denominator contains the values for the standard product A. In the CRFA method, the percentage absorption from a test product versus time is plotted with reference to a standard product. Thus, if product B is actually only 50% absorbed compared to product A, then the maximum asymptotic value for equation 14.23 will be 50%.

Area Under the Drug-Concentration Curve (AUC)

Previously, two of the three fundamental pharmacokinetic parameters that characterize a drug concentration profile in plasma, C_{max} and T_{max}, were considered. The third parameter, the area under the drug-concentration versus time curve, AUC, describes how much drug is absorbed. The area under the drug-concentration curve appears in virtually all pharmacokinetics, bioavailability, and bioequivalence studies, and this value is required by regulatory agencies worldwide as a measure of drug systemic availability (10–14). The same methods of determining AUCs and converting truncated areas to complete areas after intravenous doses apply after oral or intramuscular doses. The trapezoidal determination of areas differs in these cases only in that at time zero the drug concentration is zero, whereas in the intravenous case, the maximum concentration occurs at this time. Thus, after oral dosing, the first small area under the drug concentration curve is a triangle rather than a true trapezoid. Apart from this difference, the two calculations are identical for the intravenous, oral, and intramuscular cases.

Of possibly greater interest to the pharmacokineticist are the inferences that arise from the derived expression for the area under the drug-concentration curve. As described previously, the time course of the drug concentration in blood or plasma is given by

$$C = \frac{FD}{V}\left(\frac{k_a}{k_a - k_{el}}\right)\left(e^{-k_{el}t} - e^{-k_a t}\right) \tag{14.24}$$

Integration of this expression between limits of zero and infinite time yields the area under the drug-concentration curve:

$$AUC^{0\to\infty} = \frac{FD}{Vk_{el}} \qquad (14.25)$$

The equivalent expression for the area from zero to infinite time for the bolus intravenous case is

$$AUC^{0\to\infty} = \frac{D}{Vk_{el}} \qquad (14.26)$$

Comparing equations 14.25 and 14.26 shows that the two expressions are identical except for the value F, the fraction of dose that is systemically available.

The denominator, which is the plasma clearance, is common to both equations, but so is the absence of the absorption rate constant, k_a. This absence might be expected in the intravenous case because no absorption phase occurs. However, this phase does exist after oral or intramuscular dosing, or any dosing that involves first-order absorption, and is represented by the constant k_a in equation 14.24. However, during the integration process to achieve equation 14.25, the term k_a disappears. This disappearance shows that the area under the concentration curve is affected by the elimination rate constant but is independent of the absorption rate constant. The area under the curve is unaffected by the rate of drug absorption, but the overall shape of the curve is affected. The possible clinical impact of this situation is illustrated in Table 14.4. If the therapeutically effective drug concentration were above 69 µg/mL in this instance, then the formulation that provided the slowest release would be useless (at least in this single-dose example). This occurs despite the fact that the overall bioavailability, as reflected in the area under the curve, is identical in the three formulations.

Plasma Profiles from Formulations with Fast, Medium, and Slow Release Rates   **TABLE 14.4**

t (h)	Drug Concentration in Plasma (µg/mL)		
	$k_a = 3.0\ h^{-1}$	$k_a = 1.5\ h^{-1}$	$k_a = 0.5\ h^{-1}$
0	0	0	0
0.5	75.3	51.3	21.6
1	88.4	73.0	37.3
1.5	87.8	80.8	48.5
2	84.4	82.4	56.3
3	76.6	78.2	64.7
4	69.3	71.5	66.9
6	56.7	58.8	62.4
8	40.5	48.1	54.6
12	31.1	32.3	37.3
24	9.4	9.7	11.3
48	1.0	1.0	1.1
$AUC^{0\to\infty}$ (µg · h/mL)	1000	1000	1000

Note: Profiles constructed from equation 14.23; FD/V set at 100 µg/mL, and k_{el} at 0.1 h^{-1}.

AUC is thus independent of k_a, but dependent on k_{el}. Equation 14.25 is rearranged to equation 14.27, which shows that the overall absorption efficiency is equal to the product of k_{el} and AUC. Thus, in bioavailability and bioequivalency determinations, possible variations in k_{el} between subjects and between treatments must be considered because of their possible effect on the area under the plasma curve. For a true comparison of systemic availability, k_{el} AUC$^{0 \rightarrow \infty}$ values should be compared, as in equation 14.27, rather than just the areas.

$$\frac{FD}{V} = k_{el} \cdot \text{AUC}^{0 \rightarrow \infty} \qquad (14.27)$$

The parameter FD/V is useful even though it often cannot be further resolved to obtain estimates of F or V separately. In Chapter 13, the AUC following intravenous dosing was described by equation 13.16. This equation is rearranged to

$$\frac{D}{V} = k_{el} \cdot \text{AUC}^{0 \rightarrow \infty} \qquad (14.28)$$

The overall systemic availability from an oral or intramuscular dose compared with an intravenous standard is determined by dividing equation 14.27 by equation 14.28, as in equation 14.29

$$\frac{FD/V_{(PO)}}{D/V_{(IV)}} = \frac{k_{el} \cdot \text{AUC}^{0 \rightarrow \infty}_{(PO)}}{k_{el} \cdot \text{AUC}^{0 \rightarrow \infty}_{(IV)}} = F \qquad (14.29)$$

where PO refers to oral dose and IV refers to intravenous dose. This relationship can be used only when the same compound is given by both routes.

On the other hand, comparison of relative absorption efficiency of drug from two oral formulations, A and B, or perhaps between an intramuscular or rectal dose A and an oral dose B, is obtained from the ratio of equation 14.27 for each of the two dosages:

$$\frac{FD/V_{(A)}}{FD/V_{(B)}} = \frac{k_{el} \cdot \text{AUC}^{0 \rightarrow \infty}_{(A)}}{k_{el} \cdot \text{AUC}^{0 \rightarrow \infty}_{(B)}} = \frac{F_{(A)}}{F_{(B)}} \qquad (14.30)$$

If, for example, the ratio F_A to F_B is 0.9 after adjusting for possible differences in dose size, then the overall systemic availability from formulation A is 90% compared with formulation B. However, this comparison provides no information regarding the absolute systemic availability from either formulation. Formulation B may be 100% available or 1% available.

Absorption Lag Time

Very often, particularly with enteric-coated dosage forms and some slowly dissolving products, a time period may elapse between the time of drug administration and the time that measurable drug initially appears in plasma. This lag time is frequently difficult to detect and even more difficult to measure, principally because of the relatively small number of blood samples generally taken during the absorption phase of a blood profile, but also because of the relatively high data variability generally observed during this phase.

Interpretation of the meaning of an absorption lag time is difficult. If a lag time is not included in the interpretation of a drug-concentration profile, then prolonged absorption is reflected in a small value of k_a. If, on the other hand, a lag time is included, then prolonged absorption is reflected in a lag time and a larger value of k_a. Separating these two alternative interpretations of a drug profile is often difficult. When fitting data by nonlinear regression, many of the available computer programs tend to maximize the lag time and also maximize k_a to provide a good fit to the data. This method frequently provides a better fit to the drug-concentration profile than the fit when lag time is not used, but in doing so provides biased estimates of these two pharmacokinetic parameters. In these situations, the lag time often must be fixed at the initial estimated value when computer fitting.

Incorporating a lag time into equation 14.24 results in

$$C = \frac{FD}{V}\left(\frac{k_a}{k_a - k_{el}}\right)\left(e^{-k_{el}(t-t_0)} - e^{k_a(t-t_0)}\right) \qquad (14.31)$$

The only difference between equations 14.24 and 14.31 is that the term t in the exponential function has been replaced by $t - t_0$.

When a lag time occurs, it can be recognized in two ways: intuitively and analytically. Intuitively, if undetectable drug levels are obtained during initial sampling periods, then a lag time exists. On the other hand, if the first sampling detects drug, but at very low levels, then whether a true lag time should be included in the absorption phase is not immediately obvious. Several possible approaches exist to resolve this dilemma. One of them is described here.

Previously, methods were described to strip a drug-concentration curve based on equation 14.24. The treatment demonstrated that, if the model was correct for the data, the intercepts of the terminal elimination slope and the residual slope on the y axis would be identical and would equal $(FD/V)[k_a/(k_a - k_{el})]$. This situation was illustrated in Figure 14.1.

The situation when a lag time exists is illustrated in Figure 14.3. Interpretation of a drug profile by the method of residuals yields different intercepts from the extrapolated elimination and residual slopes at time zero, the time at which the dose was administered. In this case, the lag time, and also consequently the true value of FD/V, can be estimated from the point at which the two slopes intercept. Other ways exist for estimating the lag time, including direct computer analysis, but the graphical procedure is simple and accurate enough to provide initial estimates of this value.

Equation 14.3 described the time course of the amount of drug in the body with respect to time following an oral or intramuscular dose. The rate equation for urinary recovery of unchanged drug, dA_u/dt, is found by multiplying that equation by k_e (the rate constant for urinary excretion):

Cumulative Urinary Excretion of Unchanged Drug

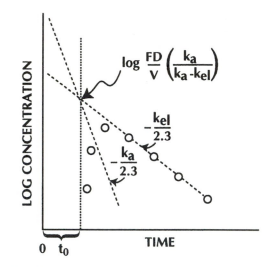

FIGURE 14.3   Graphical method to determine absorption lag time.

$$\frac{dA_u}{dt} = k_e FD\left(\frac{k_a}{k_a - k_{el}}\right)\left(e^{-k_{el}t} - e^{k_a t}\right) \tag{14.32}$$

Integration of this expression yields

$$A_u = \frac{FDk_e}{k_{el}}\left(1 - \frac{1}{k_a - k_{el}}\right)\left(k_a e^{-k_{el}t} - k_{el}e^{-k_a t}\right) \tag{14.33}$$

This equation describes cumulative urinary recovery of unchanged drug with respect to time. In this example, no absorption lag time is assumed. However, if a lag time did exist, the equation could readily be adapted simply by replacing t with $t - t_0$. Equation 14.33 can also be written as equation 14.34, where A_u^∞ is total urinary recovery of unchanged drug. This representation is analogous to equation 13.27 for the intravenous case.

$$A_u = A_u^\infty\left(1 - \frac{1}{k_a - k_{el}}\right)\left(k_a e^{-k_{el}t} - k_{el}e^{-k_a t}\right) \tag{14.34}$$

This equation shows that at time 0, the exponential terms reduce to $k_a - k_{el}$, and therefore $A_u = 0$. On the other hand, when time increases to a very large value, or infinity, $A_u = A_u^\infty$. A cumulative urinary excretion plot can be constructed between these limits. In exactly the same way as in the intravenous case, the elimination phase of this equation is controlled by the overall elimination rate constant, k_{el}, rather than k_e, and the ratio k_e/k_{el} determines the actual quantity of drug recovered by this route, that is, $A_u^\infty = FD(k_e/k_{el})$. If all of the drug is cleared unchanged in the urine, then $k_e = k_{el}$ and $A_u^\infty = FD$. On the other hand, if all of the drug is metabolized, then $k_e = 0$ and A_u will also equal zero, and no drug will be voided unchanged in the urine.

If k_a is greater than k_{el}, how might a sigma-minus plot be constructed from urinary excretion data? This situation is more complex than in the intravenous case because there are two exponential terms in equations 14.33 and 14.34. Consider equation 14.34. If k_a is greater than k_{el}, then as time postdose increases, the exponential term $e^{-k_a t}$ will approach zero more quickly than the exponential term $e^{-k_{el} t}$. When that occurs, equation 14.34 will reduce to

$$A_u = A_u^\infty \left(1 - \frac{1}{k_a - k_{el}}\right) k_a e^{-k_{el} t} \qquad (14.35)$$

Equation 14.35 is rearranged to equation 14.36, and in logarithmic form to equation 14.37.

$$A_u^\infty - A_u = A_u^\infty \frac{k_a}{k_a - k_{el}} e^{-k_{el} t} \qquad (14.36)$$

$$\log(A_u^\infty - A_u) = \log\left(A_u^\infty \frac{k_a}{k_a - k_{el}}\right) - \frac{k_{el}}{2.3} t \qquad (14.37)$$

This equation may be used to construct sigma-minus plots from urinary excretion data. If the logarithm of $A_u^\infty - A_u$, that is, the quantity of drug remaining to be excreted, is plotted against time, a slope of $-k_{el}/2.3$ and an intercept of $\log[A_u^\infty(k_a/(k_a - k_{el}))]$ are obtained. This plot is illustrated in Figure 14.4. Interpretation of the slope of the line and the intercept of the extrapolated line at zero time in this case are similar to the previous interpretation of drug-concentration data in plasma for this model (Figure 14.1).

The numerical value of the overall elimination rate constant, k_{el}, is thus obtained from the slope of the terminal linear portion of the profile. The slope is not linear at early times postdose but appears to drop off in an inverted hockey-stick shape. This effect is due to the absorption phase contributing to the profile at early times. If the value of the intercept of the straight portion of the profile, A_u^∞, and k_{el} are known, then k_a can be calculated from

$$k_a = \frac{(\text{intercept}) k_e}{\text{intercept} - A_u^\infty} \qquad (14.38)$$

The value of k_e is obtained from the ratio A_u^∞/FD if the value of F is known. The value k_m is then obtained by subtracting k_e from k_{el}. However, if the value of F is not known, neither k_e nor k_m can be calculated. For example, if only 50% of an oral dose is recovered unchanged in the urine, distinguishing whether the other 50% is metabolized, excreted by some other route, or simply not absorbed is impossible unless intravenous data are available for comparison, or some other means is available to establish systemic availability.

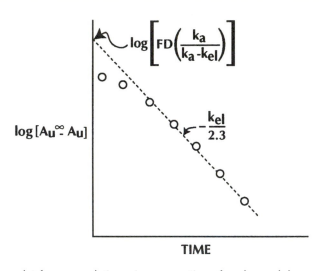

FIGURE 14.4   Sigma-minus plot from cumulative urinary excretion of unchanged drug.

If the flip-flop model occurs, that is, the absorption rate constant is smaller than the elimination rate constant, then the analytical procedure is identical to the procedure just described, but the sigma-minus expression will be expressed as

$$\log(A_\mathrm{u}^\infty - A_\mathrm{u}) = \log\left(A_\mathrm{u}^\infty \frac{k_\mathrm{el}}{k_\mathrm{el} - k_\mathrm{a}}\right) - \frac{k_\mathrm{a}}{2.3}t \qquad (14.39)$$

The terminal slope will then yield k_a, and k_el will be obtained from

$$k_\mathrm{el} = \frac{(\text{intercept})k_\mathrm{e}}{\text{intercept} - A_\mathrm{u}^\infty} \qquad (14.40)$$

The problem here is again one of establishing which rate constant is which. Again, the best way to establish this is from intravenous data, where the elimination rate constant can be obtained independent of any absorption effect.

Urinary excretion data can be used to determine drug bioavailability in much the same way as drug concentrations in plasma, and often more effectively (*15*, *16*).

From Scheme 14.1, urinary recovery of an orally administered drug is given by

$$\text{Urinary recovey} = fFD \qquad (14.41)$$

where F is the fraction of drug systemically available and f is the fraction of systemically available drug that is excreted unchanged in urine ($k_\mathrm{e}/k_\mathrm{el}$). By comparing the urinary recovery of unchanged drug from two formulations (A and B), the relationship in equation 14.42 can be obtained.

$$\frac{A_{uA}^{\infty}}{A_{uB}^{\infty}} = \frac{f_A F_A D_A}{f_B F_B D_B} \qquad (14.42)$$

Assuming that D_A equals D_B, or at least the actual administered drug doses can be accounted for, and also $f_A = f_B$, that is, the same relationship between renal and plasma clearance is obtained with the two formulations, then equation 14.42 reduces to

$$\frac{A_{uA}^{\infty}}{A_{uB}^{\infty}} = \frac{F_A}{F_B} \qquad (14.43)$$

From this equation the relative urinary excretion from the two formulations provides a direct measure of relative bioavailability. If formulation B is an intravenous injection, in other words if formulation B is 100% systemically available, then F_B can be set equal to unity, and the ratio of urinary recovery from the test formulation and the intravenous dose yields F, the absolute absorption efficiency.

Instead of measuring urinary recovery of unchanged drug, measuring recovery of a metabolite can work just as well. This approach has been used, for example, to determine griseofulvin absorption efficacy by measuring urinary recovery of the major metabolite, desmethyl griseofulvin. In this case, urinary recovery is related to available drug as in equation 14.44, where f' is the fraction of drug excreted as the metabolite (k_m/k_{el}).

$$\frac{M_{uA}^{\infty}}{M_{uB}^{\infty}} = \frac{f'_A F_A D_A}{f'_B F_B D_B} \qquad (14.44)$$

Summary

■ Absorption of drug into the systemic circulation from the GI tract is a complex phenomenon. Interpretation of absorption as a first-order process is usually a simplification.

■ The apparent absorption rate constant is a function of loss of drug from the absorption site by all processes.

■ Correct interpretation of a drug-concentration profile in plasma depends on the relative value of the absorption rate constant, which may be greater than, equal to, or less than the elimination rate constant.

■ Estimates of rate constants can be made by curve stripping procedures, or residuals.

■ Drug absorption rates can be obtained from plasma data, independent of elimination rates, using the Wagner–Nelson method for the one-compartment model or the Loo–Riegelman method for the two-compartment model. An alternative method is based on CRFA.

■ The area under the drug profile in plasma can be used to determine relative bioavailability or absolute systemic availability of an orally dosed drug.

For a drug whose absorption is delayed, incorporation of an absorption lag time is necessary to obtain accurate estimates of the absorption rate constant.

Sigma-minus plots from urinary excretion data can be used to obtain estimates of rate constants for drug absorption, excretion, and metabolism.

Problems

1. A 500-mg dose of the sulfonamide sulfamethoxazole is administered as an oral tablet to a human subject. Eighty percent of the drug is absorbed, and the balance is excreted unchanged in feces. The drug distributes into an apparently homogeneous body volume of 12 L, and has an absorption half-life of 15 min and an overall elimination half-life of 12 h. Calculate the following: (i) $AUC^{0 \to \infty}$, (ii) C_{max}, and (iii) T_{max}.

2. Recalculate the values in Problem 1 if all parameter values remained unchanged, but the elimination half-life was increased to 18 h.

3. In a controlled, crossover bioavailability study of oral dosage forms, formulation A, containing 250 mg of a drug, was tested against formulation B, containing 150 mg of the same drug. The drug is extensively metabolized in the body to one major metabolite. Urinary recovery of the metabolite was 210 mg from formulation A and 130 mg from formulation B. Calculate the ratio of systemic availability from formulation A to availability from formulation B.

4. After 400 mg of a drug is administered orally, 50% is absorbed into the circulation. The drug has a first-order absorption rate constant of 2.0 h^{-1} and is eliminated from the body in equal proportions by metabolism and by excretion in unchanged form in urine. The overall biological half-life is 3.46 h. If a sigma-minus plot were constructed from the urinary excretion data, what would be the values of (i) the rate constant from the slope of the line, and (ii) the intercept of the line on the y ordinate ($t = 0$)?

5. A patient receives a drug dose of 400 mg orally. The drug is 100% bioavailable. The drug is not appreciably bound to plasma proteins and, of the absorbed drug, 25% is metabolized and 75% is excreted as unchanged drug by the kidneys. The following drug concentrations were obtained after a single dose:

Time (h)	Concentration (µg/mL)
0.0	0.0
0.25	6.1
0.5	10.0
1	13.8
2	14.3
4	10.3
8	4.7
12	2.1
24	0.2

Calculate the following: (i) k_{el}, (ii) t, (iii) k_a, (iv) t, (v) distribution volume, and (vi) plasma clearance.

6. From the data in Problem 5, construct a plot of cumulative excretion of drug (mg) in urine. Use Cartesian coordinate graph paper and the same time intervals as for the drug concentration in plasma data. Calculate: (i) k_{el}, (ii) k_e, and (iii) k_m. Are these values the same as from the plasma data?

7. From the data in Problems 5 and 6, calculate the renal clearance. Why is the renal clearance different from the plasma clearance?

References

1. Viswanathan, C. T.; Booker, H. E.; Welling, P. G. *J. Clin. Pharm.* **1978**, *18*, 100–105.
2. Patel, R. B.; Welling, P. G. *J. Pharm. Sci.* **1982**, *71*, 529–532.
3. Li, V. J. K.; Robinson, J. R.; Lee, V. H. L. *Controlled Drug Delivery: Fundamentals and Applications*; Robinson, J. R.; Lee, V. H. L., Eds.; Marcel Dekker: New York, 1987; pp 55–56.
4. Wagner, J. G.; Nelson, E. *J. Pharm. Sci.* **1964**, *53*, 1392–1394.
5. Loo, J. C. K.; Riegelman, S. *J. Pharm. Sci.* **1968**, *57*, 918–928.
6. Wagner, J. G. *Fundamentals of Clinical Pharmacokinetics*; Drug Intelligence Publications: Hamilton, IL, 1975; pp 185–189.
7. Gibaldi, M.; Perrier, D. *Pharmacokinetics*; Marcel Dekker: New York, 1982; pp 155–161.
8. Wagner, J. G. *Fundamentals of Clinical Pharmacokinetics*; Drug Intelligence Publications: Hamilton, IL, 1975; p 171.
9. Welling, P. G.; Patel, R. B.; Patel, U. R.; Gillespie, W. R.; Craig, W. A.; Albert, K. S. *J. Pharm. Sci.* **1982**, *71*, 1259–1263.
10. Dighe, S. V.; Adams, W. P. *Pharmaceutical Bioequivalence*; Welling, P. G.; Tse, F. L. S.; Dighe, S. V., Eds.; Marcel Dekker: New York, 1991; pp 347–380.
11. McGilveray, I. J. *Pharmaceutical Bioequivalence*; Welling, P. G.; Tse, F. L. S.; Dighe, S. V., Eds.; Marcel Dekker: New York, 1991; pp 381–418.
12. Rauws, A. G. *Pharmaceutical Bioequivalence*; Welling, P. G.; Tse, F. L. S.; Dighe, S. V., Eds.; Marcel Dekker: New York, 1991; pp 419–442.
13. Walters, S.; Hall, R. C. *Pharmaceutical Bioequivalence*; Welling, P. G.; Tse, F. L. S.; Dighe, S. V., Eds.; Marcel Dekker: New York, 1991; pp 443–452.
14. Salmonson, T.; Melander, H.; Rane, A. *Pharmaceutical Bioequivalence*; Welling, P. G.; Tse, F. L. S.; Dighe, S. V., Eds.; Marcel Dekker: New York, 1991; pp 453–462.
15. Kwan, K. C.; Till, A. E. *J. Pharm. Sci.* **1973**, *62*, 1494–1497.
16. Glazko, A. J.; Kinkel, A. W.; Alegnani, W. C.; Holmes, G. L. *Clin. Pharmacol. Ther.* **1968**, *9*, 472–483.

Multiple-Dose Kinetics

15

 ost drug therapy involves repeated administration of separate doses over a variable period. In this aspect of drug therapy, a knowledge of pharmacokinetics plays a particularly important role. Factors that are important to know are how long therapy should be continued, whether a few days for an antibiotic or prolonged periods for an antidiabetic or hypotensive agent; how frequently a drug should be given to achieve the optimum blood-level profile; what type of accumulation, if any, will occur with repeated doses; and how long it will take for steady-state levels to be achieved (1). Thus, for any drug intended for long-term use it is important to examine the pharmacokinetics under multiple-dosing conditions (2). Multiple-dose studies may also provide important information not obtainable in single-dose studies. For example, Steinijans et al. (3) examined the pharmacokinetics of theophylline in subjects after single and repeated doses and showed that although single- and multiple-dose AUC (area under the drug-concentration curve) data gave similar estimates of extent of bioavailability, the peak-trough characteristics at steady state were not accurately predicted from single-dose information.

The treatment in this chapter is based on simple one-compartment and first-order kinetics and assumes that drug is given by multiple intravenous bolus injections. Despite the obvious simplifications associated with this approach, provided that the kinetics are first-order and linear, conclusions drawn from the intravenous bolus approach apply to other dosage routes that may involve repeated intravenous infusion, multiple oral or intramuscular doses, or any other means of drug administration (4, 5). Another assumption is that the drug is given at equally spaced intervals. This may not always apply in practice. The more irregular the drug dosing intervals, the more difficult will be the characterization or prediction

Drug Accumulation with Repeated Doses

of accumulation and the actual steady-state drug profile. Computer programs are available for blood-level analysis in this type of situation.

In Chapter 13, the quantity of drug in the body with respect to time following bolus intravenous injection was described by equation 13.6, where the initial amount of drug in the body, A_0, is equal to the dose, D, and k_{el} is the elimination rate constant. That equation is reproduced here as equation 15.1.

$$A = A_0 e^{-k_{el}t} = D e^{-k_{el}t} \tag{15.1}$$

If a certain time period, τ, is allowed to elapse after the dose, then the amount of drug in the body, $A_{1(min)}$, is given by

$$A_{1(min)} = D e^{-k_{el}\tau} \tag{15.2}$$

If τ is sufficiently long, that is, longer than 4.5 drug half-lives, then A will approach zero at time τ. However, if τ is not long compared to the drug half-life, then A at time τ will be less than D but greater than zero. If another dose is added at time τ, then the amount of drug in the body, $A_{2(max)}$, will equal that amount remaining from the previous dose plus the new dose:

$$A_{2(max)} = D e^{-k_{el}\tau} + D = D(1 + e^{-k_{el}\tau}) \tag{15.3}$$

At the end of another time interval τ, the quantity of drug remaining in the body, $A_{2(min)}$, will be described by

$$A_{2(min)} = D(1 + e^{-k_{el}\tau})e^{-k_{el}\tau} = D(e^{-k_{el}\tau} + e^{-2k_{el}\tau}) \tag{15.4}$$

This process can be repeated with more doses so that expressions for $A_{n(max)}$ and $A_{n(min)}$ can be written after each dose and each dosage interval, where n is the number of doses. From equations 15.3 and 15.4, the general equations 15.5 and 15.6 can be written to describe maximum and minimum quantities of drug in the body following n doses.

$$A_{n(max)} = D(1 + e^{-k_{el}\tau} + \dots + e^{-(n-1)k_{el}\tau}) \tag{15.5}$$

$$A_{n(min)} = D(e^{-k_{el}\tau} + e^{-2k_{el}\tau} + \dots e^{-nk_{el}\tau}) \tag{15.6}$$

These expressions become unwieldy as additional doses are added. To obtain simpler expressions, equations 15.5 and 15.6 can be written in shorthand form as equations 15.7 and 15.8, respectively.

$$A_{n(max)} = D \cdot X \tag{15.7}$$

$$A_{n(min)} = D \cdot X e^{-k_{el}\tau} \tag{15.8}$$

In these equations, the quantity X is equal to the entire parenthetical term in equation 15.5. If equation 15.8 is subtracted from equation 15.7, equation 15.9 is obtained.

$$A_{n(max)} - A_{n(min)} = DX(1 - e^{-k_{el}\tau}) \tag{15.9}$$

Subtracting equation 15.6 from equation 15.5 yields equation 15.10.

$$A_{n(max)} - A_{n(min)} = D(1 - e^{-nk_{el}\tau}) \tag{15.10}$$

The left side of equation 15.10 is identical to the left side of equation 15.9; therefore, equation 15.11 can be written, and this equation reduces to equation 15.12.

$$DX(1 - e^{-k_{el}\tau}) = D(1 - e^{-nk_{el}\tau}) \tag{15.11}$$

$$X = \frac{1 - e^{-nk_{el}\tau}}{1 - e^{-k_{el}\tau}} \tag{15.12}$$

Equation 15.12 is a short and manageable form of shorthand notation that can be substituted into equations 15.7 and 15.8 to obtain equations 15.13 and 15.14.

$$A_{n(max)} = \frac{D(1 - e^{-nk_{el}\tau})}{1 - e^{-k_{el}\tau}} \tag{15.13}$$

$$A_{n(min)} = \frac{D(1 - e^{-nk_{el}\tau})e^{-k_{el}\tau}}{1 - e^{-k_{el}\tau}} \tag{15.14}$$

These two expressions are general terms that describe the maximum and minimum amounts of drug in the body after n equally spaced doses with dosage interval τ.

A third expression, equation 15.15, can now be written to describe the quantity of drug in the body at any time during a dosage interval, where the term t is the time elapsed since the last dose.

$$A_{n(t)} = \frac{D(1 - e^{-nk_{el}\tau})e^{-k_{el}t}}{1 - e^{-k_{el}\tau}} \tag{15.15}$$

Three equations, 15.13, 15.14, and 15.15, have now been derived in terms of amounts of drug. These expressions can be converted to concentration terms simply by dividing both sides of the equations by the distribution volume, V, to give equations 15.16, 15.17, and 15.18.

$$C_{n(max)} = \frac{D(1 - e^{-nk_{el}\tau})}{V(1 - e^{-k_{el}\tau})} \tag{15.16}$$

$$C_{n(min)} = \frac{D(1 - e^{-nk_{el}\tau})e^{-k_{el}\tau}}{V(1 - e^{-k_{el}\tau})} \tag{15.17}$$

$$C_{n(t)} = \frac{D(1 - e^{-nk_{el}\tau})e^{-k_{el}t}}{V(1 - e^{-k_{el}\tau})} \tag{15.18}$$

These expressions describe the concentration of drug, C, in the body in blood, plasma, or serum during repeated dosing. If dosing is continued, accumulation will also continue until eventually the rate of drug elimination will equal the rate of drug administration, and steady state will be achieved. Steady state will continue indefinitely, provided the drug dosage and elimination rate remain constant. Equations 15.16–15.18 can be rewritten with n equal to ∞ as in equations 15.19–15.21.

$$C_{\infty(\text{max})} = \frac{D}{V(1 - e^{-k_{el}\tau})} \tag{15.19}$$

$$C_{\infty(\text{min})} = \frac{De^{-k_{el}\tau}}{V(1 - e^{-k_{el}\tau})} \tag{15.20}$$

$$C_{\infty(t)} = \frac{De^{-k_{el}t}}{V(1 - e^{-k_{el}\tau})} \tag{15.21}$$

Equations 15.19–15.21 describe the maximum, minimum, and intermediate drug concentrations, respectively, at steady state for a given dose, distribution volume, elimination rate constant, and dosage interval.

The pattern of drug-accumulation and steady-state $C_{\infty(\text{max})}$ and $C_{\infty(\text{min})}$ values during repeated intravenous bolus injections at intervals of τ h is illustrated in Figure 15.1.

Having been defined, $C_{\infty(\text{max})}$, $C_{\infty(\text{min})}$, and $C_{\infty(t)}$ can now be used to demonstrate some basic rules regarding multiple-dose kinetics. For example, knowing to what extent drug concentrations oscillate between maximum and minimum values at steady state is often important. How wide apart are the peaks and troughs? This information can be obtained simply by subtracting $C_{\infty(\text{min})}$ from $C_{\infty(\text{max})}$ as in equation 15.22.

$$\begin{aligned} C_{\infty(\text{max})} - C_{\infty(\text{min})} &= \frac{D}{V(1 - e^{-k_{el}\tau})} - \frac{De^{-k_{el}\tau}}{V(1 - e^{-k_{el}\tau})} \\ &= \frac{D(1 - e^{-k_{el}\tau})}{V(1 - e^{-k_{el}\tau})} = \frac{D}{V} = C_0 \end{aligned} \tag{15.22}$$

This equation shows that the difference between maximum and minimum concentrations at steady state is identical to the maximum concentration obtained after the initial dose, C_0. This identity applies whether the drug is given by intravenous injection, orally, or by any other dosage route. This relationship is not that remarkable. With each successive dose, a certain concentration of drug is added to the concentration remaining from the previous dose. The drug concentration preceding the initial dose is zero. If a maximum drug concentration of 10 μg/mL is obtained following an initial dose of a drug, then, regardless of the extent of accumulation, the maximum drug concentration in plasma will always be 10 μg/mL greater than the minimum drug concentration at steady state.

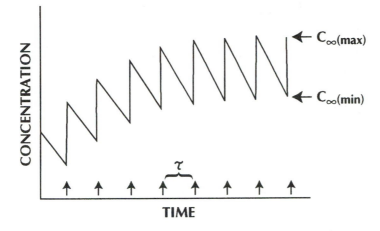

Drug concentrations during multiple intravenous injections, during the accumulation phase, and at steady state.

FIGURE 15.1

Equation 15.21 describes the drug concentration at any time t during a dosage interval at steady state. If that expression is integrated between the limits $t = 0$ and $t = \tau$ during a dosage interval at steady state, equation 15.23 is obtained.

$$\int_0^\tau \frac{De^{-k_{el}t}}{V(1-e^{-k_{el}\tau})}\,dt = \frac{D}{V(1-e^{-k_{el}\tau})}\int_0^\tau e^{-k_{el}t}\,dt = \frac{D}{Vk_{el}} = \mathrm{AUC}^{0 \to \tau} \qquad (15.23)$$

This expression describes the area under the drug-concentration curve during a dosage interval τ and is identical to the expression for the area under the drug-concentration curve from zero to infinite time following a single intravenous dose (equation 13.16). Thus, the area under the curve during a dosage interval at steady state during a multiple-dose regimen is equal to the area from zero to infinite time following a single dose, provided that the same dose is administered and the kinetics are unchanged during the multiple doses. This situation is illustrated in Figure 15.2 and applies equally to oral and intravenous dosing. The only difference is that in the oral case, the area expression must contain the parameter F, the fraction of dose absorbed into the systemic circulation.

Because of the identity between the single dose and steady-state areas, the plasma clearance, Vk_{el}, can readily be obtained from multiple dose data just as from single dose data, by dividing the dose by the appropriate area under the plasma curve. The steady-state area can be used also to determine whether any changes occur in the pharmacokinetic behavior of a drug during multiple dosing. For example, if the multiple-dose area from zero to τ is less than the single-dose area from zero to infinite time (or less than one would predict from the single-dose area), then this difference may be evidence of enzyme induction or of an increased drug distribution volume with repeated doses. If, on the other hand, the repeated-dose area is greater than the single-dose area, then this may be evidence of drug-

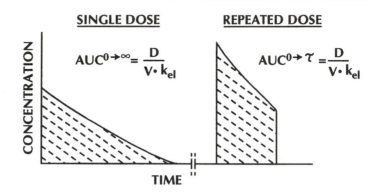

FIGURE 15.2

Areas under drug concentration vs. time profiles in plasma following a single dose, and at steady state following repeated doses.

metabolizing enzyme inhibition or of reduction in the drug distribution volume. After oral doses, drug bioavailability may be altered during multiple doses because of saturable absorption, increased gastrointestinal degradation, altered gastrointestinal physiology (perhaps due to the drug), or altered first-pass metabolism associated with enzyme induction or inhibition.

Although the steady-state area is obtained during a dosage interval, equation 15.23 does not contain the value τ. This indicates that the identity between the steady-state zero to τ area and the single-dose zero to infinity area remains the same, regardless of the dosage interval. How frequently or infrequently a drug is given is not important.

A parameter that is used frequently in multiple-dose kinetics is the mean drug concentration at steady state, $\overline{C}_\infty$. This expression is readily obtained by dividing equation 15.23 by the dosing interval τ, to yield.

$$\overline{C}_\infty = \frac{D}{Vk_{el}\tau} \tag{15.24}$$

This equation is valid for both intravenous and oral doses and has some interesting properties in the oral case (6). However, until the oral case is considered in detail, equation 15.24 may be considered a universal description of mean steady-state drug levels between $C_{\infty(max)}$ and $C_{\infty(min)}$.

Drug Accumulation Rate

Important questions that arise when considering drug accumulation with repeated doses are how fast and how much. How long does it take for drug concentrations in plasma to reach or approach steady-state values, and to what extent will levels increase during that time?

Answers to those questions will be based on equations 15.16 and 15.19 for C_{max}. Dividing equation 15.16 by equation 15.19 and canceling like terms yields

$$\frac{C_{n(max)}}{C_{\infty(max)}} = 1 - e^{-k_{el}\tau} \tag{15.25}$$

This equation can be revised and simplified to obtain a more general expression relating the frequency of drug dosing and half-life. Because k_{el} equals $\ln 2/t_{1/2}$, equation 15.25 can be written as equation 15.26, in which $t_{1/2}$ is the drug elimination half-life.

$$\frac{C_{n(\text{max})}}{C_{\infty(\text{max})}} = 1 - e^{-n \cdot \ln 2 \cdot \tau/t_{1/2}} \tag{15.26}$$

Because $e^{\ln 2} = 2$ in the same way that $10^{\log_{10} 2} = 2$, equation 15.26 can be rearranged to

$$\frac{C_{n(\text{max})}}{C_{\infty(\text{max})}} = 1 - 2^{-n\tau/t_{1/2}} \tag{15.27}$$

If the ratio $\tau/t_{1/2}$ is expressed as a single parameter, ε, then equation 15.27 can be written as

$$\frac{C_{n(\text{max})}}{C_{\infty(\text{max})}} = 1 - 2^{-n\varepsilon} \tag{15.28}$$

This expression is useful because of its general application. It is dependent only on the relationship between how often a drug is administered and the drug half-life. How can the number of doses required to reach 95% of steady state during a repeated dose regimen be determined, that is, when does $C_{n(\text{max})}/C_{\infty(\text{max})}$ equal 95%?

Consider three different situations. In the first situation, drug is administered each elimination half-life; in the second situation, once every two half-lives; and in the third situation, twice every half-life. The calculations are summarized in Figure 15.3.

The first situation occurs quite frequently. For example, doxycycline has an elimination half-life of about 10–12 h and is generally dosed twice daily. If the drug is dosed every half-life, then from Figure 15.3, $\varepsilon = 1$ and $n = 4.35$. In this case 95% of steady state is not reached at the fourth dose but is exceeded at the fifth dose.

The second situation is also common and applies to drugs such as the aminoglycosides, which have half-lives of 3–4 h and are commonly dosed at 6–8-h intervals. In this situation, $\varepsilon = 2$, and only 2.17 doses are required to reach 95% of steady state. The number of doses in this situation is only one-half the number of doses required when $\varepsilon = 1$.

The third situation is when the drug is administered twice every half-life, and applies to such drugs as digoxin, which has a half-life of about 2 days and is generally dosed once each day. In this situation, $\varepsilon = 0.5$ and the number of doses required to reach 95% of steady state increases to 8.7, which is double the value in the first situation.

The number of doses required to reach 95% of steady-state drug levels is thus directly related to dosing frequency. However, despite the different number of doses required in each of the three situations, the time taken to reach 95% of steady state

$$\frac{C_{n(max)}}{C_{\infty(max)}} = 95\%$$

SITUATION 1	SITUATION 2	SITUATION 1
$\tau = t_{1/2},\ \varepsilon = 1$	$\tau = 2(t_{1/2}),\ \varepsilon = 2$	$\tau = 0.5(t_{1/2}),\ \varepsilon = 0.5$
$1 - 2^{-n\varepsilon} = 0.95$	$2^{2n} = 20$	$2^{0.5n} = 20$
$2^{-n\varepsilon} = 0.05$	$n = 2.17$ doses	$n = 8.7$ doses
$2^{-n} = 0.05$		
$2^{n} = 20$		
$n = 4.35$ doses		

FIGURE 15.3 Calculations to obtain the number of doses required to reach 95% of steady state during a multiple-dose regimen.

is identical. If the half-life of a drug were 10 h, then in all three situations, approximately 44 h would be required to reach 95% of steady state. Therefore, just as in the intravenous infusion case discussed in Chapter 13, the time required to approach steady state during a multiple-dosage regimen is independent of the dosage frequency, or input rate, and is dependent solely on the elimination half-life of the drug. If a drug has a half-life of 24 h, then 4.4 days will be required to reach 95% of steady-state, no matter how frequently drug is administered.

The Degree of Drug Accumulation with Repeated Dosing and the Loading Dose Required To Instantaneously Achieve Steady-State Levels

The degree to which drug levels accumulate with multiple doses can be expressed in terms of the ratio $C_{\infty(max)}/C_{1(max)}$, or equation 15.19 divided by D/V, as in

$$\frac{C_{\infty(max)}}{C_{1(max)}} = R = \frac{\dfrac{D}{V(1 - e^{-k_{el}\tau})}}{\dfrac{D}{V}} = \frac{1}{1 - e^{-k_{el}\tau}} \tag{15.29}$$

In this equation, R represents $C_{\infty(max)}/C_{1(max)}$. Equation 15.29 can be rewritten as equation 15.30 after making the same substitutions as those used to obtain equation 15.28.

$$R = \frac{1}{1 - 2^{-\varepsilon}} \tag{15.30}$$

where ε again represents the ratio $\tau/t_{1/2}$. In Table 15.1, five representative values of R are described, where $\varepsilon = 0.25, 0.5, 1, 2,$ and 4 (i.e., the drug is given four times: twice, once, every second, and every fourth half-life, respectively). If a drug is given once every half-life, steady-state drug concentrations in plasma will be double those achieved after the initial dose. If the dosing frequency is increased to twice every half-life, then the accumulation factor is increased to 3.4. If the dosing frequency is reduced so that one dose is given every second half-life, the accumulation factor is reduced to only 1.3. The two additional values in the table show the more extreme cases in which dosing four times every half-life results in over sixfold accumulation, whereas dosing once every four drug elimination half-lives results in essentially zero accumulation. The latter of these two examples is a typical case in which drug levels from the initial dose decline to zero before the next dose is given. This common situation occurs with many penicillins and cephalosporins that have half-lives of approximately 1 h and are commonly dosed every 4–6 h. Drug accumulation cannot possibly occur, and repeated doses yield the same circulating drug concentrations as the first dose.

The types of drug profiles obtained in the three situations where $\varepsilon = 0.5$, 1, or 2 are shown in Figure 15.4. The time to reach steady state is identical in each case.

Just as in the infusion case described in Chapter 13, drugs that have long elimination half-lives accumulate slowly. For these compounds, it may be necessary to administer some type of loading dose in order to achieve therapeutic steady state levels quickly. By judicious borrowing from infusion concepts, and appreciating that the overall kinetics of drug accumulation and decline are identical in that situation and the one now presented, if an initial loading dose is given to provide the same quantity of drug that is in the body at steady state, then the required steady-state drug level will be achieved with the initial dose and will be maintained by subsequent doses. Table 15.1 shows that if a drug is given every elimination half-life, the accumulation factor is 2, and the required loading dose to immediately achieve steady-state levels is twice the maintenance dose. If a drug is given twice every half-life, the loading dose is 3.4 times the maintenance dose. If a drug is given once every four half-lives ($\varepsilon = 4$), then no loading dose is required because drug accumulation will be negligible.

The preceding discussions are based on multiple intravenous bolus injections. However, as discussed earlier, all of the conclusions regarding drug accumulation and its time dependency apply equally when the drug is given orally or by any other route.

Some additional aspects of multiple oral dosing need to be understood before leaving this subject. The terms for $C_{\infty(\text{max})}$, $C_{\infty(\text{min})}$, and $\overline{C}_{\infty}$ for the oral or first-order absorption case are given by equations 15.31, 15.32, and 15.33, respectively.

First-Order Absorption Case

$$C_{\infty(\text{max})} = \frac{FD}{V}\left[\frac{1}{1-e^{-k_{\text{el}}}}\right]\left[\frac{k_{\text{a}}(1-e^{-k_{\text{el}}\tau})}{k_{\text{el}}(1-e^{-k_{\text{a}}\tau})}\right]^{k_{\text{el}}/(k_{\text{el}}-k_{\text{a}})} \tag{15.31}$$

TABLE 15.1

Extent of Drug Accumulation, R, with Multiple Dosing

ε	R
0.25	6.3
0.5	3.4
1	2.0
2	1.33
4	1.07

Note: Values were calculated from equation 15.30.

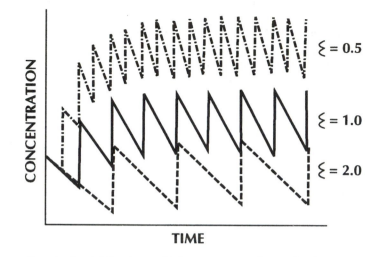

FIGURE 15.4 Drug profiles obtained following multiple intravenous doses of a drug administered at intervals of 0.5, 1, and 2 half-lives.

$$C_{\infty(min)} = \frac{FD}{V}\left[\frac{k_a}{k_a - k_{el}}\right]\left[\frac{e^{-k_{el}\tau}}{1 - e^{-k_{el}\tau}} - \frac{e^{-k_a\tau}}{1 - e^{-k_a\tau}}\right] \tag{15.32}$$

$$\overline{C}_\infty = \frac{FD}{Vk_{el}\tau} \tag{15.33}$$

The increased complexity of equations 15.31 and 15.32 is due to the presence of the absorption rate constant k_a. The first-order absorption phase has the effect of reducing $C_{\infty(max)}$ and increasing $C_{\infty(min)}$. This effect is observed by comparing the $C_{\infty(max)}$ values from equations 15.19 and 15.31 and the $C_{\infty(min)}$ values from equations 15.20 and 15.32. In the first-order absorption case a flatter drug profile (i.e., a lower $C_{\infty(max)}$ and a higher $C_{\infty(min)}$ value) is obtained with decreasing values of k_a. However, from equation 15.33 the value $\overline{C}_\infty$, the mean drug level, is independent of k_a. Except for the presence of the function F, the fraction of the dose absorbed, equation 15.33 is identical to the intravenous equation 15.24. Thus, the parameters $C_{\infty(max)}$ and $C_{\infty(min)}$ are markedly dependent on the magnitude of k_a, whereas $\overline{C}_\infty$ is unaffected. This situation is depicted in Figure 15.5. If the absorption rate constant were reduced to the point where absorption efficiency, F, was reduced, then the whole drug profile including $\overline{C}_\infty$ would be reduced. Multiplying equation 15.33 by the value τ yields the area under the drug-concentration curve shown in equation 15.34.

$$AUC^{0 \to \tau} = \frac{FD}{Vk_{el}} \tag{15.34}$$

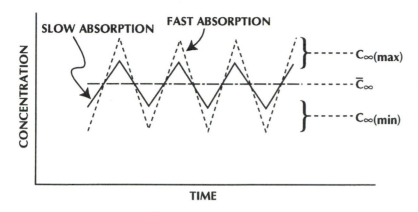

Values of $C_{\infty(max)}$, $C_{\infty(min)}$, and $\overline{C}_\infty$ for a drug that is absorbed at a fast or slow first-order rate.

FIGURE 15.5

Except for the F value, which accompanies all oral dosing, equation 15.34 is identical to equation 15.23 for the intravenous case. Just as in the case of the area from zero to infinite time after single doses, the area from time zero to τ at steady state during repeated doses is independent of the absorption rate constant.

The last parameter to consider is the time at which peak or maximum drug levels, $T_{\infty(max)}$, occur after repeated oral dosing at steady state. In the intravenous case, $T_{\infty(max)}$ occurs immediately after each dose. However, in the first-order absorption case, the value $T_{\infty(max)}$ is given by equation 15.35.

$$T_{\infty(max)} = \frac{1}{k_a - k_{el}} \ln\left[\frac{k_a(1 - e^{-k_{el}\tau})}{k_{el}(1 - e^{-k_a\tau})}\right] \qquad (15.35)$$

This equation is similar to the expression for T_{max} following a single dose (Chapter 14) except for the logarithmic function. The time $T_{\infty(max)}$, during repeated doses, is always shorter than the time T_{max} after a single dose.

Consider a drug that has an absorption rate constant, k_a, of 1 h^{-1} and an elimination rate constant, k_{el}, of 0.1 h^{-1}, and that the drug is dosed every 6 h. After the first dose, according to equation 14.19, T_{max} occurs at 2.6 h, or 2 h and 36 min after dosing. If the drug is dosed every 6 h to reach steady state, then according to equation 15.35, $T_{\infty(max)}$ is 1.7 h, or 1 h and 42 min. This value is almost 1 h earlier than after the single dose. Thus, if T_{max} occurred at 2.6 h after a single dose, and a blood sample was taken at 2.6 h after a dose at steady state to examine $C_{\infty(max)}$, then the sample would have been taken almost 1 h too late. The peak would have been missed, the sample would have been taken when drug concentrations were in the declining postpeak phase, and $C_{\infty(max)}$ would be underestimated.

The pharmacokinetic treatments just presented are based on multiple doses of drug at equally spaced dosage intervals. In the clinical environment this is not always practical, and several methods have been described to calculate pharmacokinetic parameters, and predict blood levels, for drugs given at unequal intervals (7, 8). Wagner (9) has recently described a method for both unequal doses and unequal dosage intervals. These methods are generally based on superposition principles of which the method of Ng (10) is representative.

Summary

The extent of drug accumulation during repeated doses depends on both the dose and the frequency with which it is administered. The rate of drug accumulation, on the other hand, depends exclusively on the drug elimination rate constant and is independent of the dose or dosage interval.

Whereas $C_{\infty(max)}$ and $C_{\infty(min)}$ in the first-order absorption case are both sensitive to the value of the absorption rate constant k_a, $\overline{C}_\infty$ is independent of the absorption rate constant provided the overall bioavailability is unaffected.

The time of peak drug levels occurs earlier after multiple dosing than after single dosing. This could give rise to underestimation of the degree of drug accumulation with repeated doses.

Problems

1. A drug (200 mg) is administered to a 70-kg male patient by multiple IV injections at 6-h intervals. The drug is water-soluble and has a distribution volume equivalent to that of total body water. The drug has a biological half-life of 6 h. Calculate (i) $C_{(max)}$ (initial), (ii) $C_{\infty(max)}$, (iii) $C_{\infty(min)}$, and (iv) $\overline{C}_\infty$.

2. Because of a deterioration in the patient's condition, the dose described in problem 1 has to be increased to 300 mg every 4 h. Dosing is continued until a new steady state is reached. Calculate (i) $C_{\infty(max)}$, (ii) $C_{\infty(min)}$, and (iii) $\overline{C}_\infty$.

3. A drug (50 mg) is administered to a 70-kg male patient by repeated bolus IV injections. The drug has a distribution volume of 5 L and a biological half-life of 8 h, and is injected at intervals of 8 h.

   Calculate the following:

   (i) $C_{\infty(max)}$.

   (ii) The number of doses required to reach 95% of the steady-state concentration.

   (iii) The time necessary to reach 95% of the steady-state concentration.

4. Recalculate Problem 3 if the same drug is dosed IV every 8 h to a patient with renal failure in whom the drug has a biological half-life of 20 h (volume of distribution is 5 L).

5. How could the dose be adjusted so that the same $\overline{C}_\infty$ can be obtained for the patient in Problem 4 as was obtained for the patient in Problem 3 without changing the dosage interval?

6. If the physician decided to give the same quantity of drug as in problem 3 to the patient in problem 5, at what dosing interval would it have to be given to yield the same value of $\overline{C}_\infty$ as that for the patient in Problem 3?

7. Calculate the loading dose required for the patient in Problem 6 to give the $C_{\infty(max)}$ value instantaneously with the first dose.

8. A 250-mg oral dose of a drug is administered to a patient. Pharmacokinetic analysis yielded a first-order absorption rate constant of 0.8 h^{-1} and a first-order elimination rate constant of 0.08 h^{-1}. Calculate the following:

   (i) T_{max} following the single dose.

   (ii) $T_{\infty(max)}$ at steady state if the drug was given as repeated doses at 8-h intervals.

References

1. Kruger-Thiemer, E.; Bunger, P. *Chemotherapia* **1965**, *10*, 61–74.
2. Tse, F. L. S.; Robinson, W. T.; Choc, M. G. *Pharmaceutical Bioequivalence*; Welling, P. G.; Tse, F. L. S.; Dighe, S. V., Eds.; Marcel Dekker: New York, 1991; pp 17–34.
3. Steinijans, V. W.; Sauter, R.; Jonkman, J. H. G.; Schultz, U-H.; Stricker, H.; Blume, H. *Int. J. Clin. Pharmacol. Toxicol.* **1989**, *27*, 261–266.
4. Kaplan, S. A.; Jack, M. L.; Alexander, K.; Weinfeld, R. E. *J. Pharm. Sci.* **1971**, *62*, 1789–1796.
5. Alexanderson, B. *Eur. J. Clin. Pharmacol.* **1972**, *4*, 82–91.
6. Perrier, D.; Gibaldi, M. *J. Pharmacokinet. Biopharm.* **1973**, *1*, 17–22.
7. Niebergall, P. J.; Sugita, E. T.; Schnaare, R. L. *J. Pharm. Sci.* **1974**, *63*, 100–105.
8. Howell, J. R. *J. Pharm. Sci.* **1975**, *64*, 464–466.
9. Wagner, J. G. *Pharmacokinetics for the Pharmaceutical Scientist*; Technomic: Lancaster, PA, 1993; pp 108–109.
10. Ng, P. K. *Int. J. Biomed. Comput.* **1981**, *12*, 217–226.

Metabolite Pharmacokinetics

16

The pharmacokinetic treatments in Chapters 13–15 concerned the pharmacokinetics of an administered drug. Many drug metabolites are pharmacologically inactive so that, for the most part, their pharmacokinetics is relatively unimportant; however, many exceptions exist. For example, the major metabolite of procainamide, *N*-acetyl-procainamide, has similar cardiac potency to the parent drug. Desmethyldiazepam and desmethylmethsuximide are both pharmacologically active metabolites of the parent drugs diazepam and methsuximide. Some other examples of active drug metabolites are given in Table 16.1. In these situations it is just as important to understand the pharmacokinetics of the active metabolites as that of the parent drug.

In other situations, blood or urine concentrations of parent drug may be too low to detect or measure accurately, so that attention has to be diverted to measurement of major metabolites in order to understand the pharmacokinetics of the parent compound. An example of this is the antimicotic agent griseofulvin, which is rapidly metabolized in the body to desmethylgriseofulvin. Urinary excretion of this metabolite has been used as an indicator of absorption efficiency of griseofulvin.

Thus, for a number of reasons, it is important to understand the pharmacokinetic characteristics of a metabolite in addition to or instead of those of the parent drug. From a pharmacokinetic viewpoint, formation of a metabolite from a parent drug, and subsequent further metabolism or excretion of the metabolite, can be considered in terms of catenary chain kinetics, which is a sequence of kinetic steps in the form of A→B→C, etc.

Catenary chain kinetics will be considered using a simple model of an intravenously administered drug that is partially cleared via the urine in unchanged

Pharmacokinetics of Metabolite Formation and Elimination

form and partially converted to a single metabolite that is subsequently excreted in urine. The biotransformation step from drug to metabolite is assumed in this case to obey first-order kinetics. Because metabolic processes are enzymatic, all metabolic steps are saturable at some drug (substrate) concentration. Some examples of this will be discussed in Chapter 19. However, the present argument assumes that the drug concentration is below that level where saturable, or Michaelis–Menten-type, kinetics becomes important, and the first-order kinetic approximation can be used. This assumption is reasonable because only a relatively small number of drugs exhibit saturable elimination kinetics in the therapeutic concentration range.

The model to be used is shown in Scheme 16.1. Despite the simplicity of this model, it will illustrate some inherent problems associated with analysis and interpretation of these types of data. Interpretation of the data for the parent drug presents no significant problems, as discussed in Chapter 13. To recapitulate from that chapter, the time course of the amount of drug in the body can be described by

$$A = A_0 e^{-k_{el}t} \tag{16.1}$$

where A_0 is the administered dose, t is the time elapsed since administering the drug, A is the amount of drug in the body at time t, and k_{el} is the first-order rate constant for drug elimination. The concentration of drug in the circulation is given by

$$C = \frac{D}{V} e^{-k_{el}t} \tag{16.2}$$

The overall rate constant k_{el} controls the rate of drug loss. Similarly, cumulative urinary excretion of unchanged drug was described by

$$A_u = A_u^{\infty}(1 - e^{-k_{el}t}) \tag{16.3}$$

TABLE 16.1   **Some Pharmacologically Active Metabolites**

Parent Drug	Active Metabolite
Acetylsalicylic acid	Salicylic acid
Amitriptyline	Nortriptyline
Codeine	Morphine
Diazepam	Desmethyldiazepam
Flurazepam	Desalkylflurazepam
Meperidine	Normeperidine
Metamphetamine	Amphetamine
Phenacetin	Acetaminophen
Prednisone	Prednisolone
Spironolactone	Canrenone
Trimethadione	Dimethadione

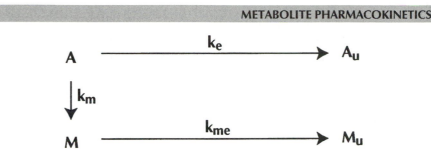

Kinetic scheme for intravenous drug administration and formation of a single metabolite; A is the amount of drug in body, M is the amount of metabolite in body, A_u is the cumulative amount of drug voided in urine, M_u is the cumulative amount of metabolite voided in urine, k_m is the first-order rate constant for formation of metabolite, and k_e and k_{me} are first-order rate constants for urinary excretion of drug and metabolite, respectively.

SCHEME 16.1

where $A_u^\infty = D(k_e / k_{el})$. This equation can be expressed in sigma-minus form as

$$\log(A_u^\infty - A_u) = \log A_u^\infty - \left(\frac{k_{el}}{2.3}\right)t \tag{16.4}$$

The overall elimination rate constant, k_{el}, controls the time dependency of urinary excretion, not k_e, which is the rate constant for urinary excretion.

Considering now the metabolite, the rate of change in the amount of metabolite in the body is given by

$$\frac{dM}{dt} = k_m A - k_{me} M \tag{16.5}$$

where M is the amount of metabolite in the body, k_m is the rate constant that controls the rate of metabolite formation, and k_{me} is the rate constant for metabolite elimination. The value of A is given by equation 16.1. Incorporating that expression into equation 16.5 and integrating yields

$$M = \frac{A_0 k_m}{k_{me} - k_{el}} (e^{-k_{el}t} - e^{-k_{me}t}) \tag{16.6}$$

This equation describes the amount of metabolite in the body with respect to time. There are subtle differences between this equation and equation 14.3, which describes the amount of drug in the body with first-order absorption and elimination. In that situation, only two rate constants, k_a and k_{el}, were involved. In this more complex situation, three constants, k_m, k_{me}, and k_{el}, are involved. The rate constant k_{el} reflects the loss of drug from the body by all processes. The rate constant for metabolite formation, k_m, plays a role in the overall scenario as a time-independent constant. Thus, the time course for the amount of metabolite in the body is controlled by the overall rate constant for loss of drug and the overall rate constant for loss of metabolite.

Analysis of Metabolite Concentrations in Plasma, and Urinary Excretion

In considering the time course of the amounts or concentrations of unchanged drug in the pharmacokinetic model with first-order absorption and elimination, three possible situations were established in which $k_e > k_{el}$, $k_a < k_{el}$, and $k_a = k_{el}$. The same situation frequently occurs with metabolites. That is, the overall rate constant for loss of drug, k_{el}, may be greater than, equal to, or less than the overall rate constant for elimination of metabolite, k_{me}.

As described earlier, the primary objective of drug metabolism is to produce a more water-soluble and hence more rapidly excretable derivative of the parent drug (1). Therefore, k_{me} would be expected to be greater than k_{el} in most cases. However, sometimes the metabolite has a longer intrinsic elimination half-life than the parent drug, and k_{el} is greater than k_{me}.

In equation 16.6, for the case where k_{el} is greater than k_{me}, as time increases, the term $e^{-k_{el}t}$ will approach zero before the term $e^{-k_{me}t}$, and the equation will reduce to equation 16.7.

$$M = \frac{A_0 k_m}{k_{me} - k_{el}} e^{-k_{me}t} \qquad (16.7)$$

which is rearranged to equation 16.8 in order to keep the exponential term positive.

$$M = \frac{A_0 k_m}{k_{el} - k_{me}} e^{-k_{me}t} \qquad (16.8)$$

Thus, the profile of the amount of metabolite in the body after an intravenous dose of parent drug can be interpreted as in Figure 16.1. The value k_{me} is obtained from

$$k_{me} = \frac{\text{intercept} (k_{el} - k_{me})}{A_0} \qquad (16.9)$$

This method of interpretation is straightforward, but it occurs only when $k_{el} > k_{me}$.

The second situation, where the rate constant for metabolite elimination is greater than that for the parent drug, is illustrated in Figure 16.2. In this case, $k_{me} > k_{el}$. As time increases, the term $e^{-k_{me}t}$ in equation 16.6 will approach zero before the exponential term $e^{-k_{el}t}$, and the equation will then reduce to

$$M = \frac{A_0 k_m}{k_{me} - k_{el}} e^{-k_{el}t} \qquad (16.10)$$

This expression shows that the terminal linear portion of the metabolite profile in plasma is controlled by k_{el} (i.e., the rate constant for overall loss of parent drug from the body). In this case, the value of k_{me} is obtained from residuals, and the value of the common intercept of the terminal and residual slopes at time zero is $\log[A_0 k_m(k_{me} - k_{el})]$. The value k_m is then obtained from

$$k_m = \frac{\text{intercept} (k_{me} - k_{el})}{A_0} \qquad (16.11)$$

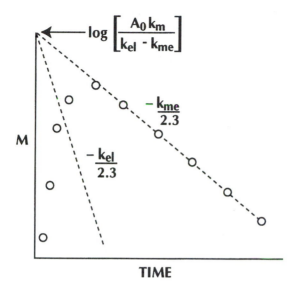

Time dependency of the amount of metabolite in the body following intravenous bolus injection of parent drug. The profile is interpreted in terms of equation 16.6, with $k_{el} > k_{me}$.

FIGURE 16.1

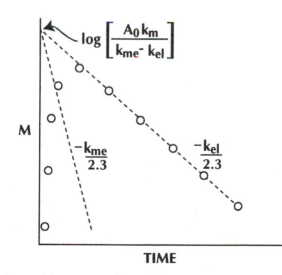

Time dependency of the amount of metabolite in the body following intravenous bolus injection of parent drug. The profile is interpreted in terms of equation 16.6, with $k_{me} > k_{el}$.

FIGURE 16.2

A number of factors can be learned by comparing the two situations represented by Figures 16.1 and 16.2. First, there are two possible ways to analyze metabolite data. Unless additional information is available, distinguishing the correct from the incorrect method is not possible. That is, does the terminal linear slope of the metabolite profile represent k_{el} or k_{me}? This dilemma is similar to the flip-flop model that was discussed in Chapter 14. The additional information necessary to resolve this problem can be obtained from the elimination slope for unchanged drug. This plot will yield k_{el} directly, so that the two constants can then be distinguished from each other.

A second factor to be learned by comparing Figures 16.1 and 16.2 is that the observed terminal linear portion of the metabolite profile can have a slope that is less than that of the parent drug (when $k_{el} > k_{me}$), but it cannot have a slope that is greater than that of the parent drug. If k_{me} is greater than k_{el}, then the terminal slope of the metabolite profile is controlled by k_{el}. If the terminal slope for the metabolite appears to be steeper than that of the parent drug, then probably something is wrong with the assay.

Measurement of metabolites in urine is very common because metabolites frequently occur at much higher concentrations in urine than in the systemic circulation. Appropriate expressions for urinary excretion of unchanged drug were given in equation 13.28 and in sigma-minus form in equation 13.30.

Analogous expressions for urinary metabolite data are based on the rate equation:

$$\frac{dM_u}{dt} = k_{me}M = \frac{k_{me}A_0k_m}{k_{me} - k_{el}}(e^{-k_{el}t} - e^{-k_{me}t}) \tag{16.12}$$

where M_u is the amount of metabolite excreted in urine. This equation can be solved for M_u:

$$M_u = \frac{A_0k_m}{k_{el}}\left[1 - \frac{1}{k_{me} - k_{el}}(k_{me}e^{-k_{el}t} - k_{el}e^{-k_{me}t})\right] \tag{16.13}$$

Because the function $A_0(k_m/k_{el}) = M_u^\infty$, equation 16.13 can be rewritten as

$$M_u = M_u^\infty\left[1 - \frac{1}{k_{me} - k_{el}}(k_{me}e^{-k_{el}t} - k_{el}e^{-k_{me}t})\right] \tag{16.14}$$

The time dependency of this equation can be examined in the same way as that of equation 16.6. Depending on the relative magnitude of k_{el} and k_{me}, one or the other of the two exponential terms in equation 16.14 approaches zero with increasing time, so that during the terminal linear portion of the cumulative urinary excretion profile, equation 16.14 may reduce to

$$M_u = M_u^\infty\left[1 - \frac{1}{k_{me} - k_{el}}(-k_{el}e^{-k_{me}t})\right] \tag{16.15}$$

or

$$M_u = M_u^\infty \left[1 - \frac{1}{k_{me} - k_{el}} (-k_{me} e^{-k_{el}t}) \right] \qquad (16.16)$$

Expressed in sigma-minus form, these equations become, respectively, equations 16.17 and 16.18.

$$\log(M_u^\infty - M_u) = \log\left[M_u^\infty \left(\frac{k_{el}}{k_{el} - k_{me}} \right) \right] - \frac{k_{me}}{2.3} t \qquad (16.17)$$

$$\log(M_u^\infty - M_u) = \log\left[M_u^\infty \left(\frac{k_{me}}{k_{me} - k_{el}} \right) \right] - \frac{k_{el}}{2.3} t \qquad (16.18)$$

These two equations present the same quandary as occurred with the plasma data, reflected in equations 16.6, 16.8, and 16.10. If $\log(M_u^\infty - M_u)$ is plotted versus time, a plot is obtained that is at first curved, but then becomes linear as the contribution of either $e^{-k_{el}t}$ or $e^{-k_{me}t}$ in equation 16.14 drops out, and the log–linear relationships described by equations 16.17 or 16.18 are obtained. However, without additional information, it is not possible to say which of these equations is appropriate for a given set of data.

Resolving the Pharmacokinetics of Metabolite and Parent Drug

The problem of resolving the pharmacokinetics of metabolite and parent drug is best illustrated with a numerical example. Suppose 100 mg of a drug is administered by rapid bolus intravenous injection to a patient and metabolite voided in urine is collected during a 50-h postdose period. The quantity of metabolite remaining to be excreted in urine, expressed in equivalents of unchanged drug, is given in Table 16.2. A plot of $\log(M_u^\infty - M_u)$ versus time would yield a curve similar to that shown in Figure 16.3.

Quantity of Metabolite Recovered in Urine

TABLE 16.2

Time (h)	Cumulative M_u (mg)	$M_u^\infty - M_u$ (mg)
0	0	80
0.5	1.1	78.9
1	3.1	76.9
2	10.5	69.5
5	35.3	44.7
10	62.2	17.8
15	73.4	6.6
20	77.5	2.5
25	79.1	0.9
30	79.7	0.3
40	80.0	0
50	80.0	0

Note: All values were obtained during a 50-h period following a single 100-mg bolus intravenous injection of the parent drug.

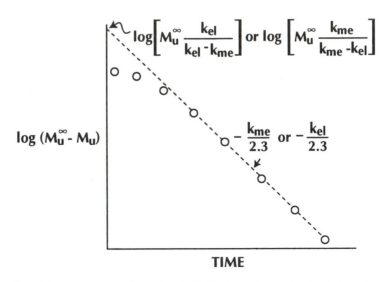

FIGURE 16.3 Plot of $\log(M_u^\infty - M_u)$ vs. time from data in Table 16.2. Interpretation is based on equations 16.17 and 16.18.

These data can now be interpreted in two different ways depending on whether k_{el} or k_{me} is the larger rate constant. If k_{el} is larger, equation 16.17 is appropriate; if k_{me} is larger, equation 16.18 is appropriate. The calculations involved and the results obtained using these different approaches are shown in Table 16.3. The half-life of the terminal log–linear slope for the metabolite is 3.4 h. If equation 16.17 is assumed, then $k_{me} = 0.693/3.4 = 0.2$ h^{-1}. If, on the other hand, equation 16.18 is assumed, then this value is assigned to k_{el}. The intercept of the terminal portion of the log–linear curve on the ordinate at time zero is 140 mg. From this value and known values of k_{me} or k_{el}, k_{el} is calculated to be 0.47 h^{-1} from equation 16.17, whereas k_{me} is calculated to have this value from equation 16.18. Further substitution yields values of 0.38 h^{-1} and 0.09 h^{-1} for k_m and k_e, respectively, from equation 16.17, and 0.16 h^{-1} and 0.04 h^{-1}, respectively, from equation 16.18. Thus, two completely different sets of parameter values are obtained depending on the assumption of the relative values of k_{el} and k_{me}.

Unless some additional data are available, it is not possible to say which of these two sets of results is correct. However, this Gordian knot can be untied by two methods. The first method is to identify the value of k_{el} from the elimination slope of the plasma profile or urinary excretion data for the parent drug. If the terminal elimination rate constant for the metabolite is approximately equal to that of unchanged drug, then equation 16.18 probably applies because k_{el} is most likely smaller than k_{me}. If, on the other hand, the terminal elimination rate constant for the metabolite is smaller than that for the unchanged drug, then equation 16.17 applies because k_{me} is most likely smaller than k_{el}. Another method is to dose the metabolite, if it is available, and thereby establish k_{me} independent of any influence from the parent drug.

It is evident from this brief treatment that pharmacokinetic characterization of metabolite formation and excretion is difficult. The example used here is based on formation of a single metabolite by a nonsaturable mechanism following a single intravenous dose of a drug. In practice, the situation is more complex and more difficult to resolve. For example, the drug may be given orally, so that the drug absorption constant must be taken into account. Drugs may form more than one metabolite, and one metabolite may be transformed to another in the body. Some metabolic pathways may be parallel, and some may be more saturable than others (2–4). For complete characterization of metabolite kinetics, it is preferable to dose the metabolite so that its pharmacokinetics can be examined in the absence of other substances, particularly parent drug (5, 6).

As indicated at the beginning of this chapter, the catenary chain model, while providing insight into the kinetics of metabolite formation and elimination, is a relatively simple model. Drug metabolism may be more complex than this. Oral or intramuscular administration introduces the complexity of an additional first- or zero-order absorption rate constant for parent drug, which makes the pharmacokinetics more complicated. A drug may be metabolized by a number of parallel and competing pathways, some of which may be saturable. Also, as described in Chapter 9, metabolism pathways may be subject to induction or inhibition, giving rise to changes in the mathematical values of kinetic parameters.

Despite its simplifying assumptions, the catenary chain model serves to illustrate the complexity of the dynamics of drug-metabolite relationships, and the pitfalls that may be encountered in analyzing such data. Detailed descriptions of

Pharmacokinetic Parameters

TABLE 16.3

Equation 16.17	Equation 16.18
$t_{1/2} = 3.4$ h	$t_{1/2} = 3.4$ h
$k_{me} = 0.2$ h^{-1}	$k_{el} = 0.2$ h^{-1}
$M_u^\infty = 80$ mg	$M_u^\infty = 80$ mg
$M_u^\infty \left(\dfrac{k_{el}}{k_{el} - k_{me}} \right) = 140$ mg	$M_u^\infty \left(\dfrac{k_{me}}{k_{me} - k_{el}} \right) = 140$ mg
$\dfrac{k_{el}}{k_{el} - k_{me}} = \dfrac{140}{80} = 1.75$	$\dfrac{k_{me}}{k_{me} - k_{el}} = \dfrac{140}{80} 1.75$
$k_{el} = 0.47$ h^{-1}	$k_{el} = 0.47$ h^{-1}
$k_m = \dfrac{k_{el} M_u^\infty}{D} = \dfrac{0.47 \times 80}{100} = 0.38$ h^{-1}	$k_m = \dfrac{k_{el} M_u^\infty}{D} = \dfrac{0.2 \times 80}{100} = 0.16$ h^{-1}
$k_e = k_{el} - k_m = 0.47 - 0.38 = 0.09$ h^{-1}	$k_e = k_{el} - k_m = 0.2 - 0.16 = 0.04$ h^{-1}

Note: All parameters are based on urinary excretion data from Table 16.2.

more complex models of drug metabolism are given in recent reviews by Wagner (7) and Rowland and Tozer (8).

The use of hepatic clearance models to predict metabolite formulation and elimination was recently described by St-Pierre and co-workers (9).

Summary

Metabolite formation and excretion can be described in terms of catenary chain kinetics.

The primary objective of drug metabolism is to generate more water-soluble and rapidly excreted compounds. However, in some instances the metabolite may be excreted at a slower rate than parent drug.

Because of the close relationship between the kinetic behavior of parent drug and its metabolite(s), care must be exercised when pharmacokinetic parameter values are assigned to metabolite data. This is particularly important when the rate constant for metabolite elimination is faster than the overall elimination rate constant for parent drug.

Problems

1. A drug (10 mg) is dosed to a 70-kg male patient by rapid IV injection. The drug does not bind to plasma proteins, distributes into a volume equivalent to that of blood, and has a plasma clearance of 50 mL/min. The drug is cleared from the body as unchanged drug and as one metabolite. Both compounds are voided quantitatively into the urine. A sigma-minus plot of metabolite in urine data yielded a rate constant of 0.2 h^{-1}, and total urinary excretion of metabolite accounted for 25% of the dose. Calculate the following: (i) k_e (ii) k_m, and (iii) k_{me}.

2. A drug (100 mg) is dosed orally to a patient. Only 20% of the dose is recovered in the feces. Of this, one-half is recovered as unchanged drug and may be assumed not to have been absorbed. Sixty percent of the original dose appears in urine as unchanged drug and the balance as urinary metabolites. A sigma-minus plot of unchanged drug excreted in urine yielded an apparent elimination rate constant of 0.3 h^{-1}, and the intercept of the linear portion of the plot, extrapolated to time zero, was 75 mg. Calculate the following: (i) k_e, (ii) k_m, and (iii) the drug absorption rate constant, k_a.

3. One-half of an intravenously dosed drug is voided unchanged in urine, and the balance of the drug is metabolized to inactive metabolites. The decline in plasma levels of unchanged drug is first-order and has a $t_{1/2}$ of 4 h. What is the value of the first-order rate constant obtained from a sigma-minus plot of unchanged drug appearing in urine?

4. A drug, which gives rise to a single metabolite that is excreted in urine, has a biological half-life of 6 h. When the metabolite is administered alone, a biological half-life of 2 h results. What would the observed biological half-life of the metabolite be if measured following administration of parent drug?

5. What would the observed metabolite half-life in plasma be following administration of parent drug if the drug half-life were 4 h and the metabolite half-life, when administered alone, were 6 h?

References

1. Taylor, J. A. *Clin. Pharmacol. Ther.* **1972,** *13,* 710–718.
2. Levy, G. *J. Pharm. Sci.* **1965,** *54,* 959–967.
3. Levy, G.; Vogel, A. W.; Amsel, L. P. *J. Pharm. Sci.* **1969,** *58,* 503–504.
4. Hewick, D. S.; McEwen, J. *J. Pharm. Pharmacol.* **1973,** *25,* 458–465.
5. Tse, F. L. S.; Welling, P. G. *Biopharm. Drug Disp.* **1980,** *1,* 221–223.
6. Tse, F. L. S.; Welling, P. G. *Res. Commun. Subst. Abuse* **1980,** *1,* 185–195.
7. Wagner, J. G. *Pharmacokinetics for the Pharmaceutical Scientist;* Technomic Publishing Company, Inc.: Lancaster, PA, 1993; pp 269–280.
8. Rowland, M.; Tozer, T. N. *Clinical Pharmacokinetics: Concepts and Applications,* 2nd ed.; Lea and Febiger: Philadelphia, PA, 1989; pp 347–375.
9. St-Pierre, M. V.; Lee, P. I.; Pang, K. S. *J. Pharmacokinet. Biopharm.* **1992,** *20,* 105–145.

The Two-Compartment Open Model with Intravenous or Oral Administration

17

The pharmacokinetic treatment in preceding chapters has been based on the one-compartment model. That is, the drug, after introduction into the body, is assumed to distribute rapidly into an apparently homogeneous volume.

For many drugs the simple one-compartment approach is inadequate to describe observed blood level versus time profiles. Drug may not rapidly distribute into one apparently homogeneous volume, but it may distribute more slowly into some tissues than others, and variable lengths of time may elapse before equilibrium is reached between drug concentrations, or amounts, in different body tissues and fluids. This situation is illustrated in Scheme 17.1. In the scheme, the central compartment represents plasma water and any other tissues or fluids into which the drug rapidly equilibrates. The other compartments in the scheme represent either single tissues or groups of tissues into which drug equilibrates at a slower rate than those of the central compartment. This model is complex and cannot be characterized simply by measuring drug concentrations in plasma or urinary excretion. However, a less complex version of the model is obtained by combining all of the tissues outside the central compartment and summing individual rate constants. Scheme 17.2, which represents the two-compartment open pharmacokinetic model, is thus obtained. This particular version of the model represents the situation where drug is introduced into the central compartment (of which the blood, plasma, or serum is generally representative); distributes between the central compartment and the second, or peripheral, compartment; and is eliminated from the central compartment (1–2). This is the most common model used to describe circulating drug levels, particularly after bolus intravenous administration.

Several variations of this model have been described. For example, drug could be introduced into or eliminated from the tissue and plasma compartments

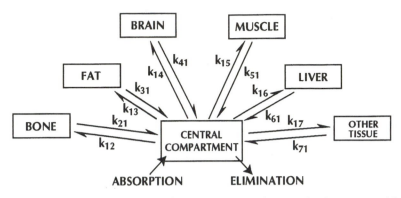

SCHEME 17.1   Distribution of drug between a central compartment and various body tissues and fluids.

simultaneously, or drug could conceivably be eliminated exclusively from the second, or tissue, compartment (3). However, the model in Scheme 17.2 is the most common and makes good anatomical sense. Intravenous dosage always delivers drug directly into the plasma compartment. Similarly, drug elimination occurs principally via the liver or kidney, and these organs are intimately associated with, if not part of, the plasma compartment because of rapid blood flow through these organs and rapid equilibrium of substances between these organs and blood. Each compartment has three names. The plasma compartment is also called the central or first compartment. Similarly, the tissue compartment is also known as the peripheral or second compartment. These names are largely self-explanatory and are synonymous within each compartment.

The Two-Compartment Model with Bolus Intravenous Injection

As shown in Chapter 13, in the one-compartment model drug equilibrates rapidly into those fluids and tissues that constitute the distribution volume. Thus, a plot of drug concentration versus time following intravenous dosage yields a simple monoexponential curve, and a plot of the logarithm of drug concentration versus time yields a straight line. With the two-compartment model, a distinctive curve is obtained that readily distinguishes the two-compartment model from the one-compartment model.

The sequence of events in the two-compartment model, as depicted in Scheme 17.2, is as follows: After drug is introduced into the first compartment by bolus intravenous injection, initial and rapid distribution into those fluids and tissues that constitute the first compartment occurs. At the same time, drug starts to disperse into the less accessible and more slowly equilibrating tissues and fluids that constitute the second compartment. Of course, drug is simultaneously being eliminated from the first compartment. Because drug is not only being eliminated (excreted or metabolized) but also being taken up by second-compartment tissues and fluids during the early postdose period, rapid net loss of drug from the first compartment, and a rapid decline in plasma concentrations, occurs. This period of rapidly falling drug concentrations due to the combined effects of tissue uptake and elimination is called the α phase. After a certain time period depending on the

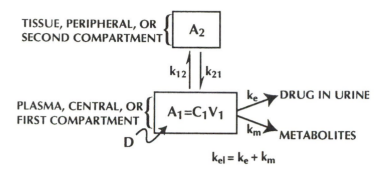

Two-compartment open model with rapid intravenous injection.

SCHEME 17.2

magnitude of "microscopic" rate constants between the compartments, k_{12} and k_{21}, drug will equilibrate between the various tissues and fluids, and net loss of drug from the first compartment due to distribution will no longer occur. Even though equilibrium is reached between the two compartments at the end of the α phase, the concentration of drug is not necessarily equal in the two compartments. The concentration of drug in tissues may be less or greater than that in the first compartment, but the concentrations will nonetheless be at equilibrium. Because the second compartment usually comprises several organs and tissues, the concentration of drug in this compartment is unlikely to be homogeneous.

Once equilibrium is reached between the two compartments, the rate of drug loss from the bloodstream is reduced. This period of more slowly declining blood levels is the β phase. The type of drug profile that occurs in this situation is shown in Figure 17.1, which shows rapid decline of drug concentrations during the α phase and relatively slow decline during the β phase.

Having established the kinetic model in Scheme 17.2, and having described the type of blood level curve that occurs in Figure 17.1, it is now possible to derive the appropriate equations that describe the drug profile in plasma.

The rates of change in the amounts of drug in the first and second compartments, and also the cumulative amount of drug voided in urine with respect to time, are given by equations 17.1 through 17.3.

Interpretation of Drug-Concentration Profiles in Plasma To Obtain Parameter Estimates

$$\frac{dA_1}{dt} = k_{21}A_2 - (k_{12} + k_{el})A_1 \qquad (17.1)$$

$$\frac{dA_2}{dt} = k_{12}A_1 - k_{21}A_2 \qquad (17.2)$$

$$\frac{dA_u}{dt} = k_e A_1 \qquad (17.3)$$

These equations can be solved simultaneously to obtain

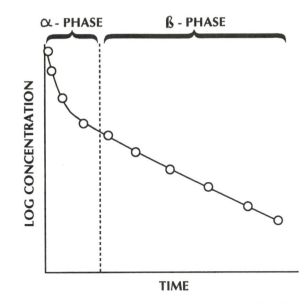

FIGURE 17.1 Blood levels of a drug that obeys two-compartment model kinetics following bolus intravenous injection.

$$A_1 = \left(\frac{D}{\alpha - \beta}\right)\left[(k_{21} - \beta)e^{-\beta t} - (k_{21} - \alpha)e^{-\alpha t}\right] \qquad (17.4)$$

$$A_2 = \left(\frac{Dk_{12}}{\alpha - \beta}\right)(e^{-\beta t} - e^{-\alpha t}) \qquad (17.5)$$

$$A_u - A_u^{\infty}\left[1 - \frac{k_{el}}{\alpha - \beta}\left(\frac{k_{21} - \beta}{\beta}e^{-\beta t} - \frac{k_{21} - \alpha}{\alpha}e^{-\alpha t}\right)\right] \qquad (17.6)$$

which represent the quantity of drug in the first compartment, the quantity of drug in the second compartment, and the quantity of drug cumulatively excreted unchanged in urine, respectively. The parameters α and β are complex rate constants that are related to the "microscopic" constants k_{12}, k_{21}, and k_{el} as shown in equations 17.7 and 17.8.

$$\alpha = 0.5\{(k_{12} + k_{21} + k_{el}) + [(k_{12} + k_{21} + k_{el})^2 - 4k_{21}k_{el}]^{1/2}\} \qquad (17.7)$$

$$\beta = 0.5\{(k_{12} + k_{21} + k_{el}) - [(k_{12} + k_{21} + k_{el})^2 - 4k_{21}k_{el}]^{1/2}\} \qquad (17.8)$$

The only difference in the right sides of these two equations is the sign separating the parenthetical term and the root function. These equations appear complex, but careful examination will show that the combined equations 17.7 and 17.8 can be written

$$X = \frac{-b \pm \sqrt{b^2 - 4ac}}{2a} \tag{17.9}$$

This equation represents the two possible values of X in the quadratic equation:

$$aX^2 + bX + c = 0 \tag{17.10}$$

Equation 17.9 thus represents the two possible roots of equation 17.10. In exactly the same way, the values α and β are the two possible roots of a quadratic function that one reaches when obtaining simultaneous solutions for the rate equations 17.1, 17.2, and 17.3.

The values α and β are therefore intimately related to the values of k_{12}, k_{21}, and k_{el}, and both α and β are influenced by all three rate constants. If k_{12} is very high, then drug will distribute rapidly into peripheral tissue. Detecting a distribution phase in this case will be difficult, and the drug will tend to exhibit simple one-compartment model kinetics. The lower k_{12} becomes relative to the other microscopic constants, the slower the distribution phase, and the more two-compartment character is evident in the drug plasma profile.

Relationships between the values of α and β and the magnitude of the elimination rate constant, k_{el}, are interesting. For example, if values of k_{12} and k_{21} are held constant, and the value of the elimination rate constant is varied over a fairly wide range, then for very high values of k_{el}, with very fast intrinsic elimination, the value of α will approach k_{el}. For very low values of k_{el} relative to the other microscopic constants, α approaches the sum of k_{12} and k_{21}. Because of the negative sign in equation 17.8, the value of β is affected differently by changes in k_{el}. For instance, at high values of k_{el}, when elimination is rapid relative to k_{12} and k_{21}, the value of β approaches k_{21}. At very low values of k_{21}, β approaches zero.

Having established the basic equations associated with this model, the next task is to obtain estimates of the rate constants and other parameters from the drug profile in plasma. Equation 17.4 can be rewritten in concentration form as

$$C = \frac{D}{V_1(\alpha - \beta)}\left[(k_{21} - \beta)e^{-\beta t} - (k_{21} - \alpha)e^{-\alpha t}\right] \tag{17.11}$$

where V_1 is the volume of the first compartment. Equation 17.11 can be expressed in shorthand notation as

$$C = Ae^{-\alpha t} + Be^{-\beta t} \tag{17.12}$$

In this equation, the values A and B relate to the parameters in equation 17.4, as shown in equations 17.13 and 17.14.

$$A = \frac{D(k_{21} - \alpha)}{V_1(\alpha - \beta)} \tag{17.13}$$

$$B = \frac{D(k_{21} - \beta)}{V_1(\alpha - \beta)} \qquad (17.14)$$

Graphical Estimation of Kinetic Parameters

Graphical estimates of pharmacokinetic parameters associated with this model can be obtained by plotting drug concentration data as shown in Figure 17.1, and stripping the curve to obtain estimates of the four constants, A, B, α, and β in equation 17.12. This procedure is demonstrated in Figure 17.2. According to equations 17.7 and 17.8, the rate constant α must be greater than β; therefore, as time increases after dosing, the term $Ae^{-\alpha t}$ will approach zero before the term $Be^{-\beta t}$, and equation 17.12 will reduce to

$$C = Be^{-\beta t}, \quad \text{or} \quad \log C = \log B - \left(\frac{\beta}{2.3}\right)t \qquad (17.15)$$

The swiftly declining portion of the drug profile is the α phase. The portion in which only equation 17.15 contributes to the curve, the more slowly declining portion, is the β phase.

The value β is thus obtained from the terminal linear portion of the log concentration versus time curve, and the intercept of this linear portion extrapolated back to time zero is the concentration term B. The values of A and α are obtained by the method of residuals or curve stripping as described in Chapter 14 for the one-compartment model with first-order absorption. In this case, however, residual values are obtained by subtracting the values on the extrapolated portion of the slope from the actual data values obtained during the α phase. Residual values thus obtained are plotted as shown in Figure 17.2. The slope of the residual line gives the value α and the intercept at time zero gives the value A. Proof that the slope and intercepts of the residual line represent α and A is given by

$$R = Ae^{-\alpha t} + Be^{-\beta t} - Be^{-\beta t} \qquad (17.16)$$

in which the right side of equation 17.15 is subtracted from the right side of equation 17.12. The logarithmic form of the resulting residual function is given by

$$\log R = \log A - \left(\frac{\alpha}{2.3}\right)t \qquad (17.17)$$

Numerical values have thus been obtained for the concentration terms A and B and the rate constants α and β. Using these terms to find numerical values for the microscopic constants and other values associated with the model in Scheme 17.2 is a relatively simple process.

The first relationship to note is that the initial drug concentration at time zero, C_0, which is the concentration obtained when the drug has distributed within the first compartment but has not yet started to enter the second compartment, is equal to $A + B$, as in

$$C_0 = A + B \qquad (17.18)$$

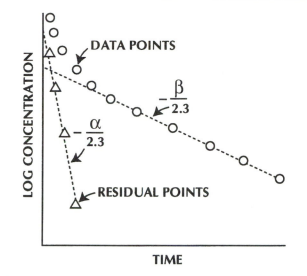

FIGURE 17.2

Method to obtain initial estimates of the constants A, B, α, and β in equation 17.12 from the blood profile of a drug that obeys two-compartment model kinetics after bolus intravenous injection.

As $C_0 = D/V_1$, equation 17.18 can be rewritten as

$$\frac{D}{V_1} = A + B \qquad (17.19)$$

which is rearranged to give the value V_1 in terms of the dose and two intercepts, as in

$$V_1 = \frac{D}{A + B} \qquad (17.20)$$

Estimates for the rate constants k_{12}, k_{21}, and k_{el} are obtained in a sequential manner, by solving for one, and then using that value to solve for the next. The first constant to solve for is k_{21}, by means of equation 17.13, which is rearranged for convenience as

$$k_{21} - \alpha = \frac{AV_1(\beta - \alpha)}{D} \qquad (17.21)$$

Equation 17.21 is rearranged further to

$$k_{21} = \alpha + \frac{AV_1(\beta - \alpha)}{D} \qquad (17.22)$$

Equation 17.20 can be substituted into equation 17.22, which is then rearranged to equation 17.23.

$$k_{21} = \frac{A\beta + B\alpha}{A + B} \qquad (17.23)$$

In this equation, the unknown constant, k_{21}, is expressed in terms of four known values: A, α, B, and β.

The next constant to solve for is k_{el} by using the relationships in equations 17.7 and 17.8. If α is multiplied by β, then similar multiplication of the right side of equations 17.7 and 17.8 and appropriate cancellation of like terms yield the simple relationship

$$\alpha\beta = k_{21}k_{el} \qquad (17.24)$$

Equation 17.24 is rearranged to

$$k_{el} = \frac{\alpha\beta}{k_{21}} \qquad (17.25)$$

which describes k_{el} in terms of known values. The known values on the right side of equation 17.25 include the microscopic constant k_{21}.

The third microscopic constant, k_{12}, is obtained by adding α and β in terms of the relationships in equations 17.7 and 17.8. Addition of these two equations yields equation 17.26, which rearranges to equation 17.27.

$$\alpha + \beta = k_{12} + k_{21} + k_{el} \qquad (17.26)$$

$$k_{12} = \alpha + \beta - k_{21} - k_{el} \qquad (17.27)$$

Recognizing the difference between k_{el} and β is important. From the model in Scheme 17.2, k_{el} is the intrinsic rate constant for loss of drug from the body, or more correctly from the first compartment. However, the actual elimination slope obtained from drug concentration data is given by β, and the overall drug elimination half-life, or biological half-life, is given by

$$t_{\frac{1}{2}(\beta)} = \frac{0.693}{\beta} \qquad (17.28)$$

In exactly the same way, the half-life of α, or the fast disposition phase of the drug profile, is given by

$$t_{\frac{1}{2}(\alpha)} = \frac{0.693}{\alpha} \qquad (17.29)$$

Derivation of AUC$^{0\to\infty}$, Plasma Clearance, and Renal Clearance

The area under the drug-plasma curve in this model can be calculated by means of the trapezoidal rule in the same way as in the one-compartment model because the area is a model-independent parameter. In the present model, the "end correction" that may be necessary to convert a truncated area to an area of infinite time is given by

$$\text{end correction} = \frac{C_t}{\beta} \tag{17.30}$$

The area under the curve from zero to infinite time can be expressed in pharmaco-kinetic terms by integrating equation 17.11 or 17.12 between the limits of zero and infinite time. Integration of equation 17.11 yields equation 17.31, while integration of equation 17.12 yields equation 17.32.

$$\text{AUC}^{0 \to \infty} = \frac{D}{V_1 k_{el}} \tag{17.31}$$

$$\text{AUC}^{0 \to \infty} = \frac{A}{\alpha} + \frac{B}{\beta} \tag{17.32}$$

Proving that equations 17.31 and 17.32 are identical is an interesting exercise.

In the one-compartment model, plasma clearance following an intravenous dose is obtained by dividing the dose by the area under the drug-concentration curve from zero to infinite time. The identical situation occurs with the two-compartment model. As two possible expressions have been written for the area under the curve, equations 17.31 and 17.32, two expressions can also be written for plasma clearance. Equation 17.31 yields equation 17.33, and equation 17.32 yields equation 17.34.

$$Cl_p = \frac{D}{\left(\dfrac{D}{V_1 k_{el}}\right)} = V_1 k_{el} \tag{17.33}$$

$$Cl_p = \frac{D}{\dfrac{A}{\alpha} + \dfrac{B}{\beta}} \tag{17.34}$$

The expression $V_1 k_{el}$ in equation 17.33 is similar to the clearance expression $V k_{el}$ in the one-compartment case. Renal clearance is readily obtained by dividing the total quantity of drug recovered in urine by the area under the plasma curve, or alternatively by multiplying the plasma clearance by the ratio k_e/k_{el}, as in equation 17.35.

$$Cl_r = \frac{A_u^\infty}{\left(\dfrac{A}{\alpha}\right)\left(\dfrac{B}{\beta}\right)} = V_1 k_{el} \frac{k_e}{k_{el}} = V_1 k_e \tag{17.35}$$

Volume of Distribution at Equilibrium

Just as the term V is used to represent the overall distribution volume of a drug that obeys one-compartment model kinetics, so is the term V_{dss} used to describe the volume that a drug obeying two-compartment model kinetics would occupy if the distribution volume at equilibrium were homogeneous. The overall drug distribution volume is derived in the following way. At steady state, the clearance of drug between the first and second compartments of the two-compartment model should be equal:

$$V_1 k_{12} = V_2 k_{21} \tag{17.36}$$

which rearranges to

$$V_2 = \frac{V_1 k_{12}}{k_{21}} \tag{17.37}$$

Because V_{dss}, or the total distribution volume, is equal to $V_1 + V_2$, the distribution volume can be described in terms of V_1 and the microscopic constants k_{12} and k_{el}, as in equation 17.38.

$$V_{dss} = V_1 + V_1 \frac{k_{12}}{k_{21}} = V_1 \left(1 + \frac{k_{12}}{k_{21}} \right) \tag{17.38}$$

This expression is used most often to describe the overall drug distribution volume with the two-compartment model.

Other methods have been used to describe the overall distribution volume of a drug that obeys two-compartment model kinetics. The two methods most commonly used lead to the overall distribution volumes $V_{d(extrap)}$ and $V_{d\beta}$. The second of these, $V_{d\beta}$, is also known as $V_{d(area)}$ (4).

The term $V_{d(extrap)}$ is derived from the assumption that drug is homogeneously distributed in the body during the β phase of the drug profile. If the curve during the β phase is extrapolated back to time zero to obtain point B as in Figure 17.2, then an estimate of the drug concentration in the equilibrium condition before any drug loss from the body is obtained:

$$V_{d(extrap)} = \frac{D}{B} \tag{17.39}$$

From equation 17.14, B is related to the dose, V_1, and various rate constants as in

$$B = \frac{D(k_{21} - \beta)}{V_1 (\alpha - \beta)} \tag{17.40}$$

As the volume term $V_{d(extrap)}$ is equal to the dose divided by the intercept B (equation 17.39), equation 17.40 can be substituted into equation 17.39 to obtain the new equation:

$$V_{d(extrap)} = \frac{V_1 (\alpha - \beta)}{k_{21} - \beta} \tag{17.41}$$

This equation is the standard form of the relationship among $V_{d(extrap)}$, V_1, and rate constants associated with the two-compartment model.

The term $V_{d\beta}$, or $V_{d(area)}$, is derived from the relationship:

$$AUC^{0\to\infty} = \frac{D}{V_{d\beta}\beta} \qquad (17.42)$$

This equation is analogous to

$$AUC^{0\to\infty} = \frac{D}{Vk_{el}} \qquad (17.43)$$

that was used to relate area, dose, and plasma clearance for the one-compartment model in Chapter 13. In Equation 17.42 the denominator $V_{d\beta}\beta$ is the plasma clearance. Another expression for plasma clearance with the two-compartment model is $V_1 k_{el}$, as in

$$\text{Plasma clearance} = V_1 k_{el} \qquad (17.44)$$

Equating the two different terms for plasma clearance for the two-compartment model, equations 17.33 and 17.44, yields equation 17.45.

$$V_1 k_{el} = V_{d\beta}\beta \qquad (17.45)$$

which is rearranged to

$$V_{d\beta} = V_1 \frac{k_{el}}{\beta} \qquad (17.46)$$

which gives the overall volume term, $V_{d\beta}$, or $V_{d(area)}$, in terms of V_1, k_{el}, and β. Because $\alpha\beta = k_{21} k_{el}$ (equation 17.24), equation 17.46 can be written in the alternative form:

$$V_{d\beta} = V_1 \frac{\alpha}{k_{21}} \qquad (17.47)$$

Thus, three expressions have been derived to describe the overall drug distribution volume at equilibrium, V_{dss}, $V_{d(extrap)}$, and $V_{d\beta}$. Note that the term V_{dss} does not contain an elimination rate constant in the form of β or k_{el}, but is dependent only on V_1 and the distribution constants k_{12} and k_{21}. The other volume terms are dependent on k_{el} or β and are thus elimination rate constant-dependent. To illustrate the dependency of $V_{d(extrap)}$ and $V_{d\beta}$ on drug elimination rate, consider the relationship between these values and the elimination rate-independent volume V_{dss} as k_{el} decreases in value. As k_{el} or β approaches zero (and recall that if k_{el} is zero then β must also equal zero), so do the terms $V_{d(extrap)}$ and $V_{d\beta}$ approach the value of V_{dss}. Both $V_{d(extrap)}$ and $V_{d\beta}$, although used extensively in the literature, overestimate the overall distribution volume relative to the value V_{dss}. This relationship is illustrated in equations 17.48 and 17.49.

$$V_{d(extrap)} = \frac{V_1(\alpha - \beta)}{(k_{21} - \beta)} \overset{(\beta \to 0)}{=} V_1 \frac{\alpha}{k_{21}} = V_1 \frac{k_{12} + k_{21}}{k_{21}} = V_1\left(1 + \frac{k_{12}}{k_{21}}\right) \qquad (17.48)$$

$$V_{d\beta} = V_1 \frac{\alpha}{k_{21}} \overset{(k_{el} \to 0)}{=} V_1 \frac{k_{12} + k_{21}}{k_{21}} = V_1\left(1 + \frac{k_{12}}{k_{21}}\right) \qquad (17.49)$$

The degree of overestimation depends on how fast the drug is eliminated relative to the values of the distribution constants k_{12} and k_{21} (5).

Kinetics of Tissue Distribution

For drugs that obey two-compartment model kinetics after bolus intravenous administration, drug concentrations in blood, plasma, or serum decline initially at a fast rate because of combined tissue uptake and elimination, and decline subsequently at a slower rate.

While drug is drawn out of plasma during the α distribution phase, it accumulates in tissue so that, as the amount of drug in plasma decreases, the amount of drug in tissue increases. This process continues until equilibrium is reached between the plasma and tissue compartments. The time at which equilibrium is achieved will mark the end of the α phase. During the β phase, drug levels in both compartments will decline at the same rate, although the amount of drug in the two compartments may be different depending on the relative affinity of drug for various central and peripheral tissues. This situation is illustrated in Figure 17.3. The quantity of drug in the second compartment is represented in this figure as being greater than that in the first compartment during the β phase, but this is not necessarily the case.

The quantities of drug in the two compartments were described previously by equations 17.4 and 17.5. Although the two equations are obviously different, they have identical time dependencies, containing the rate constants α and β, so that the rate of drug increase in tissue during the α phase following intravenous dosage will be the mirror image of the rate of decline in drug concentrations in plasma. During the β phase, drug levels in tissue and plasma will decline at the same rate according to the first-order rate constant β. During the α phase, the relative amounts of drug in the two compartments are changing rapidly with respect to each other. However, during the β phase an equilibrium situation exists and the ratio of the quantity of drug in tissue to the quantity of drug in plasma should be constant.

This ratio can be established from the β phase segments of equations 17.4 and 17.5. During the β phase of a drug profile in tissue or plasma, the term $e^{-\alpha t}$ has reduced to zero so that equations 17.4 and 17.5 collapse to equations 17.50 and 17.51, respectively.

$$A_1 - \left(\frac{D}{\alpha - \beta}\right)(k_{21} - \beta)e^{-\beta t} \qquad (17.50)$$

$$A_2 = \frac{Dk_{12}}{\alpha - \beta}e^{-\beta t} \qquad (17.51)$$

Dividing equation 17.51 by equation 17.50 and canceling like terms yields equation 17.52.

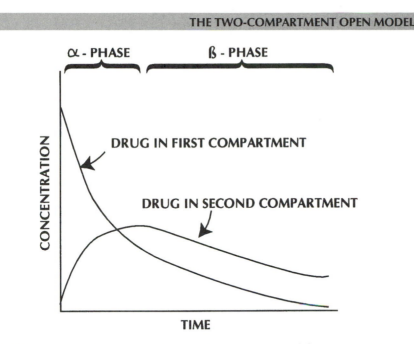

α - PHASE β - PHASE

CONCENTRATION

DRUG IN FIRST COMPARTMENT

DRUG IN SECOND COMPARTMENT

TIME

Drug concentrations in the first and second compartments of the two-compartment open model following bolus intravenous injection into the central compartment.

FIGURE 17.3

$$\frac{A_2}{A_1} = \frac{k_{12}}{k_{21} - \beta} = \frac{T}{P} \qquad (17.52)$$

This equation indicates that the ratio of drug in tissue to drug in plasma, the T/P ratio, is given by k_{12} divided by the difference between k_{21} and β (i.e., the equilibrium ratio of drug in the two compartments is a function not only of the relative magnitude of the two rate constants, k_{12} and k_{21}, but also of the magnitude of β). Imagine the situation where k_{12} and k_{21} are both 0.1 h^{-1} and β is only 0.01 h^{-1}. The T/P ratio will then be 0.1/ 0.09 = 1.1. However, if β is 0.05 h^{-1}, then the T/P ratio will be 0.1/0.05 = 2.0. If β were 0.08 h^{-1}, then T/P = 0.1/0.02 = 5.0. Thus the ratio of the quantity of drug in tissue to that in the plasma compartment is influenced by the overall drug elimination rate.

The dependency of T/P on the drug elimination rate is similar to the elimination rate dependency of $V_{d(extrap)}$ and $V_{d\beta}$ described previously. Failure to recognize this dependency can lead to false conclusions regarding the influence (or lack of influence) of drug elimination on drug disposition in the body.

The equation for cumulative urinary excretion of drug with this model was given as equation 17.6. Obtaining good estimates of pharmacokinetic parameters from urinary data with this model is difficult because of the model complexity and the difficulty of obtaining sufficiently frequent urine samples, even with water loading. However, given that satisfactory urine collection can be obtained, one way to analyze urinary excretion data is to construct sigma-minus plots in the same way as described previously for the one-compartment model (Chapter 14). During the

Obtaining Parameter Estimates from Urinary Excretion Data

postdistributive phase, that is, when the value of $e^{-\alpha t}$ approaches zero, equation 17.6 reduces to

$$A_{u} = A_{u}^{\infty}\left[1 - \left(\frac{k_{el}}{\alpha - \beta}\right)\left(\frac{k_{21} - \beta}{\beta}\right)e^{-\beta t}\right] \tag{17.53}$$

Because $\alpha\beta = k_{21}k_{el}$ (equation 17.24), equation 17.53 can be rearranged to equation 17.54.

$$\log(A_{u}^{\infty} - A_{u}) = \log\left\{A_{u}^{\infty}\left[\frac{k_{el}(k_{21} - \beta)}{\beta(\alpha - \beta)}\right]\right\} - \frac{\beta}{2.3}t \tag{17.54}$$

$$\log(A_{u}^{\infty} - A_{u}) = \log\left(A_{u}^{\infty}\frac{\alpha - k_{el}}{\alpha - \beta}\right) - \frac{\beta}{2.3}t \tag{17.55}$$

This equation shows that if the logarithm of the amount of drug remaining to be excreted, $\log(A_{u}^{\infty} - A_{u})$, is plotted against time, the situation illustrated in Figure 17.4 is obtained. Using the relationship $k_{21}k_{el} = \alpha\beta$, the terminal linear portion, the plot can also be described by equation 17.55. The rate constant β is obtained from the slope of the line and $\log[A_{u}^{\infty}(\alpha - k_{el})/(\alpha - \beta)]$ from the intercept at zero time. If residuals are taken in the same way as with the drug profile in plasma (i.e., by subtracting values on the extrapolated β slope from the actual values at early sampling times), a residual slope can be constructed that will have a slope of $-\alpha/2.3$ and an intercept of $\log\{A_{u}^{\infty}[1 - (\alpha - k_{el})/(\alpha - \beta)]\}$.

Knowing the values of the slopes and intercepts of the two lines, k_{el} can be solved as in equation 17.56.

$$k_{el} = \alpha - \left(\text{intercept}\frac{\alpha - \beta}{A_{u}^{\infty}}\right) \tag{17.56}$$

After k_{el} is obtained, k_{21} can be obtained from equation 17.57, and k_{12} from equation 17.58.

$$k_{21} = \frac{\alpha\beta}{k_{el}} \tag{17.57}$$

$$k_{12} = \alpha + \beta - k_{el} - k_{21} \tag{17.58}$$

No volume calculations can be done with these data because urinary excretion of drug provides no information regarding concentrations of drug within the body compartments.

Although the preceding calculations are straightforward on a purely theoretical basis, good data are required for correct and accurate parameter estimation; for example, accurate, frequent, and complete urine collections, and also accurate and specific assays. Accurate and specific assays are particularly important in urine

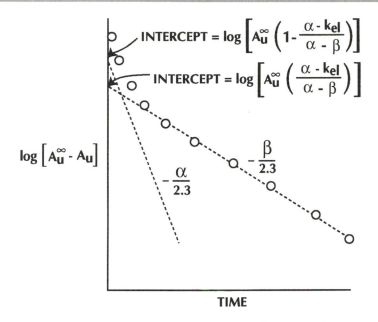

$$\text{INTERCEPT} = \log\left[A_u^\infty\left(1 - \frac{\alpha - k_{el}}{\alpha - \beta}\right)\right]$$

$$\text{INTERCEPT} = \log\left[A_u^\infty\left(\frac{\alpha - k_{el}}{\alpha - \beta}\right)\right]$$

$\log\left[A_u^\infty - A_u\right]$

$-\dfrac{\alpha}{2.3}$

$-\dfrac{\beta}{2.3}$

TIME

Plot of the logarithm of the amount of drug remaining to be excreted in urine vs. time, illustrating a method of obtaining graphical estimates of pharmacokinetic parameters.

FIGURE 17.4

analysis compared to plasma analysis because of the greater possibility of metabolite interference.

The two-compartment model with first-order drug input is presented in Scheme 17.3, and a typical drug-in-blood profile is shown in Figure 17.5. This model is similar to the model for the bolus intravenous case, except that a first-order absorption rate constant is included. With this model, drug is assumed to be absorbed from the absorption site at a first-order rate. During the absorption phase and also extending beyond that phase, the drug distributes from the blood or plasma and all other fluids associated with the first compartment into the less accessible, slowly equilibrating peripheral tissues that constitute the second compartment. When equilibrium is established then, as in the intravenous case, drug levels in the circulation and in tissue decline at a first-order rate controlled by the rate constant β. As shown in the figure, the three phases of the blood profile can be described as the absorption, α, and β phases.

Before describing the mathematics and methods of obtaining parameter values for this model, some words of caution are appropriate. This model contains many traps for the unwary, and analysis must be done with cognizance that several different methods can be used to analyze the data and to assign kinetic parameter values.

Detecting two-or multicompartment model kinetics following an oral dose of a drug is difficult, although such models may be clearly evident after intravenous dosing. This difficulty is due to the absorption and α phases being of similar dura-

The Two-Compartment Model with First-Order Drug Input

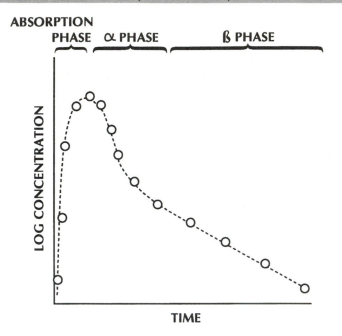

FIGURE 17.5 Blood concentration versus time profile of a drug that obeys two-compartment model kinetics following first-order drug input.

tion; therefore, the α or distribution phase will be obscured by the absorption phase. This condition not only prevents two-compartment model analysis but also throws further doubt on the significance of the absorption rate constant, k_a. A high probability of a flip-flop model also exists, not only between absorption and elimination rate constants as discussed earlier for the one-compartment model (Chapter 14), but also between the k_a and α rate constants. Thus, incorrect values could be assigned to these two constants. Finally, because of the increased complexity of this model, more data points are required for accurate parameter characterization. Assigning numerical values to the various rate constants associated with the model is unrealistic unless sufficient data points exist to define each of the three phases of the drug profile (6, 7).

From Scheme 17.3, the rates of change in the amounts of drug in the first and second compartments, and also the cumulative amount of drug voided in urine with respect to time, are given by equations 17.59 through 17.61

$$\frac{dA_1}{dt} = k_a X_a + k_{21}A_2 - (k_{12} + k_{el})A_1 \qquad (17.59)$$

$$\frac{dA_2}{dt} = k_{12}A_1 - k_{21}A_2 \qquad (17.60)$$

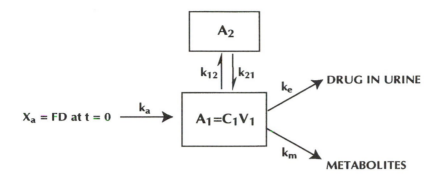

Two-compartment open model with first-order drug input.   **SCHEME 17.3**

$$\frac{dA_u}{dt} = k_e A_1 \qquad (17.61)$$

These equations can be solved simultaneously to obtain equations 17.62 through 17.64.

$$A_1 = FDk_a\left[\frac{k_{21} - \alpha}{(k_a - \alpha)(\beta - \alpha)}e^{-\alpha t} + \frac{k_{21} - \beta}{(k_a - \beta)(\alpha - \beta)}e^{-\beta t} + \frac{k_{21} - k_a}{(\alpha - k_a)(\beta - k_a)}e^{-k_a t}\right] \quad (17.62)$$

$$A_2 = FDk_a k_{21}\left[\frac{e^{-\alpha t}}{(k_a - \alpha)(\beta - \alpha)} + \frac{e^{-\beta t}}{(k_a - \beta)(\alpha - \beta)} + \frac{e^{-k_a t}}{(\alpha - k_a)(\beta - k_a)}\right] \qquad (17.63)$$

$$A_u = A_u^\infty\left\{1 - k_a k_{el}\left[\frac{k_{21} - \alpha}{\alpha(k_a - \alpha)(\beta - \alpha)}e^{-\alpha t} + \frac{k_{21} - \beta}{\beta(\alpha - \beta)(k_a - \beta)}e^{-\beta t} \right.\right.$$
$$\left.\left. + \frac{k_{12} - k_a}{k_a(\alpha - k_a)(\beta - k_a)}e^{-k_a t}\right]\right\} \qquad (17.64)$$

which represent the quantity of drug in the first compartment, the quantity of drug in the second compartment, and the quantity of drug cumulatively excreted unchanged in the urine, respectively (8). In these equations, the constants α and β have identical relationships to k_{12}, k_{21}, and k_{el} that were established in the intravenous case (equations 17.7 and 17.8). Equation 17.62, the amount of drug in the first compartment, can be converted to concentration form by dividing by the volume of the first compartment, V_1:

$$C = \frac{FDk_a}{V_1}\left[\frac{k_{21} - \alpha}{(k_a - \alpha)(\beta - \alpha)}e^{-\alpha t} + \frac{k_{21} - \beta}{(k_a - \beta)(\alpha - \beta)}e^{-\beta t} + \frac{k_{21} - k_a}{(\alpha - k_a)(\beta - k_a)}e^{-k_a t}\right] \quad (17.65)$$

equations 17.62–17.65 are similar. They all have the same time relationships, and they all contain the same three exponential functions, α, β, and k_a.

Drug Concentration in the First Compartment

In the intravenous case, the equation for C, equation 17.11, was written in short-hand notation as equation 17.12. Equation 17.65 can similarly be written in the form of

$$C = Ae^{-\alpha t} + Be^{-\beta t} + C'e^{-k_a t} \tag{17.66}$$

The symbol C' is used for the third concentration constant in order to differentiate it from the concentration in plasma term, C. The three constants, A, B, and C', represent the groups of kinetic parameters in equation 17.65 as shown in equations 17.67, 17.68, and 17.69, respectively.

$$A = \frac{FDk_a(k_{21} - \alpha)}{V_1(k_a - \alpha)(\beta - \alpha)} \tag{17.67}$$

$$B = \frac{FDk_a(k_{21} - \beta)}{V_1(k_a - \beta)(\alpha - \beta)} \tag{17.68}$$

$$C = \frac{FDk_a(k_{21} - k_a)}{V_1(\alpha - k_a)(\beta - k_a)} \tag{17.69}$$

The task at hand is to take a set of drug concentration versus time data and use it to obtain graphical estimates of all of the pharmacokinetic parameters associated with this model. Consider the drug concentration data in Table 17.1. The three sequential steps in the analysis of the data are shown in Figure 17.6.

The graphical procedure is as follows. The data are first plotted on semilogarithmic graph paper, and the terminal linear slope is assigned the value $-\beta/2.3$, and the extrapolated intercept at zero time is B. The first residual is then obtained by subtracting the values on the extrapolated portion of the β line from the actual data points during the pre-β portion of the curve. In this example, four data points were used to construct the β slope (because those points were on a straight line), leaving 10 data points including time zero to construct the first residual. In the table, the first residual column is designated $C-\hat{C}$. The first residual has both negative and positive values depending on whether the data points fall above or below the extrapolated β line, as shown in Figure 17.6a. The positive values obtained from the first residual are then plotted on semilogarithmic graph paper, and the terminal linear descending portion of that curve is shown in Figure 17.6b. The last three residual points are linear, and a line is therefore constructed through these points and extrapolated back to time zero as in Figure 17.6b. This line has a slope of $-\alpha/2.3$ (assuming that $\alpha < k_a$), and the extrapolated intercept at zero time is A.

To obtain the remaining two parameters, C' and k_a, the values on the extrapolated slope of the terminal portion of the first residual curve are subtracted from the values on the early nonlinear portion of the first residual curve to obtain the second residual. This procedure is described in Table 17.1 and shown in Figure

17.6b. In the table, the extrapolated first residual line is depicted as $\hat{C}$. The slope of the second residual, $C-\tilde{C}$, is composed entirely of negative values, reflecting the opposite direction of this phase of the drug profile to that of the α and β phases, and also the negative value of the intercept of the second residual slope at zero time. To obtain the values of C' and k_a, the absolute values of the residuals in Table 17.1 are plotted, as shown in Figure 17.6c. These values should yield a straight line because this final slope is monoexponential, the α and β components now having been removed. The line has a slope of $-k_a/2.3$, and an intercept of C'.

The numerical values obtained from the data in Table 17.1 are summarized in

$$C = 116e^{-1.33t} + 27e^{-0.06t} - 143e^{-3.01t} \tag{17.70}$$

The value C' is negative because of the negative values of the second residual slope in the table. Two good tests of whether the analysis is accurate are (i) to make sure that α, β, and k_a have different values from one another and increase in the order $\beta < \alpha < k_a$ (assuming no flip-flop situation), and (ii) to verify that the concentration of drug in the bloodstream should be zero at zero time because no drug has been absorbed. When $t = 0$ in equation 17.66, equation 17.71 is obtained.

$$C = A + B + C' \tag{17.71}$$

Therefore, if the three constants A, B, and C' are added (C' is negative), a value should be obtained that equals or approaches zero.

Drug Concentration Data Analysis by the Method of Residuals

TABLE 17.1

		First Residual		Second Residual		
Time	Concentration	$\hat{C}(27.1e^{-0.6t})$	$C - \hat{C}$(residual)	C	$\tilde{C}(116e^{-1.33t})$	$C - \tilde{C}$
0.0	0.00	27.10	−27.10	−27.10	116.0	−143.1
0.1	24.53	26.94	−2.41	−2.41	101.55	−103.96
0.2	40.51	26.78	+13.73	13.73	88.91	−75.18[a]
0.5	58.49	26.30	+32.19	32.19	59.66	−27.47
1.0	52.28	25.52	+26.27	26.76	30.70	−3.94
1.5	41.07	24.77	+16.30	16.30	15.78	
2.0	32.73	24.04	+8.69	8.69	8.11	
3.0	25.02	22.63	+2.39[b]			
4.0	21.95	21.32	+0.63			
6.0	18.95	18.91	+0.04			
8.0	16.77					
10.0	14.87[c]					
20.0	8.16					
40.0	2.46					

Note: Drug concentrations in plasma data were obtained during a 40-h period for a drug that obeys two-compartment model kinetics with first-order absorption and elimination.
[a]Linear k_a phase, intercept $(C) = 143$, $t_{1/2} = 0.23$, and $k_a = 3.01$.
[b]Linear α phase, intercept $(A) = 116.0$, $t_{1/2} = 0.52$, and $\alpha = 1.33$.
[c]Linear β phase, intercept $(B) = 27.1$, $t_{1/2} = 11.5$, and $\beta = 0.06$.

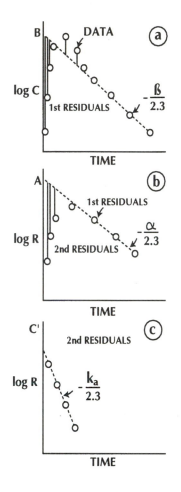

FIGURE 17.6   Method of obtaining residuals from the drug profile in Figure 17.5. (a) Terminal β slope and first residuals obtained from Figure 17.5, (b) slope of first residuals and method of obtaining second residuals, and (c) slope of second residuals.

The three concentration terms and three rate constants in equation 17.70 can now be used to obtain values of the microscopic constants, k_{12}, k_{21}, k_{el}. These calculations are done with equations 17.67, 17.68, and 17.69. As a first step, it is convenient to remove the term k_a from the right side of these equations. To do this, each equation is divided by k_a to obtain the new functions P, Q, and R as in equations 17.72, 17.73, and 17.74.

$$P = \frac{A}{k_a} = \frac{FD}{V_1}\left[\frac{k_{21}-\alpha}{(k_a-\alpha)(\beta-\alpha)}\right] = \frac{116}{3.01} = 38.54 \tag{17.72}$$

$$Q = \frac{B}{k_a} = \frac{FD}{V_1}\left[\frac{k_{21} - \beta}{(k_a - \beta)(\alpha - \beta)}\right] = \frac{27.1}{3.01} = 9.00 \qquad (17.73)$$

$$R = \frac{C'}{k_a} = \frac{FD}{V_1}\left[\frac{k_{21} - k_a}{(\alpha - k_a)(\beta - k_a)}\right] = \frac{143}{3.01} = -47.51 \qquad (17.74)$$

From equations 17.72 and 17.73, the relationships in equations 17.75 and 17.76, respectively, can be derived.

$$\frac{FD}{V_1} = P(k_a - \alpha) + Q(k_a - \beta) \qquad (17.75)$$

$$V_1 = \frac{FD}{P(k_a - \alpha) + Q(k_a - \beta)} \qquad (17.76)$$

Equation 17.76 provides an expression for the volume of the central compartment in this model. However, as in the oral absorption case for the one-compartment model, solving for V_1 is not always possible unless the value for F, the fraction of administered drug that is absorbed into the systemic circulation, is known.

To solve for the three microscopic rate constants, a little additional information is needed. The area under the plasma profile can be obtained by integrating equation 17.66 between the limits of zero and infinite time to obtain

$$\text{AUC}^{0 \to \infty} = \frac{A}{\alpha} + \frac{B}{\beta} + \frac{C'}{k_a} \qquad (17.77)$$

The same area can be obtained, however, by integrating equation 17.65 between the same limits, and canceling like terms to obtain

$$\text{AUC}^{0 \to \infty} = \frac{D}{V_1 k_{el}} \qquad (17.78)$$

Equations 17.77 and 17.78 can be combined to form equation 17.79.

$$V_1 k_{el} = \frac{D}{\dfrac{A}{\alpha} + \dfrac{B}{\beta} + \dfrac{C'}{k_a}} \qquad (17.79)$$

Substituting from equations 17.72, 17.73, and 17.74 for A, B, and C' in equation 17.79 yields

$$V_1 k_{el} = \frac{D}{\dfrac{Pk_a}{\alpha} + \dfrac{Qk_a}{\beta} + R} \qquad (17.80)$$

Taking reciprocals of both sides of this equation, using equation 17.79 to solve for V_1, and using the relationship in equation 17.81 yield equation 17.82.

$$k_{el} = \frac{\alpha\beta}{k_{21}} \qquad (17.81)$$

$$k_{21} = \frac{\beta k_a P + \alpha k_a Q + \alpha\beta R}{P(k_a - \alpha) + Q(k_a - \beta)} \qquad (17.82)$$

This equation provides the first of the three microscopic rate constants. Once k_{21} is solved for, the other two rate constants are readily obtained from equations 17.81 and 17.83.

$$k_{12} = \alpha + \beta - k_{21} - k_{el} \qquad (17.83)$$

The numerical values for these calculations are given in equations 17.84, 17.85, and 17.86.

$$k_{21} = \frac{(0.06)(3.01)(38.54) + (1.33)(3.01)(9.0) - (1.33)(0.06)(47.51)}{38.54(3.01 - 1.33) + 9.0(3.01 - 0.06)} = 0.43 \quad (17.84)$$

$$k_{el} = \frac{(1.33)(0.06)}{0.43} = 0.185 \qquad (17.85)$$

$$k_{12} = 1.33 + 0.06 - 0.43 - 0.185 = 0.775 \qquad (17.86)$$

Some Model-Independent Parameters for a Drug that Obeys Two-Compartment Model Kinetics

In Chapter 1, three approaches to pharmacokinetics were described: the compartment model approach, the physiological model approach, and the model-independent approach. The one-compartment model, being the simplest way in which drug disposition can be described, is really a model-independent method.

The use of true model-independent kinetics becomes particularly useful when dealing with compounds that obey multicompartment kinetics, for example, the two-compartment model. The two-compartment model with intravenous injection will be used to illustrate the model-independent approach.

The blood-level curve in Figure 17.1 was previously described in shorthand notation by equation 17.12. If further interpretation of this equation (i.e., as in equation 17.11) is avoided, then model-independent kinetics is being applied. The drug-level curve is being described by a biexponential equation that has two coefficients (or concentration terms), A and B, and two rate constants, α and β. The volume term in equation 17.20, the terminal half-life in equation 17.28, the area under the drug-concentration curve in equation 17.32, and the plasma clearance in equation 17.34 were obtained from the four parameters A, B, α, and β. In other words, these equations were based on a simple biexponential equation and are therefore completely model-independent.

Thus, a great deal of information can be obtained about a drug by the model-independent approach. Whether the model-dependent or model-independent approach is used depends entirely on the type of study, the questions being asked, the quality of the data, and the curiosity of the investigator.

Although calculating pharmacokinetic parameters using a model-independent approach is quite simple, one parameter, the overall drug distribution volume at steady state, is more difficult to obtain. This parameter is useful because it describes, in a model-independent manner, the distribution characteristics of a drug that has equilibrated into tissues and body fluids.

The first solution to this problem was described by Wagner (9), who showed that the value V_{dss} can be obtained from

$$V_{dss} = \text{dose} \frac{\Sigma \dfrac{A_i}{\lambda_i^2}}{\left(\Sigma \dfrac{A_i}{\lambda_i}\right)^2} \tag{17.87}$$

where A is a concentration term and λ is a rate constant. For the biexponential intravenous case (equation 17.12), equation 17.87 can be written as equation 17.88.

$$V_{dss} = \text{dose} \frac{\dfrac{A}{\alpha^2} + \dfrac{B}{\beta^2}}{\left(\dfrac{A}{\alpha} + \dfrac{B}{\beta}\right)^2} \tag{17.88}$$

An alternative method to calculate V_{dss} was described by Benet and Galeazzi (10). This approach is both model- and rate constant-independent. The derivation of the method is not included here, but the final expression is

$$V_{dss} = \text{dose} \frac{\text{AUMC}^{0 \to \infty}}{(\text{AUC}^{0 \to \infty})^2} \tag{17.89}$$

This equation yields V_{dss} as a function of the administered dose, the area under the moment curve (AUMC), and the square of the area under the blood-level curve. The AUMC is given by

$$\text{AUMC}^{0 \to \infty} = \int_0^\infty tC \cdot dt \tag{17.90}$$

which is simply the area under the curve of the product of time and plasma concentration from zero to infinite time. The AUMC can be considered to be the sum of the residence times of all drug molecules in the body from the time of administration to the time when all drug molecules have left the body.

As indicated earlier in this chapter, plasma drug profiles following intravenous dosing frequently exhibit biphasic elimination profiles consistent with a two-compartment kinetic model, whereas this is less common after oral or intramuscular dosing because of the overlap of distribution and absorption phases. Some drugs whose plasma profiles have recently been described in terms of bi-or multiphasic kinetics, after intravenous or oral doses, include amiodarone in subjects with normal and impaired renal function (11), the new, third generation cephalosporin cefadozime in healthy volunteers (12) and in the elderly (13), the 5α-reductase inhibitor epristeride in healthy subjects (14), the antidementia drug E 2020 in healthy subjects (15), fluconazole in immune-compromised children (16), recombinant hirudin in healthy volunteers (17), the novel inotropic agent OPC-18790 in healthy male subjects (18), the Class III antiarrhythmic agent sematilide in patients with renal failure (19), the individual components of the new antibiotic agent teicoplanin in healthy volunteers (20), the antifungal agent terbinafine in healthy subjects (21), the loop diuretic torsemide in patients with congestive heart failure (22), and the cardiac agent zatebradine in healthy subjects (23).

Summary

For many drugs, particularly after bolus intravenous administration, a multicompartment kinetic model may provide a more accurate description of drug plasma profiles than a one-compartment kinetic model. The multicompartment model is frequently represented by a two-compartment kinetic model.

With the two-compartment kinetic model, drug-concentration profiles in plasma after bolus intravenous administration are characterized by an α phase of fast decline and a β phase of slower decline.

Initial graphical estimates can be used to calculate values of microscopic disposition constants associated with the two-compartment model.

A variety of methods have been described to calculate the overall equilibrium distribution volume for a drug that obeys two-compartment model kinetics. Only one of these, V_{dss}, is a true constant. Other constants are influenced by the relative magnitudes of microscopic elimination and disposition constants, and may yield biased values.

For a drug that obeys two-compartment model kinetics, the ratio of drug in the tissue compartment to that in plasma is a function of the distribution rate constants and drug elimination rate.

Estimates of pharmacokinetic parameters can be obtained from urinary excretion data by use of sigma-minus plots.

Plasma drug profiles after oral administration of a drug that obeys two-compartment model kinetics are triexponential. The curves are characterized by absorption, α, and β phases.

Estimates of kinetic parameters can be obtained by standard curve-stripping procedures. Values of microscopic rate constants may be derived from initial estimates of components of the triexponential function.

Values of many pharmacokinetic parameters, including the equilibrium distribution volume, can be obtained by model-independent methods.

Problems

1. A drug (200 mg) was administered to a patient by bolus intravenous injection. The decline in log plasma concentrations of unchanged drug was biphasic, and the half-life of the terminal linear elimination phase was 6.9 h. By curve stripping, the following values of microscopic rate constants were found: k_{12} = 0.6 h^{-1}, k_{21} = 0.4 h^{-1}, and k_{el} = 0.3 h^{-1}. The concentration of unchanged drug in plasma immediately after injection (allowing 1–2 min for mixing) was 40 µg/mL. Total excretion of unchanged drug in urine accounted for 55% of the dose; the rest of the dose was metabolized. Calculate the following: (i) $V_d\beta$, (ii) $V_{d(extrap)}$, (iii) V_{dss}, (iv) plasma clearance, and (v) renal clearance.

2. Given that a drug obeys two-compartment model kinetics, and that the disposition rate constants, k_{12} and k_{21}, have values of 0.37 h^{-1} and 0.20 h^{-1}, respectively, calculate the shortest possible biological half-life of the drug.

3. Following a rapid intravenous dose of 250 mg of cefotaxime to a patient with normal renal function, plasma levels of unchanged drug declined in a biphasic manner, and the following parameter values were obtained: A = 25 µg/mL, B = 7.5 µg/mL, α = 3.5 h^{-1}, β = 0.6 h^{-1}. What is the plasma clearance of cefotaxime?

4. From the data in Problem 3, calculate the terminal-phase elimination half-life of cefotaxime.

5. Are plasma concentrations of a drug that obeys two-compartment model kinetics following an oral dose (i) monophasic, (ii) biphasic, or (iii) triphasic?

6. Mathematical analysis of plasma levels of a drug following a single oral dose yielded the expression: $C = 59e^{-1.2t} + 130e^{-0.32t} - 189e^{-3.1t}$. What is the drug terminal elimination half-life?

7. From the data in Problem 6, what is the area under the plasma curve from zero to infinite time?

8. If the drug dose leading to the plasma levels described in Problem 6 was 100 mg, and the drug was 80% available to the systemic circulation, what is the plasma clearance?

9. Suppose the drug described in Problem 6 was cleared 50% as unchanged drug via the kidneys and 50% by hepatic metabolism. What are the renal clearance and the hepatic clearance?

References

1. Gibaldi, M.; Perrier, D. *Pharmacokinetics*; Marcel Dekker: New York, 1982; pp 45–111.
2. Wagner, J. G. *Pharmacokinetics for the Pharmaceutical Scientist*; Technomic: Lancaster, PA, 1993; pp 15–43.

3. Wagner, J. G. *J. Pharmacokinet. Biopharm.* **1975,** *3*, 457–478.

4 Riegelman, S.; Loo, J. C. K.; Rowland, M. *J. Pharm. Sci.* **1968,** *57*, 128–133.

5 Jusko, W. J.; Gibaldi, M. *J. Pharm. Sci.* **1972,** *61*, 1270–1273.

6. Benet, L. Z. *J. Pharm. Sci.* **1972,** *61*, 536–541.

7. Kaplan, S. A.; Jack, M. L.; Alexander, K.; Weinfeld, R. E. *J. Pharm. Sci.* **1973,** *62*, 1789–1796.

8. Welling, P. G.; Lee, K. P.; Patel, J. A.; Walker, J. A.; Wagner, J. G. *J. Pharm. Sci.* **1971,** *60*, 1629–1634.

9. Wagner, J. G. *Pharmacokinet. Biopharm.* **1976,** *4*, 443–467.

10. Benet, L. Z.; Galeazzi, R. L. *J. Pharm. Sci.* **1979,** *68*, 1071–1074.

11. Ujhelyi, M. R.; Klamerus, K. J.; Vadiei, K.; O'Rangers, E.; Izard, M.; Neefe, L.; Zimmerman, J. J.; Chow, M. S. S. *J. Clin. Pharmacol.* **1996,** *36*, 122–130.

12. Bryskier, A.; Procyk, T.; Tremblay, D.; Lenfant, B.; Fourtillan, J. B. *J. Antimicrob. Chemother.* **1990,** *26* (Suppl C), 65–70.

13. Nilsen, O. G.; Rennemo, F.; Rennemo, R.; Lenfant, B. *J. Antimicrob. Chemother.* **1990,** *26* (Suppl), 71–75.

14. Benincosa, L. J.; Audet, P. R.; Lundberg, D.; Zariffa, N.; Jorkasky, D. K. *Biopharm. Drug. Dispos.* **1996,** *17*, 249–258.

15. Ohnishi, A.; Mihara, M.; Kamakura, H.; Tomono, Y.; Hasegawa, J.; Yamazaki, K.; Morishita, N.; Tanaka, T. *J. Clin. Pharmacol.* **1993,** *33*, 1086–1091.

16. Seay, R. E.; Larson, T. A.; Toscano, J. P.; Bostrom, B. C.; O'Leary, M. C.; Uden, D. L. *Pharmacotherapy* **1995,** *15*, 52–58.

17. Cardot, J-M. A.; Lefèvre, G. Y.; Godbillon, J. A. *J. Pharmacokinet. Biopharm.* **1994,** *22*, 147–154.

18. Ohniski, A.; Toyoki, T.; Ohno, T.; Takeshige, Y.; Fujita, T.; Kodama, K.; Mishima, M.; Hirayama, A.; Kitani, M.; Miyamoto, G.; Odomii, M.; Tanaka, T. *J. Clin. Pharmacol.* **1994,** *34*, 131–143.

20. Bernareggi, A.; Banese, A.; Cometti, A.; Buniva, G.; Rowland, M. *J. Pharmacokinet. Biopharm.* **1990,** *18*, 525–544.

21. Kovarik, J. M.; Mueller, E. A.; Zehender, H.; Denouël, J.; Caplain, H.; Milleroux, L. *Antimicrob. Agents Chemother.* **1995,** *39*, 2738–2741.

22. Kramer, W. G.; Smith, W. B.; Ferguson, J.; Serbas, T.; Grant, A. G.; Black, P. K.; Brater, D. C. *J. Clin. Pharmacol.* **1996,** *36*, 265–270.

23. Roth, W.; Bauer, E.; Heinzel, G.; Cornelissen, P. J.; van Tol, R. G.; Jonkman, J. H.; Zuiderwijk, P. B. *J. Pharm. Sci.* **1993,** *82*, 99–106.

Physiological Pharmacokinetic Models

18

The focus in the preceding chapters has been on classical compartment-based pharmacokinetic models. The simplifying assumptions in these models are that drugs or metabolites distribute into one or more body compartments, and that movement of compound between compartments, and also elimination from the body, can be described by one or more first-order or saturable processes.

These models are simple approximations of the true anatomical situation and tend to serve a descriptive function, usually with regard to circulating drug or metabolite profiles or urinary excretion. They do not attempt to address the more specific and often intractable problems that relate to the actual time course of drug or metabolite disposition in particular body organs and tissues.

One approach to address these problems is the physiological model, in which the body is divided into compartments based on true anatomical regions or volumes such as blood, heart, liver, and central nervous system. The time course of drug or metabolite levels in the various "physiological" organs or compartments is calculated on the basis of blood flow rate through each particular region, diffusion of drug between blood and tissue, and the relative affinity of drug for blood and the various tissues and organs.

Because of the large number of parameters involved in these studies, physiological models are necessarily more complicated than classical compartment models. Physiological models have been developed for very few drugs and are really justified only to investigate specific tissue localization of drugs such as anticancer agents, or to examine detailed mechanisms of drug metabolism or excretion by the liver, kidney, lung, or other organs.

Physiological models have some advantages and some disadvantages compared with compartment models. The principal advantages of the physiological model are:

Description of Physiological Pharmacokinetic Model

1. Pharmacokinetic parameters are realistic because they are based on observed or predicted physiological values.

2. Parameter values can be altered to allow for changes in physiological function.

3. Total, free, and bound compound concentration profiles can be predicted for selected tissues and regions of the body.

4. Insight can be obtained regarding specific organ elimination mechanisms.

5. Parameter values can be "scaled" for different animal species and humans.

The advantages of the physiological model approach may be summarized briefly in that this model provides more information based on specific tissue localization and handling of compounds, and is particularly useful for some drugs.

The principal disadvantages of the physiological model approach are as follows:

1. The models are complicated; their mathematics are often difficult to solve.

2. It is difficult to validate a physiological model in animals, and virtually impossible in humans, because large numbers of tissue samples are required at different time intervals after dosing.

3. It is difficult to obtain tissue samples that are free of blood.

4. Model development requires a large data base for each drug or drug group in a particular species.

5. In vitro testing is frequently required to establish or validate model parameters.

6. Despite their complexity, physiological models often contain many simplifying assumptions that are difficult to verify regarding diffusion of drug into tissues, complete mixing within organs, and the relative degree of intravascular and extravascular drug binding.

The disadvantages may be summarized briefly in that considerable information is needed to develop the model, the model still contains many assumptions, certainly of a more microscopic nature than the comparatively gross assumptions associated with the compartmental approach, the model is often difficult and sometimes impossible to validate, and species differences may also cause complications.

On a more general and positive note, the development and validation of physiological models, whether successful or not, are likely to improve understanding of the rates and mechanisms of drug absorption, distribution, metabolism, and excretion, whether the work is carried out in vivo or in situ with animal models, or in vitro in isolated organ systems. At the very worst, this approach increases understanding of the processes involved. At best, the approach may facilitate more rigorous characterization of drug pharmacokinetic and pharmacodynamic properties than can be obtained by other compartment-model systems.

Before discussing some basic aspects of physiological pharmacokinetic models, it is important to revisit the concept of organ clearance, a fundamental component for this type of model.

Organ Clearance

Perhaps the best way to understand organ clearance is to consider a single organ that is well perfused by the bloodstream and is capable of eliminating a drug. This simple model, with or without an elimination component, provides the basic building block of physiological pharmacokinetic models. The model is shown in Scheme 18.1. The symbol Q represents blood flow rate through the organ, and C_i and C_0 represent drug concentrations in blood entering and leaving the organ, respectively. If an eliminating organ is considered, then $C_i > C_0$.

The rate at which blood introduces drug to an organ is equal to the blood flow rate multiplied by the drug concentration:

$$\text{Rate at which drug enters organ in blood} = QC_i \qquad (18.1)$$

Because Q has units of volume per unit time, and C_i has units of mass per volume, the product of Q and C_i has units of mass per unit time. If drug is eliminated by the organ, then the rate that drug leaves the organ is similarly equal to the blood flow rate, Q, multiplied by the drug concentration, C_0:

$$\text{Rate at which drug leaves organ in blood} = QC_0 \qquad (18.2)$$

The difference between the rate that drug enters the organ and the rate that drug leaves the organ gives the rate of drug elimination by the organ:

$$\text{Drug elimination rate} = QC_i - QC_0 = Q(C_i - C_0) \qquad (18.3)$$

This type of expression, flow rate multiplied by concentration difference, is commonly used to denote organ clearance, which has units of mass per unit time. This expression is useful because it indicates how much of the drug the organ can eliminate in a certain time period. However, in Chapter 13 clearance was described in terms of the volume of plasma cleared of drug per unit time. Clearance due to particular organs was also defined in terms of the volume of plasma cleared of drug by that particular organ, for example, renal clearance and hepatic clearance. Describing organ clearance with the same units in the physiological model is possible. To do this, the organ extraction ratio, E, needs to be defined. This ratio compares the rate of drug elimination with the rate that drug enters the organ:

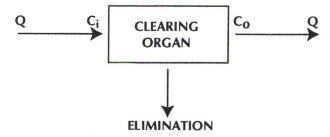

Physiological flow model for a well-perfused organ that is capable of drug elimination.   **SCHEME 18.1**

$$E = \frac{Q(C_i - C_0)}{QC_i} = \frac{C_i - C_0}{C_i} \qquad (18.4)$$

The organ extraction ratio is a dimensionless quantity. This ratio indicates how efficiently the organ removes drug from the bloodstream. If removal is inefficient and removes essentially no drug, then C_0 will equal C_i, and the extraction ratio is zero. On the other hand, if removal is efficient and removes all of the drug as it passes through the organ, then C_0 is zero, and the extraction ratio is unity. For most eliminating organs, the extraction ratio lies somewhere between these two extremes. For example, an extraction ratio of 0.45 indicates that 45% of the blood flowing through the organ is completely cleared of the drug. Thus, the organ clearance in this case will be 45% of the blood flow rate through the organ. This relationship is described by

$$Cl_{or} = \frac{Q(C_i - C_0)}{C_i} = QE \qquad (18.5)$$

In this case, the organ clearance, Cl_{or}, has units of volume of blood cleared of drug per unit time. Thus, to express organ clearance in the same units as blood or plasma clearance established in Chapter 13, the blood flow rate must be multiplied by the extraction ratio. If all of the individual organ clearances are added together, the total body clearance or plasma clearance is obtained. The terms body clearance and plasma clearance are generally used synonymously.

Blood Flow Rate-Limited Transport

A physiological pharmacokinetic model consists of body tissues or organs linked by blood flow in a pattern that approximates or simulates the true anatomical and physiological situation. An example of this type of model is shown in Scheme 18.2. The choice of organs to be included in a model depends on anatomical and physiological reality and on the characteristics and site of action of the drug to be examined. Typically, organs where the compound exerts its pharmacological activity, organs that contain or accumulate large quantities of drug, and organs involved in drug elimination might be included.

When a drug in the bloodstream reaches an organ, the drug cannot enter the cells of that organ directly, but must pass through the capillary membranes into interstitial fluid, and then through the membrane(s) separating interstitial fluid from the organ or tissue cells. This process is illustrated in Scheme 18.3. It is often assumed that drug transport across these membranes is fast, that drug distribution into a specific organ is limited only by blood flow rate into that organ, and also that the concentration of drug in the emergent or venous blood from a particular organ is in equilibrium with the drug concentration in the intracellular fluid in that organ. This description is the essence of the blood flow rate- or perfusion-limited model, which permits organs to be represented by a simple, single compartment as illustrated in Scheme 18.3. This approach is suitable for any molecule that readily and rapidly crosses membranes. These molecules include only partially ionized or un-ionized fat-soluble compounds.

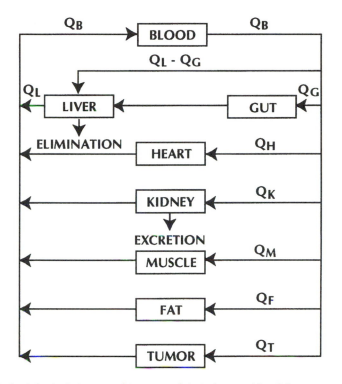

A general physiological pharmacokinetic model. Q denotes blood flow rate to a region. Subscripts are as follows: B, blood; L, liver; G, gut; H, heart; K, kidney; M, muscle; F, fat; and T, tumor. Elimination occurs from the liver and kidney.

SCHEME 18.2

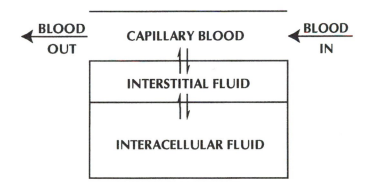

Model of drug transport between blood and intracellular fluid; this is the blood flow rate-limited model.

SCHEME 18.3

For ionized or water-soluble compounds, the assumption of very rapid equilibrium across physiological membranes may be inappropriate. For these types of compounds, membrane transport will probably be slow, and rapid equilibrium cannot be assumed. This situation is recognized in the membrane-limited model that will be discussed later in this chapter.

Most whole body physiological models are complex and are similar to the model shown in Scheme 18.2. However, in order to understand how the appropriate mathematical expressions are derived, one can consider two small subunits, one consisting of blood and a noneliminating organ such as muscle, and the other consisting of blood and an eliminating organ such as the kidney. These subunits are illustrated in Scheme 18.4. In Model A of this scheme, the rates of change in the total drug concentration in the blood pool, C_B, and also in the muscle, C_M, may be represented by the difference in drug concentration in the blood that flows through each tissue or fluid volume. The rates of change of total drug in blood and muscle are represented by equations 18.6 and 18.7, respectively.

$$V_B \frac{dC_B}{dt} = Q_M C_0 - Q_M C_i = Q_M(C_0 - C_i) \tag{18.6}$$

$$V_M \frac{dC_M}{dt} = Q_M C_i - Q_M C_0 = Q_M(C_i - C_0) \tag{18.7}$$

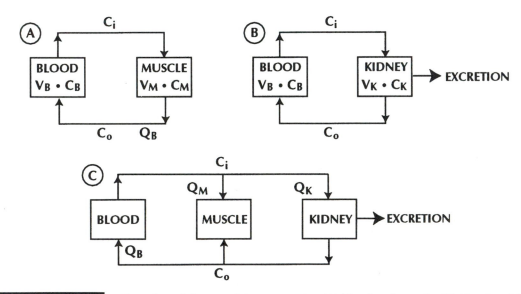

SCHEME 18.4 Physiological flow models representing (A) blood and muscle, (B) blood and kidney, and (C) blood, muscle, and kidney. Q_x is blood flow rate through organ X, and V is volume of the organ. Subscripts are as follows: B, blood; M, muscle; and K, kidney. C_i is drug concentration in arterial (afferent) blood, and C_0 is drug concentration in venous (efferent) blood.

In these equations, V_B and V_M are the volumes of the blood and muscle compartments, respectively; Q_M represents blood flow through muscle; and C_i and C_0 are the entering, or afferent, and exiting, or efferent, blood drug concentrations. In these models, drug concentration in entering blood, C_i, is equivalent to arterial drug concentration, C_B, and the drug concentration in exiting blood, C_0, is equal to C_M if no binding in blood or muscle occurs or if binding in blood and muscle is identical, which is unlikely. If unequal binding occurs, or some other situation exists that prevents C_0 from being exactly equal to C_M, then

$$C_0 = \frac{C_M}{R_M} \tag{18.8}$$

where R_M is the partition coefficient for distribution of drug from blood into tissue at equilibrium. Again the assumption is made that blood leaving the tissue is in equilibrium with drug within the tissue. Clearly, if $R_M = 2$, then the total drug concentration in tissue is double that in blood, and $C_M = C_M/2$. Equation 18.8 can also be expressed as

$$R_M = \frac{C_M}{C_0} \tag{18.9}$$

The concentration of total drug in a tissue compartment is equal to the free concentration multiplied by the free fraction. That is, if the free concentration is 10 μg/mL and the drug is 50% free, then the total concentration is 20 μg/mL. In that case, R_M can be described in terms of the relative free fractions of drug in blood, f_B, and tissue, f_M, by

$$R_M = \frac{f_B}{f_M} \tag{18.10}$$

Assume that a drug is extensively bound to muscle protein, and the free fraction in muscle is only 20%, while the drug is bound to only a small extent in blood, and the free fraction in blood is 80%. In this case, R_M is equal to 0.8/0.2, which equals 4. In other words, the concentration of total drug in the muscle compartment is four times greater than efferent, or exiting blood, at equilibrium. Taking this into account, equations 18.6 and 18.7 can be rewritten as equations 18.11 and 18.12.

$$V_B \frac{dC_B}{dt} = Q_M \frac{C_M}{R_M} - Q_M C_B = Q_M \left(\frac{C_M}{R_M} - C_B \right) \tag{18.11}$$

$$V_M \frac{dC_M}{dt} = Q_M C_B - Q_M \frac{C_M}{R_M} = Q_M \left(C_B - \frac{C_M}{R_M} \right) \tag{18.12}$$

In these equations, drug concentrations in afferent and efferent blood are more specifically identified in relation to drug concentrations in tissue and partitioning of drug from blood into tissue.

When drug distribution into an eliminating organ is considered, the elimination process has to be taken into account. Therefore, in Model B of Scheme 18.4, the equivalent equations to 18.6 and 18.7 for blood and kidney tissue, respectively, are equations 18.13 and 18.14.

$$V_B\left(\frac{dC_B}{dt}\right) = Q_K C_0 - Q_K C_i = Q_K(C_0 - C_i) \tag{18.13}$$

$$V_K\left(\frac{dC_K}{dt}\right) = Q_K C_i - Q_K C_0 - Cl_K C_K \tag{18.14}$$

where V_K is the volume of the kidney, Q_K is blood flow through the kidney, and Cl_K is clearance of drug by the kidney. If the drug is partially bound in the kidney, and only free drug is cleared, then the term $Cl_K C_K$, which represents the rate of drug removal by the kidney, would be replaced by $Cl'_K C'_K$, where the primes denote the free drug clearance and the free drug concentration, respectively. Equations 18.13 and 18.14 can also be converted to equations 18.15 and 18.16 in terms of total drug concentrations in blood and kidney compartments.

$$V_B\left(\frac{dC_B}{dt}\right) = Q_K\left(\frac{C_K}{R_K} - C_B\right) \tag{18.15}$$

$$V_K\left(\frac{dC_K}{dt}\right) = Q_K\left(C_B - \frac{C_K}{R_K}\right) - Cl_K C_K \tag{18.16}$$

Thus, expressions have been derived that describe the rates of change in drug concentrations in blood and tissue for the situations in Models A and B of Scheme 18.4. These expressions can now be combined to obtain overall expressions for the more complete situation in Model C of Scheme 18.4, that is, for the combined Models A and B. Drug in blood, muscle, and kidney is represented by the rate equations 18.17, 18.18, and 18.19, respectively.

$$V_B\left(\frac{dC_B}{dt}\right) = Q_M\frac{C_M}{R_M} + Q_K\frac{C_K}{R_K} - Q_B C_B \tag{18.17}$$

$$V_M\left(\frac{dC_M}{dt}\right) = Q_M\left(C_B - \frac{C_M}{R_M}\right) \tag{18.18}$$

$$V_K\left(\frac{dC_K}{dt}\right) = Q_K\left(C_B - \frac{C_K}{R_K}\right) - Cl_K C_K \tag{18.19}$$

Equation 18.17, the rate equation for drug in blood, has thus been expanded to include input terms from both muscle and kidney.

Equations 18.17–18.19 are thus a complete set of rate equations for Model C of Scheme 18.4. From this relatively simple example, construction of a similar set of equations for more complex models, such as the one in Scheme 18.2, is not difficult (*1*). The liver compartment in that model, with its dual blood input from the systemic and splanchnic circulations, is representative of the true anatomical situation, and the rate of change of total drug concentration in liver in that model is given by

$$V_L\left(\frac{dC_L}{dt}\right) = C_B(Q_L - Q_G) + Q_G\frac{C_G}{R_G} - Q_L\frac{C_L}{R_L} - Cl_L C_L \qquad (18.20)$$

In these examples, drug distribution into body tissues and fluids is assumed to be limited by blood flow rate. This limit is generally the case, but not always. Some compounds, such as water-soluble compounds, methotrexate, and actinomycin D, have tissue uptake characteristics that are not consistent with the simple blood flow model (*2–4*). Typically, relationships between concentrations of these compounds in blood and in certain tissues are not simple or linear. These drugs behave as though there is some barrier preventing rapid equilibrium. One way of interpreting these types of situations incorporates the concept of membrane-limited transport.

Membrane-Limited Transport

A typical model for membrane-limited transport of drug between blood and organ tissue is shown in Scheme 18.5. This model is different from the one in Scheme 18.3 because the membrane barrier between interstitial fluid and intracellular fluid is now rate-limiting.

The net flux or drug transport across this limiting membrane may be controlled by passive diffusion or by a saturable process (*4, 5*) (See Chapter 19.)

Whenever drug metabolism or drug movement from one place to the other, such as across membranes into cells, is dependent upon an enzymatic or active transport process, then a likelihood of Michaelis–Menten, or saturable kinetics,

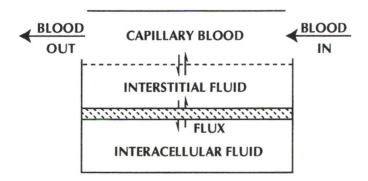

Model of drug transport between blood and intracellular fluid; this is the membrane-limited transport model.

SCHEME 18.5

exists. In Michaelis–Menten kinetics, the constants that characterize a saturable reaction are V_m (or V_{max}), which is the maximum rate of a reaction, and K_m, the Michaelis constant, which is the substrate concentration at which the rate is one-half the maximum value, V_m. Thus, at high drug concentrations (the actual concentration depending on the efficiency of the particular process), the enzyme or transporting agent can become saturated. Although first-order kinetics may be observed at low drug concentrations, this process may become less efficient at high drug concentrations as the rate of a particular event becomes limited by availability of enzyme or carrier. When this occurs, the kinetics become or approach zero-order.

For diffusion transport, which has been shown to occur, for example, with actinomycin D uptake by dog testes, net flux across the limiting membrane is given by

$$\text{Flux} = K(C_E - C_I) \tag{18.21}$$

where K is the membrane permeability coefficient, and C_E and C_I represent extracellular and intracellular drug concentrations, respectively. If drug movement across the membrane is saturable, then equation 18.21 needs to be expanded to include the Michaelis–Menten functions:

$$\text{Flux} = K(C_E - C_I) + \left(\frac{V_M C_E}{K_M + C_E} \right) - \left(\frac{V_M C_I}{K_M + C_I} \right) \tag{18.22}$$

Michaelis–Menten functions have been included in both directions because active transport may occur for drug movement both into and out of the cell.

Incorporating these membrane-limited transport expressions into overall physiological models is difficult. More parameters have been introduced for which initial estimates have to be obtained, and the complexity and uncertainty of the model are increased. This approach has nonetheless been used successfully to describe methotrexate kinetics. Figure 18.1 illustrates the close agreement that has been obtained between predicted and observed methotrexate levels in bone marrow and plasma of rats using a membrane-limited model (4).

Experimental Considerations

Discussion so far has been limited to the theory relating to physiological pharmacokinetic models. But how are the models used in practice? The most common approach is to set up a model that is thought to be most appropriate for the system to be studied; substitute estimates for the constants involved, for example, blood flow rates, organ volumes, and binding parameters, and solve the equations simultaneously as a function of time to establish theoretical concentration versus time plots for the sites of interest. Introduction of drug into the model is usually done as a program step. For example, an intravenous bolus injection is an initial condition in the blood pool, whereas an intramuscular injection is introduced as an initial condition in muscle. Similarly, intrathecal injection is an initial condition in cerebrospinal fluid.

Once the model is validated (if this can be done), it can be used to predict drug disposition in humans under a variety of dosing conditions provided that

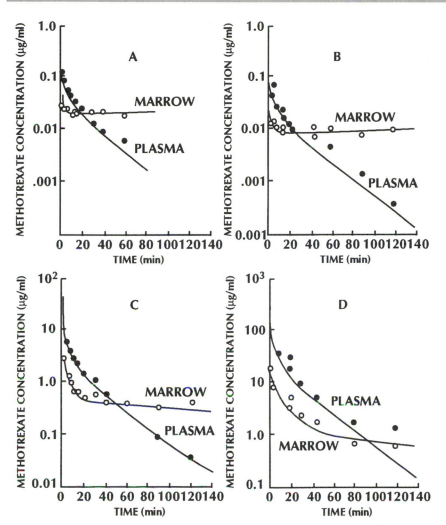

Plasma and average bone-marrow concentrations of methotrexate in the rat. Curves represent model simulations. Data points were obtained from one rat each time. Key (mg/kg IV): A, 0.05; B, 0.25; C, 2.5; and D, 25. (Reproduced with permission from reference 4.)

FIGURE 18.1

animal data can be successfully applied to humans for that particular drug and situation (6, 7).

Blood flow rates and tissue volumes used are generally average values for a particular species. Most of these, under normal conditions, are available in the literature. However, drugs are often given under disease conditions, which may influence these values, and the drug itself may also affect physiological conditions because of its pharmacological actions. These factors can perturb a model and lead to difficulty in interpretation. Some typical tissue volumes and blood flow rates in humans are given in Table 18.1 (8).

This very brief introduction to physiological kinetic modeling has shown that this approach to pharmacokinetics is complex and experimentally demanding. As indicated at the beginning of this chapter, physiological modeling is not of general application in most cases, but it may be used to solve particular drug distribution questions.

Despite the difficulties associated with the use of physiological kinetic models, they continue to be used to monitor disposition of some exogenous and endogenous substances, where specific tissue concentrations are of interest. Typical studies have recently examined mechanisms of action of lovastatin (9), interactions of lipoprotein lipase with the active site inhibitor tetrahydrolipstatin (Orlistat) (10), splanchnic insulin dynamics and secretion in abdominal obesity (11), and analysis of triiodothyronine kinetics in hypothyroid patients treated with triiodothyronine (12). Most of the studies use the SAAM computer software (13) to handle the complex kinetic models. Charnick et al. (14) have recently discussed the use of physiologically based pharmacokinetic modeling as a tool for drug development (14).

Summary

Physiological pharmacokinetic models are based on true anatomical regions and volumes, blood flow rate through these regions, and transport of drug between blood and tissue.

Physiological pharmacokinetic models have advantages and disadvantages compared with conventional compartment models. Advantages are associated with more accurate characterization of drug disposition. Disadvantages are associated with model complexity and difficulty in validation.

Physiological models are particularly useful for compounds such as anticancer agents when drug location in specific organs is important.

Drug transport from blood to organs or tissues may be limited by blood flow rate or by the rate at which drug diffuses between blood and tissue.

TABLE 18.1 **Percent of Body Volume and Blood Flow Rate in Some Organs and Tissues**

Organ or Tissue	Percent of Body Volume	Blood Flow Rate (mL/min)
Blood	7	(5000)[a]
Muscle (resting)	42.0	750
Kidneys	0.4	1100
Liver	2.3	1350
Heart	0.5	200
Lungs	0.7	5000
Fat	10.0	200
Brain	2.0	700

Note: Values are under basal conditions in a standard 70-kg man.

[a]Total blood volume.

Source: Adapted with permission from reference 8.

Diffusion- or membrane-limited transport may be controlled by limited passive diffusion or by saturable transport processes.

Predictability of physiological models is influenced by species differences and disease states that can affect physiological conditions.

Problems

1. The blood concentration of drug entering the kidneys via the renal arteries is 10 µg/mL. The blood concentration of drug leaving the kidneys via the renal veins is 5 µg/mL. If blood flow to the kidneys is 1200 mL/min, what are (i) the kidney extraction ratio and (ii) the renal clearance?

2. If the drug concentration in blood entering the kidneys is 10 µg/mL and the drug concentration in blood leaving kidneys is zero, what is the kidney extraction ratio for the drug?

3. Blood flows through a noneliminating muscle tissue at a rate of 450 mL/min. Drug concentration is 9 µg/mL in afferent blood and 4 µg/mL in efferent blood. The partition coefficient for drug distribution from blood to muscle tissue is 1.8. Calculate the concentration of drug in muscle tissue.

4. What would the concentration of drug in muscle tissue be if the drug in Problem 3 were 80% bound to blood proteins?

5. A drug that does not bind to plasma proteins or enter red cells is cleared by hepatic metabolism. A patient with a hematocrit of 0.44 has a hepatic blood flow of 1.5 L/min. If the hepatic extraction ratio for the drug is 0.4, and the drug level in afferent blood is 10 µg/mL, what are (i) the hepatic drug clearance, (ii) the concentration of drug in efferent blood, and (iii) the concentration of drug in efferent plasma?

References

1. Himmelstein, K. J.; Lutz, R. J. *J. Pharmacokinet. Biopharm.* **1979,** 7, 127–145.
2. Lutz, R. J.; Galbraith, W. M.; Dedrick, R. L.; Shrager, R.; Mellett, L. B. *J. Pharmacol. Exp. Ther.* **1971,** 200, 469–478.
3. Bischoff, K. B.; Dedrick, R. L.; Zaharko, D. S.; Longstreth, J. A. *J. Pharm. Sci.* **1971,** 60, 1128–1133.
4. Dedrick, R. L.; Zaharko, D. S.; Lutz, R. J. *J. Pharm. Sci.* **1973,** 62, 882–890.
5. Mintun, M.; Himmelstein, K. J.; Schroder, R. L.; Gibaldi, M.; Shen, D. D. *J. Pharmacokinet. Biopharm.* **1980,** 8, 373–409.
6. Dedrick, R. L. *J. Pharmacokinet. Biopharm.* **1973,** 1, 435–461.
7. Dedrick, R. L.; Bischoff, K. B. *Fed. Proc.* **1980,** 39, 54–59.
8. Rowland, M.; Tozer, T. *Clinical Pharmacokinetics: Concepts and Applications*, 2nd ed.; Lea and Febiger: Philadelphia, PA, 1989; p 132.
9. Aguilar-Salinas, C. A.; Barrett, P. H.R.; Kelber, J.; Delmez, J.; Schonfeld, G. *J. Lipid Res.* **1995,** 36, 188–199.
10. Lookene, A.; Skottova, N.; Olivecrona, G. *Eur. J. Biochem.* **1994,** 222, 395–403.
11. Sonnenberg, G. E.; Hoffman, R. G.; Mueller, R. A.; Kissebah, A. H. *Diabetes* **1994,** 43, 468–477.

12. Zaninovich, A. A.; el Tamer, E.; el Tamer, S.; Noli, M. I.; Hays, M. T. *Thyroid* **1994,** *4*(3), 285–93.

13. Berman, M.; Weiss, M. F.; *SAAM Manual*, DHEW Publication #(NIH) 78-180.; U. S. Government Office: Washington, DC, 1978.

14. Charnick, S. B.; Kawai, R.; Nedelman, J. R.; Lemaire, M.; Niederberger, W.; Sato, H. *J. Pharmacokinet. Biopharm.* **1995,** *23*, 217–229.

Nonlinear Pharmacokinetics

19

espite the wide diversity of classical and physiological pharmacokinetic models, all of them, with the exception of the membrane-limited transport model discussed in Chapter 18, incorporate the assumption that drug elimination from the body is a first-order process and the rate constant for drug elimination is a true constant, independent of drug concentration. In these cases, the percentage of body drug load that is cleared per unit time is constant, and the drug has an elimination half-life.

Fortunately, first-order elimination, or at least apparent first-order elimination, is common. This greatly simplifies dosage design and adjustment, prediction of drug accumulation, bioavailability assessment, dose–response relationships, and many other aspects of pharmacokinetics.

In fact, very few drugs are eliminated from the body by mechanisms that are truly first-order in nature. True first-order elimination applies only to compounds that are eliminated by mechanisms that do not involve enzymatic or transport processes. This applies only to drugs that are cleared from the body by urinary excretion, and among those, only drugs that enter the renal tubules by glomerular filtration. All other processes require some form of energy-consuming metabolism or transport mechanism and are therefore saturable. These processes apply to drugs that are metabolized in the liver or any other organ in the body, cleared in the urine by active secretion, or cleared in the bile by a similarly active process. Even compounds like riboflavin, bethanidine, cephaloridine, and cephapirin, which may undergo capacity-limited reabsorption from the distal renal tubule back into the circulation, are included in this group (1). Saturable elimination is therefore a potential condition for the great majority of compounds. Of course,

Saturable Processes

drugs may be excreted in saliva, sweat, or even to a small extent in bile by passive processes; but these elimination routes generally account for only a small proportion of total eliminated drug and are often negligible.

Why is it that although the vast majority of drugs and other compounds are cleared from the body by saturable processes, most drugs exhibit first-order, or apparent first-order, elimination kinetics? The answer of course is that in most cases concentrations of drug in the bloodstream, or more correctly at the site of elimination, are well below those required to saturate the processes involved.

One notable exception to this explanation is ethyl alcohol. Ethyl alcohol is cleared from the body by oxidative metabolism at an apparent zero-order rate that is equivalent to about 12 oz of beer or 1 oz of liquor per hour (2, 3). The opinion is often expressed that zero-order elimination is a unique property of ethyl alcohol and differentiates it from most other drugs that are cleared either entirely at a first-order rate, or at least at such a rate at low drug concentrations.

However, alcohol loses much of this uniqueness if one considers how much compound the body is being required to cope with. One bottle of eight-proof beer is equivalent to about 8 g of pure alcohol. The molecular weight of ethyl alcohol is 46, which is about one-quarter the molecular weight of aspirin, and much less than that of many drugs that are in the 200–500 molecular weight range. Thus, if an equivalent dose on a "per molecule" basis is considered, then the 8-g dose of ethyl alcohol is equivalent to 32 g of aspirin, or 77 g of tetracycline. Ingestion of this quantity of any drug that is eliminated by a saturable pathway would yield apparent zero-order kinetics, or possibly no kinetics at all. In fact, administration of a sufficiently small dose of ethyl alcohol such that circulating levels are low relative to K_m for this compound yields first-order elimination kinetics (4).

Compounds such as phenytoin, salicylate, theophylline, and probenecid exhibit saturable kinetics within the therapeutic range (5–8). Phenytoin kinetics are usually interpreted as though its elimination proceeds via a single saturable pathway. Salicylate, on the other hand, forms a number of metabolites and formation of only two of these, salicyluric acid and salicylphenolic glucuronide, appears to be saturable. Thus, overall elimination of salicylate consists of parallel saturable and nonsaturable pathways (9, 10).

As mentioned earlier, a constant percentage of a drug that is eliminated by first-order kinetics is cleared per unit time, and the drug has a discrete concentration-independent elimination half-life. For drugs that are eliminated by zero-order kinetics or by saturable pathways, however, the same quantity of drug is cleared per unit time, and this quantity is drug concentration-independent. Thus, a different percentage of the drug is cleared per unit time depending on the concentration, and the drug does not have a constant, characteristic elimination half-life.

The difference between simple first-order elimination and saturable, leading ultimately to zero-order, elimination may have a profound effect on drug concentrations, duration of drug activity, and the time course and extent of drug accumulation with repeated or continuous doses. Saturable hepatic metabolism may also markedly affect drug absorption because of altered first-pass metabolism.

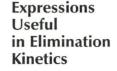

As mentioned briefly in Chapter 18, saturable elimination of a compound, whether it be drug or metabolite, is generally described in the form of Michaelis–Menten kinetics.

For a drug that obeys one-compartment model kinetics, and whose elimination is first-order, the kinetic model following intravenous injection is given in Scheme 19.1, and the rate equation for loss of drug from the circulation with respect to time is given by equation 19.1. The symbols for this model were described in Chapter 13.

$$\frac{dC}{dt} = -k_{el}C \tag{19.1}$$

For a drug that is eliminated by a single saturable process, the kinetic model is given in Scheme 19.2, and the rate equation for loss of drug from the circulation is given by equation 19.2.

$$\frac{dC}{dt} = -\left(\frac{V_m C}{K_m + C}\right) \tag{19.2}$$

where V_m is the maximum rate of elimination and K_m is the Michaelis constant.

The term on the right side of equation 19.2 is a form of the familiar Michaelis–Menten expression for a saturable reaction based on a single substrate mechanism. Typical drug profiles resulting from the models in Schemes 19.1 and 19.2 are shown on linear and semilogarithmic scales in Figure 19.1. The profiles in Figure 19.1A describe first-order loss of drug. The profiles in Figure 19.1B, however, describe Michaelis–Menten elimination. When the concentration in B is plotted on a linear scale, the decline in levels is initially linear or zero-order, but it becomes curved or first-order at lower levels to yield the familiar hockey-stick

Expressions Useful in Elimination Kinetics

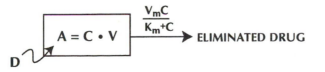

One-compartment open model with first-order elimination after bolus intravenous injection.

SCHEME 19.1

One-compartment open model with a single route of saturable elimination after bolus intravenous injection.

SCHEME 19.2

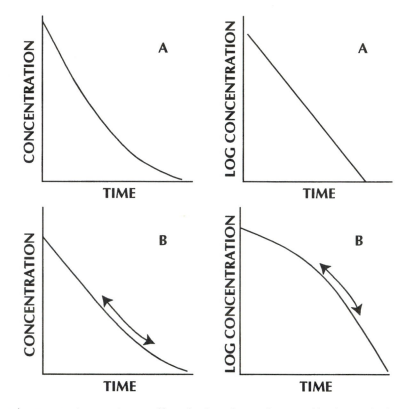

FIGURE 19.1   Blood concentration vs. time profiles of a drug that is eliminated by first-order kinetics (A), and by Michaelis–Menten kinetics (B). In (A), the profile is convex-curved when plotted on a linear scale and linear when plotted on a semilogarithmic scale (see Figure 13.1). In (B) the profile is initially linear at high drug concentrations, when elimination is essentially zero-order, and then curvilinear on a linear scale. On a semilogarithmic scale, the profile is initially convex-curved and then linear. The arrows indicate the transition concentration range between apparent zero-order and first-order elimination kinetics.

shape. When the data are plotted on a semilogarithmic scale, the decline is initially curvilinear but becomes linear at lower concentrations.

Why does Figure 19.1B yield apparent zero-order kinetics at high drug levels and apparent first-order kinetics at low drug levels? Consider the two situations where the drug concentration C is much greater or much less than the Michaelis constant K_m. In the situation where C is much greater than K_m, equation 19.2 becomes

$$\frac{dC}{dt} = -\left(\frac{V_m C}{C}\right) = -V_m \qquad (19.3)$$

In this equation, the rate of elimination becomes essentially zero-order. However, in the situation where C is much smaller than K_m, equation 19.2 becomes

$$\frac{dC}{dt} = -\left(\frac{V_m C}{K_m}\right) = -kC \tag{19.4}$$

The units of V_m/K_m are concentration per time per concentration, which cancels to units of reciprocal time, i.e., a first-order rate constant. Thus, in equation 19.4, V_m/K_m can be expressed as a single apparent first-order rate constant, k.

In the region where the values of drug concentration and K_m are similar, equation 19.2 is operative. This region is the transition state between apparent first-order and apparent zero-order elimination kinetics. This region is identified by the arrows in Figure 19.1B.

Equation 19.2 can be expanded for various types of models. For example, the one-compartment model with first-order input and saturable elimination is represented by equation 19.5, and the same model with zero-order drug input is given by equation 19.6.

$$\frac{dC}{dt} = \frac{k_a F D}{V} e^{-k_{el} t} - \frac{V_m C}{K_m + C} \tag{19.5}$$

$$\frac{dC}{dt} = k_0 V - \frac{V_m C}{K_m + C} \tag{19.6}$$

where F is the fraction of the dose, D, absorbed, and k_0 is the zero-order rate constant for drug input. If a drug is eliminated by both parallel saturable and nonsaturable pathways, then the rate equation after intravenous injection is given by

$$\frac{dC}{dt} = -k_{el} C - \frac{V_m C}{K_m + C} \tag{19.7}$$

If the drug obeys two-compartment model kinetics and is also cleared by both saturable and nonsaturable pathways, then referring back to Chapter 17, the rate equation for the amount of drug in the central compartment of that model after bolus intravenous injection is given by

$$\frac{dA_1}{dt} = k_{21} A_2 - (k_{12} + k_{el}) A_1 - \frac{V_m A_1}{K_m + A_1} \tag{19.8}$$

where A_1 and A_2 are the quantity of drug in the first and second compartments, respectively.

Writing the appropriate rate equations for the various situations is not difficult. However, obtaining initial estimates for actual values of V_m and K_m, the Michaelis constants, is more difficult and requires good data. Several methods can be found in references 9–11. Two are described here.

Obtaining Estimates of V_m and K_m from Plasma-Level Data

The first method of obtaining estimates of V_m and K_m from plasma data requires determination of the rate of change in plasma drug concentrations between successive sampling times during the postabsorptive and postdistributive phase of a plasma profile. Thus the rate of change in plasma concentration, together with the drug concentration at the midpoint of each sampling period, C_m, can be incorporated into a number of expressions to solve for V_m and K_m.

A typical expression is

$$\frac{1}{\left(\dfrac{\Delta C}{\Delta t}\right)} = \frac{1}{V_m} + \left(\frac{K_m}{V_m}\right)\left(\frac{1}{C_m}\right) \tag{19.9}$$

This equation is the Lineweaver–Burk expression, which is a linear form of the Michaelis–Menten equation. In this equation, $\Delta C/\Delta t$ and C_m represent the decline in drug concentrations during a time interval and drug concentration at the midpoint of the time interval, respectively. A plot of $1/(\Delta C/\Delta t)$ versus $1/C_m$ will thus yield a slope of K_m/V_m and an intercept of $1/V_m$, as in Figure 19.2.

In the second method, estimates of V_m and K_m are obtained directly from log C versus t data. Equation 19.2 is rearranged to

$$-dC - \left(\frac{K_m \cdot dC}{C}\right) = V_m dt \tag{19.10}$$

Integration of equation 19.10 yields

$$-C - K_m \ln C = V_m t + i \tag{19.11}$$

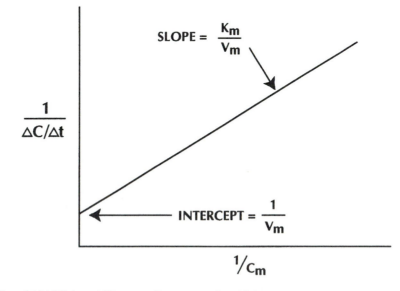

SLOPE = $\dfrac{K_m}{V_m}$

$\dfrac{1}{\Delta C/\Delta t}$

INTERCEPT = $\dfrac{1}{V_m}$

$^1/C_m$

FIGURE 19.2   Plot of $1/(\Delta C/\Delta t)$ vs. $1/C_m$ according to equation 19.9.

where i is the constant of integration. Solving for i at t equals zero, where C equals C_0, yields equation 19.12.

$$i = -C_0 - K_\mathrm{m} \ln C_0 \qquad (19.12)$$

Substituting for i in equation 19.11 gives

$$t = \left(\frac{C_0 - C}{V_\mathrm{m}} \right) + \left(\frac{K_\mathrm{m}}{V_\mathrm{m}} \right) \ln\left(\frac{C_0}{C} \right) \qquad (19.13)$$

Conversion of equation 19.13 to logarithms to the base 10, and solving for log C yields

$$\log C = \left(\frac{C_0 - C}{2.3 K_\mathrm{m}} \right) + \log C_0 - \left(\frac{V_\mathrm{m} t}{2.3 K_\mathrm{m}} \right) \qquad (19.14)$$

Consider log C versus t data following a bolus intravenous dose of a drug that obeys one-compartment model kinetics. The terminal log–linear portion of the log C versus t plot, that is, the region where apparent first-order kinetics occur, is a straight line described by

$$\log C = \log C_0^\star - \frac{V_\mathrm{m} t}{2.3 K_\mathrm{m}} \qquad (19.15)$$

where $C_0^\star$ is the extrapolated intercept of C on the y axis, as shown in Figure 19.3.

At low plasma concentrations in the log–linear region, equation 19.15 is identical to equation 19.14; therefore, the two expressions can be set equal:

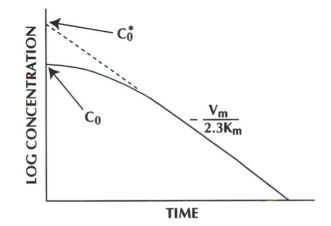

Estimates of V_m and K_m from a plot of log C vs. time after a bolus intravenous injection of a drug that obeys one-compartment model kinetics and undergoes saturable elimination.

FIGURE 19.3

$$\log C_0^\star - \frac{V_m t}{2.3 K_m} = \frac{C_0 - C}{2.3 K_m} + \log C_0 - \frac{V_m t}{2.3 K_m} \qquad (19.16)$$

Canceling the common term $V_m t / 2.3 K_m$ and rearranging yield

$$\log\left(\frac{C_0^\star}{C_0}\right) = \frac{C_0 - C}{2.3 K_m} \qquad (19.17)$$

Because equation 19.16 occurs only during the terminal linear phase of a drug profile, that is, when C is much less than C_0, the approximation that $C_0 - C = C_0$ can be applied. Equation 19.17 is then rearranged to

$$K_m = \frac{C_0}{2.3 \log \dfrac{C_0^\star}{C_0}} = \frac{C_0}{\ln \dfrac{C_0^\star}{C_0}} \qquad (19.18)$$

Because both C_0 and $C_0^\star$, the actual and extrapolated y intercepts, can be measured, K_m can be calculated from equation 19.18, and V_m from equation 19.19.

$$V_m = -2.3 \,(\text{slope})\, K_m \qquad (19.19)$$

The preceding are two of several possible ways to calculate Michaelis constants from plasma data. One method uses the rates of change in plasma concentrations, which may be based on either oral or intravenous data. The other method is based on direct estimates from plots of log plasma concentration versus time data, and is restricted to intravenous data.

A word of caution is appropriate regarding the fitting of saturable data to kinetic models, especially computer-fitting procedures. Equations 19.2, 19.5, and 19.6 are written as rate equations and have not been solved to yield expressions for drug concentration as a function of time because there is no analytical solution to these rate equations. Therefore, the rate equations must be used during curve-fitting procedures, and the necessary integration to obtain parameter values is carried out by means of numerical integration subroutines.

Influence of Saturable Kinetics on Drug Elimination, Area Under the Drug-Concentration Curve, and First-Pass Effect

Elimination Half-Life

While blood concentrations of a drug are in the saturable range, the drug will not have a true half-life. The half-life will change continuously with drug concentration. The higher the drug concentration, the smaller the percentage cleared per unit time, and the longer the apparent half-life becomes. This phenomenon is particularly important with toxic drug overdoses, when levels of drug are most likely to be in the saturable range.

Although saturable elimination may thus profoundly affect elimination of a drug that is cleared by one saturable process, a lesser effect may occur for elimination of a drug that is cleared by parallel saturable and nonsaturable pathways. As drug concentrations are increased in this case, the apparent elimination half-life will tend to stabilize at a value larger than that observed at low drug concentrations. The apparent elimination half-life will nonetheless become concentration-independent because the nonsaturable elimination component becomes dominant and controls the new half-life, with negligible contribution from the saturable component. The effect of drug concentrations on the half-lives of drugs that are cleared by nonsaturable, saturable, and combined saturable and nonsaturable pathways is shown in Figure 19.4.

An important consequence of the increase in apparent drug elimination half-life with increasing drug concentration is that of increased and prolonged drug accumulation with repeated dosing. Levy and Tsuchiya (9) showed that a twofold increase in salicylate dosage from 0.5 to 1 g given every 8 h could result in a more than sixfold increase in steady-state salicylate levels and an increase in the time to reach steady state from 2 to 7 days.

As long as a drug obeys linear kinetics, and provided the kinetics are dose-independent, the AUC will be directly proportional to the systemically available dose. This relationship is shown in equation 19.20 for the one-compartment model after bolus intravenous injection.

Area Under the Drug-Concentration Curve

$$AUC^{0\to\infty} = \frac{D}{Vk_{el}}$$
(19.20)

For drugs that obey saturable kinetics, however, this relationship may not be the case. The AUC after intravenous injection of a drug that is eliminated by a single saturable pathway is given by

$$AUC^{0\to\infty} = \left(\frac{C_0}{V_m}\right)\left(\frac{C_0}{2} + K_m\right)$$
(19.21)

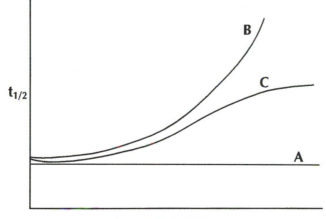

LOG DOSE

Changes in apparent elimination half-life with increasing dose for drugs that are cleared from the body by A, nonsaturable; B, saturable; and C, parallel saturable and nonsaturable pathways.

FIGURE 19.4

When the initial drug level, C_0, is much less than K_m, equation 12.21 reduces to equation 19.22.

$$\text{AUC}^{0\to\infty} = \frac{C_0 K_m}{V_m} = \frac{D \cdot K_m}{V \cdot V_m} = \frac{D}{V \cdot k} \tag{19.22}$$

In this situation, the AUC is again directly proportional to dose, just as in the first-order elimination case. However, if C_0 is much greater than K_m, then equation 19.21 becomes

$$\text{AUC}^{0\to\infty} = \frac{C_0^2}{2V_m} = \frac{D^2}{2V^2 V_m} \tag{19.23}$$

Quite a different situation is obtained in this case because the AUC is proportional to the square of the dose and inversely proportional to the square of the distribution volume. However, the distribution volume should remain constant and should not influence drug levels from different doses.

Thus, the more drug concentrations lie within the saturable range, the greater the tendency for AUC values to increase disproportionately with increasing dose. The example represented by equation 19.23 is based on simple kinetics associated with intravenous dosing and the one-compartment kinetic model. However, the problem of disproportionate increases in AUC values with small dose increments is independent of both the dosage route and the kinetic model.

First-Pass Metabolism

Most drugs are metabolized predominantly in the liver, which is the site where saturation of metabolism is most likely to occur. Because saturation occurs at high drug concentrations, the time at which saturation is most likely to occur is during drug absorption via the splanchnic circulation and the portal vein. Drug concentrations are generally much higher at this time (i.e., during the first pass through the liver) than after they have entered the general circulation. This can lead to saturation of drug-metabolizing enzymes during absorption and hence greater bioavailability of unchanged drug. Because saturation occurs at high drug concentrations, a rapidly absorbed drug is likely to undergo less first-pass metabolism than a more slowly absorbed drug. This situation is a potentially serious problem for sustained-release dosage forms. If a drug undergoes saturable metabolism, then the slower the release, and hence the lower the drug concentrations in the splanchnic circulation, and the greater the opportunity for first-pass metabolism. Therefore, accurate determination of the relative bioavailability of slow-release dosage forms compared to conventional dosage forms or oral solutions is important.

An interesting example of possible saturation of hepatic metabolizing enzymes during the first pass is provided by the anticancer compound 5-fluorouracil. This drug is extensively metabolized in the body and has a very short elimination half-life of about 10 min. With such a short half-life, the drug would be expected to be extensively metabolized during the first pass after oral dosing, resulting in low bioavailability. In fact, 5-fluorouracil is absorbed into the systemic circulation reasonably well after oral doses, and the AUC from equivalent oral doses is approximately 30% of that after intravenous injection (*12*). So it appears (and remains to be proven) that the high oral dose of approximately 1 g of 5-fluorouracil is capable of saturating hepatic enzymes during the first pass, resulting in more efficient absorption of unchanged drug than might be expected.

Some compounds that have recently been shown to obey nonlinear, saturable pharmacokinetics include aminoglycosides (*13*), the anticancer agent paclitaxel (Taxol) (*14*, *15*), the antidementia agent tacrine (*16*), and the antidepressant nefazodone (*17*). In the case of nefazodone, whereas oral doses increased in the ratio 1:2:4 in healthy volunteers, $AUC^{0 \to \tau}$ values increased in the proportion 1:4.2:16.8. A major metabolite, hydroxynefazodone, also exhibited nonlinear kinetics, whereas a minor metabolite, *m*-chlorophenylpiperazine, exhibited linear kinetics. Nonlinear kinetics of nefazodone may be related to saturable first-pass metabolism and also metabolic clearance from the general circulation.

Summary

Most drugs are eliminated from the body by mechanisms that are potentially saturable. True first-order, nonsaturable elimination occurs only with drugs that are excreted in urine via passive glomerular filtration and do not undergo active reabsorption.

Some drugs exhibit saturable elimination at therapeutic blood concentrations.

A drug that is cleared by zero-order or Michaelis–Menten kinetics does not exhibit a characteristic, constant elimination half-life. The apparent half-life changes continuously with drug concentration in plasma.

Saturable first-pass metabolism may lead to increased absorption efficiency of drug from a rapid-release oral formulation compared with a slow-release oral formulation.

Saturable elimination may introduce bias into systemic availability estimates of drug from fast- and slow-release formulations, and from large versus small oral doses.

Problems

1. A drug that undergoes saturable elimination has a V_m of 5 µg/mL·h and a K_m of 10 µg/mL. What are (i) the apparent first-order elimination rate constant and (ii) the apparent elimination half-life of the drug when circulating drug levels are below the saturation range?

2. Following bolus intravenous injection of a drug that obeys one-compartment kinetics and exhibits saturable elimination, the observed maximum drug concentration in plasma obtained immediately after dosing was 8 µg/mL. The drug concentration in plasma obtained by extrapolating the terminal log–linear portion of the drug concentration curve back to zero time was 13.5 µg/mL, and the apparent half-life from the log–linear elimination phase was 1.5 h. Calculate V_m and K_m from these data.

3. A person who weighs 150 lb will attain a blood alcohol concentration (BAC) of approximately 0.025 percent upon complete absorption of 12 oz of beer (4% ethanol, 8 proof) or 1 oz of 100-proof (50% ethanol) liquor. During the zero-order elimination range, the average rate of ethanol elimination from blood is 0.018%/h.

A 150-lb person started drinking at 6:00 p.m. and continued until 10:00 p.m. During that time the person drank six 12-oz bottles of beer. What was the person's BAC at 11:00 p.m.?

4. How many ounces of 100-proof (50% ethanol) liquor would a 150-lb person have to drink between 6:00 and 10:00 p. m. to have a blood alcohol content of 0.1 mg% at 11:00 p.m.?

5. A single 100-mg oral tablet dose of a drug that is eliminated by a single saturable pathway yielded a peak plasma drug concentration of 15 µg/mL at approximately 2 h postdose. Reformulation into a fast-release suspension resulted in a peak plasma level of 25 µg/mL from the same dose size, occurring at 0.5 h. Calculate (i) the approximate increase in AUC that results from the fast-release suspension compared to the slower release tablet formulation and (ii) to what level the fast-release dosage would have to be reduced to yield the same AUC as the slower release formulation. K_m is 0.5 µg/mL.

References

1. Jusko, W. J.; Levy, G. *J. Pharm. Sci.* **1970,** *59*, 765–772.
2. Lundquist, F.; Wolthers, H. *Acta Pharmacol. Toxicol.* **1958,** *14*, 265–289.
3. Winek, C. L. *Trial* **1983,** *19*, 38.
4. Welling, P. G.; Lyons, L. L.; Elliot, R.; Amidon, G. L. *J. Clin. Pharmacol.* **1977,** *E7*, 199–206.
5. Arnold, K.; Gerber, N. *Clin. Pharmacol. Ther.* **1970,** *11*, 121–134.
6. Dayton, P. G.; Cucinell, C. A.; Weiss, M.; Perell, J. M. *J. Pharmacol. Exp. Ther.* **1967,** *158*, 305–316.
7. Levy, G. *J. Pharm. Sci.* **1965,** *54*, 959–967.
8. Selen, A.; Amidon, G. L.; Welling, P. G. *J. Pharm. Sci.* **1982,** *71*, 1238–1241.
9. Levy, G.; Tsuchiya, T. *New Engl. J. Med.* **1972,** *287*, 430–432.
10. Levy, G.; Tsuchiya, T.; Amsel, P. *Clin. Pharmacol. Ther.* **1972,** *13*, 258–268.
11. Wagner, J. G. *Fundamentals of Clinical Pharmacokinetics*; Drug Intelligence: Hamilton, IL, 1975; pp 247–284.
12. Phillips, T. A.; Howell, A.; Grieve, R. J.; Welling, P. G. *J. Pharm. Sci.* **1980,** *69*, 1428–1431.
13. Czock, D.; Giehl, M. *Int. J. Clin. Pharmacol. Ther.* **1995,** *33*, 537–539.
14. Creaven, P.; Raghavan, D.; Pendyala, L.; Perez, R.; Loewen, G.; Meropol, N.; Levine, E.; Hicks, W. *Semin. Oncol.* **1995,** *5* (Suppl 12), 13–16.
15. Sparreboom, A.; Van Tellingen, O.; Nooijen, W. J.; Beijnan, J. H. *Cancer Res.* **1996,** *6*, 2112–2115.
16. Parnetti, L. *Clin. Pharmacokinet.* **1995,** *29*, 110–129.
17. Kaul, S.; Shukla, U. A.; Barbhaiya, R. H. *J. Clin. Pharmacol.* **1995,** *35*, 830–839.

Applications of Pharmacokinetics in Drug Discovery and Development

ᐟ

Pharmacokinetics and Toxicokinetics

20

Pharmacokinetics and toxicokinetics, together with the closely related disciplines of pharmacodynamics and toxicodynamics, are important components of drug discovery and development (*1*). Pharmacokinetics has played an ever-increasing role in discovery and development during the last 30 years and it is now a critical and highly interactive discipline, contributing to knowledge regarding drug disposition and activity throughout nonclinical and clinical development.

Toxicokinetics is of more recent origin but is rapidly becoming an extremely important discipline in pharmaceutical research and development. The study of toxicokinetic–toxicodynamic relationships is in its infancy, but it is also evolving rapidly as technology improves and understanding of this critical relationship increases. The objective of this chapter is to compare the disciplines of pharmacokinetics and toxicokinetics from both technical and philosophical perspectives.

Pharmacokinetics as an end unto itself, is essentially redundant. The major emphasis in drug disposition studies now is to obtain relationships and correlations among drug absorption, distribution, metabolism, and excretion (ADME) and pharmacologic or therapeutic events (*2*). These types of studies occur in both nonclinical and clinical environments. Appreciation of the importance of pharmacokinetic–pharmacodynamic relationships has provided added interest and impetus to this area of research and has projected these disciplines into a central and critical position in drug discovery and development. This is particularly so in the case of clinical studies. In Phase 1, Phase 2, and Phase 3 clinical studies, blood level determinations, and examination of the relationships between these values and observed pharmacologic or therapeutic end points, are included in most study design protocols. Much of the initiative for this approach has come from the availability of appropriate mixed-effect modeling computer programs (*3*), and also from

Pharmacokinetics and Pharmacodynamics

the high level of interest shown by regulatory agencies in this novel marriage of disciplines (*4*).

Toxicokinetics and Toxicodynamics

The discipline of toxicokinetics is of far more recent origin than pharmacokinetics. Blood drug levels were originally determined during toxicology trials merely to ensure that the drug was administered and was to some extent systemically available. Safety assessments were based on dose–effect relationships. This situation has changed. Toxicokinetic studies are now a necessary and central component of toxicology studies, and it is now required that the ADME of a drug be well characterized in toxicity species at toxicologic dose levels (*5*). The very high doses generally used in toxicology studies compared to those used in nonclinical pharmacology or clinical studies result in considerable technical differences between pharmacokinetic and toxicokinetic disciplines, and these are described later in this chapter.

The discipline of toxicodynamics, and the relationship between toxicokinetics and toxicodynamics (drug level–effect relationships), is of even more recent origin, and few cases of clearly established toxicokinetic and toxicodynamic relationships have been described. Nonetheless, this area is advancing rapidly, and awareness of the relationships between the time course of circulating drug or metabolite(s) and toxic events is setting a new dimension in the conduct and interpretation of toxicology studies. Such relationships have recently been reviewed for several therapeutic areas including anticancer compounds (*6*), anti-atherosclerotic agents (*7*), drugs acting on the central nervous system (CNS) (*8*), and antibacterial agents (*9*).

Technical Differences Between Pharmacokinetics and Toxicokinetics

Toxicokinetics is a unique expansion of the science of pharmacokinetics. The major difference between the two disciplines is that doses used in toxicokinetics studies are generally much higher than doses used in pharmacokinetics studies. In toxicology studies tolerance to a drug is examined at high dosage levels, and it is unrealistic to assume that drug or metabolite levels could be handled in the same way from these high doses as from therapeutic or pharmacologic doses. Toxicokinetics may thus be technically different from pharmacokinetics in many ways (*10*).

Solubility

The high doses used in toxicokinetics often give rise to drug solubility problems. These may occur during dosage-form preparation and administration and also in terms of drug solubility in the gastrointestinal (GI) tract. More seriously perhaps, exceeding drug solubility limits could give rise to precipitation in biological fluids, organs, and tissues, giving rise to toxicity that may not be associated with the intrinsic pharmacologic or toxicologic effects of the drug. Crystalline deposition in kidney tubules occurs with many compounds at toxicologic dose levels, but not at therapeutic doses.

Stability

Compound stability may be greatly influenced by the concentrations and amounts of substances used in toxicology. Traditionally in toxicology and toxicokinetic studies, drugs may be administered to animals mixed with feed, and this is a realistic approach in terms of economy of resources. Dosing compounds by gavage in toxicology studies is labor intensive and expensive. However, fine dispersion of drug with feed, and sometimes storage over considerable periods in this finely dispersed form, can give rise to drug degradation problems.

While most drugs are absorbed by passive processes, so that the quantity of drug administered should not influence intrinsic absorption rate or efficacy, the amounts of drug administered during toxicology studies are almost certain to induce some changes in absorption. The changes may result from limited solubility in the GI tract giving rise to slower or less efficient absorption, or from pharmacologic or toxicologic effects of the compound on absorption mechanisms and on the general condition of the GI tract.

Absorption

As discussed in Chapters 16 and 19, gastrointestinal and hepatic clearance (presystemic clearance) of drug during absorption are enzyme-dependent and therefore saturable processes. First-pass presystemic drug clearance will inevitably saturate at toxicological doses, giving rise to changes in drug and metabolite systemic availability. There are many examples of this in the literature (*11*). Bioavailability of compounds may thus be markedly influenced by dose size, and for this reason it cannot be assumed that bioavailability at toxicologic dose levels will be the same as or even similar to that at pharmacologic or therapeutic dose levels. High drug doses may also give rise to induction of hepatic drug-metabolizing enzymes, which may have toxicokinetic as well as toxicologic sequelae (*12*).

First-Pass Effect or Presystemic Clearance

Binding of compounds to plasma proteins and other tissues is generally reversible, and always saturable (*13*) (Chapter 8). Thus, considerable changes may occur in drug binding to plasma proteins, and also to tissues, at toxicologic doses. This binding can in turn influence drug distribution and penetration into tissues, giving rise to different plasma concentration–effect relationships at toxicokinetic doses compared to pharmacokinetic doses.

Protein Binding

As with other enzyme-dependent systems, metabolism of compounds is substrate concentration-dependent (*14*) (Chapter 9). Thus, metabolic pathways and metabolic efficiency may differ at toxicologic doses compared to therapeutic and pharmacologic doses. This is important when comparing the metabolism of a compound in toxicology species to humans for toxicology species validation.

Metabolism

Renal excretion comprises both saturable and nonsaturable mechanisms and can be markedly influenced by circulating drug concentrations, giving rise to changes in renal excretion efficiency and clearance of drug from the body. This is particularly true for any compound that is actively excreted into the renal tubules or actively reabsorbed from the tubules back into the bloodstream.

Renal Excretion

The high concentrations of drugs used in toxicology studies are, by definition, likely to be toxic to the host. Depending on the site and nature of toxic events, this may have a traumatic effect on physiological feedback that may affect drug absorption, distribution, metabolism, or excretion.

Physiological Feedback

Drug Interactions

Drug interactions are frequently concentration-dependent, and different types of interactions may occur in toxicity studies compared to pharmacokinetic studies. This is particularly relevant to toxicokinetic studies of drug combinations and of enantiomeric compounds.

As illustrated by the preceding examples, and as described in Chapter 19, most aspects of drug disposition in the body may be affected by saturable processes in one way or another, and these may affect the toxicokinetics and toxicity profile of an administered compound. Thus, from many perspectives, toxicokinetics cannot be considered to be the same as pharmacokinetics, but as representing an extension of the latter discipline to examine the behavior of drugs at toxicological doses.

Philosophical Differences Between Pharmacokinetics and Toxicokinetics

Apart from the above technical differences, another major difference between pharmacokinetics and toxicokinetics lies in their overall philosophy and objectives. As toxicokinetic studies are carried out exclusively in animals, it is appropriate to compare the objectives of pharmacokinetic and toxicokinetic studies in these species.

The objective of nonclinical pharmacokinetic studies is to obtain pharmacokinetic and pharmacodynamic data to provide a link between nonclinical and human studies. The nonclinical–clinical link is strengthened if relationships can be established between pharmacokinetic and pharmacodynamic end points in the nonclinical species (*15*). Typically, pharmacologic and pharmacodynamic end points may be used not only to provide links between nonclinical and clinical studies, but also to better define the activity of the drug. Quantifiable pharmacodynamic end points have been described for various therapeutic areas, including anti-infectives (minimum inhibitory concentration), cardiovascular agents (blood pressure), the CNS (anticonvulsant effect), immunomodulators (inosine levels), diabetes (blood sugar), antipsychotic agents (behavior), analgesic agents (pain threshold), and cancer chemotherapy (tumor regression).

While nonclinical–clinical pharmacokinetic–pharmacodynamic links are useful, and sometimes very important, they are seldom critical during a drug development program. Whether or not nonclinical information gives sufficient information to accurately predict drug behavior in humans may have little influence on the final decision to go forward into a clinical program, and which direction such a program might take.

The situation is different in the case of toxicokinetics. The objective of toxicokinetic studies is to attempt to quantify the pharmacokinetics of the compound in toxicology species at toxicologic doses and to establish appropriate drug "exposure" in these species (*16*). A further objective is to establish, with priority, metabolic profiles of new chemical entities in the toxicity species relative to humans. The objective of linking toxicokinetics and toxicodynamics is to establish relationships between drug exposure and toxic events and to predict toxicity and tolerance in humans on the basis of nonclinical information. By far the most critical parameter generated by toxicodynamic–toxicokinetic studies is the safety margin. This parameter relates the maximum tolerated concentration (C_{max}, C_{ss}, or AUC) in the toxicity species to the minimum effective concentration in humans. This safety margin may well be critical to decisions whether or not to proceed with a clinical program. Whereas pharmacokinetic–pharmacodynamic studies are often facilitated by means of well-established end points, toxicokinetic–toxicodynamic studies are extremely difficult because of the poor and frequently unpredictable end points in experimental animals, and also because of the capricious interspecies differences in organ and tissue sensitivity among animals and between animal species and humans (17–20). Thus in pharmacokinetics and pharmacodynamics, prediction from nonclinical data to the human situation is relatively straightforward, but often not critical to a drug development program. In toxicokinetics and toxicodynamics, on the other hand, prediction is critical for the nonclinical–clinical transition, but often extremely difficult to achieve.

Conclusions

Pharmacokinetics and toxicokinetics differ in terms of the goals, technology, and philosophical emphasis of the two disciplines. Each discipline plays an important role in drug discovery and development, but the roles are different. Both are now recognized as critical but different components of the process. Each discipline gives rise to different alliances. Pharmacokinetics studies require extensive collaboration with disciplines such as pharmacology, clinical pharmacology, and clinical development. Toxicokinetics studies, on the other hand, demand close alliances between pharmacokinetic–toxicokinetic scientists and toxicologists. Interactions between toxicokinetics and both clinical pharmacology and clinical development are based largely on clinical predictions associated with safety assessment and therapeutic margins. These latter interactions are not as continual as the interactions between pharmacokinetics and the clinical disciplines. These various multi-interdisciplinary alliances emphasize the global impact of pharmacokinetics and drug metabolism across the broad spectrum of pharmaceutical R&D.

Summary

Toxicokinetic studies are necessary components of toxicology programs.

Because of the different dosage levels used, toxicokinetic and pharmacokinetic profiles of a compound are likely to be different.

Nonclinical pharmacokinetic–pharmacodynamic correlations are often relatively straightforward to establish, but may not directly impact drug development decisions.

Nonclinical toxicokinetic–toxicodynamic relationships are extremely difficult to establish, but may have considerable impact on drug development decisions.

References

1. Peck, C. C.; Barr, W. H.; Benet, L. Z.; Collins, J.; Desjardins, R. E.; Furst, D. E.; Harter, J. G.; Levy, G.; Ludden, T.; Rodman, J. H.; et al. *Clin. Pharm. Ther.* **1992,** *51*, 465–473.
2. Yasuda, S. U.; Schwartz, S. L.; Wellstein, A.; Woosley, R. L. *Integration of Pharmacokinetics, Pharmacodynamics and Toxicokinetics in Rational Drug Development.* Yacobi, A.; Skelly, J. P.; Shah, V. P.; Benet, L. Z., Eds.; American Association of Pharmaceutical Sciences: New York, 1993; pp 225–238.
3. Sheiner, L. B. *Clin. Pharmacol. Ther.* **1989,** *46*, 605–615.

4. Peck, C. C. *Population Approach in Pharmacokinetics and Pharmacodynamics, FDA View. New Strategies in Drug Development and Clinical Evaluation, Population Approach*; Rowland, M.; Aarons, L., Eds.; Commission of European Communities: Brussels, Belgium, 1992; pp 157–168.

5. De la Iglesia, F. A. *Drug Toxicokinetics*; Welling, P. G.; de la Iglesia, F. A., Eds.; Marcel Dekker: New York, 1993; pp 381–402.

6. Whitfield, L. R.; Pegg, D. G. *Drug Toxicokinetics*; Welling, P. G.; de la Iglesia, F. A., Eds.; Marcel Dekker: New York, 1993; pp 267–299.

7. Susick, R. L.; Woolf, T. *Drug Toxicokinetics*; Welling, P. G.; de la Iglesia, F. A., Eds.; Marcel Dekker: New York, 1993; pp 305–324.

8. Wright, D. S.; Ross, S. E.; MacDonald, J. R. *Drug Toxicokinetics*; Welling, P. G.; de la Iglesia, F. A., Eds.; Marcel Dekker: New York, 1993; pp 325–361.

9. Smyth, R. D.; Smyth, R. J. *Drug Toxicokinetics*; Welling, P. G.; de la Iglesia, F. A., Eds.; Marcel Dekker: New York, 1993; pp 363–380.

10. Welling, P. G. *Pharmacokinetics of Drugs*; Welling, P. G.; Balant, L., Eds.; Springer-Verlag: Heidelberg, Germany, 1994; pp 3–19.

11. Balant, L. P.; McAinsh, J. *Concepts in Drug Metabolism*; Jenner, P.; Testa, B., Eds.; Marcel Dekker: New York, 1980; pp 331–371.

12. Cayen, M. N.; Black, H. E. *Drug Toxicokinetics*; Welling, P. G.; de la Iglesia, F. A., Eds.; Marcel Dekker: New York, 1993; pp 69–83.

13. Rowland, M.; Tozer, T. N. *Clinical Pharmacokinetics: Concepts and Applications*; Lea and Febiger: Philadelphia, PA, 1980; p 42.

14. La Du, B. N., Jr. *Drug Toxicokinetics*; Welling, P. G.; de la Iglesia, F. A., Eds.; Marcel Dekker: New York, 1993; pp 221–244.

15. Levy, G. *Integration of Pharmacokinetics, Pharmacodynamics, and Toxicokinetics in Rational Drug Development*; American Association of Pharmaceutical Scientists: New York, 1993; pp 7–13.

16. Sjöberg, P. *Drug Inf. J.* **1994,** *28,* 151–157.

17. Munro, A. *Drug Inf. J.* **1994,** *28,* 259–262.

18. Smith, D. A.; Humphrey, M. J.; Charuel, C. *Xenobiotica* **1990,** *20,* 1187–1199.

19. Smith, D. A. *Drug Metab. Rev.* **1990,** *23,* 355–373.

20. Smith, D. A. *Eur. J. Drug Metab. Pharmacokinet.* **1993,** *18,* 31–39.

Role of Pharmacokinetics in Drug Discovery and Development

21

The evolution of pharmacokinetics is described briefly in Chapter 1. The contribution of this discipline to pharmaceutical research and development (R&D) has been influenced by evolution of the awareness of its importance as a determinant of drug action and also by developments in analytical technology. Rapid advances in the appreciation of pharmacokinetic concepts and in analytical technology have thrust this discipline into a central role in pharmaceutical R&D.

Prior to 1960, pharmacokinetic and drug metabolism information comprised a very small proportion of regulatory submissions. Not only was there little appreciation of the importance of this type of information to the understanding of drug action, but there were also limited means of providing such information even if it had been required.

Times have changed! Current submissions for drug marketing approval contain a considerable quantity of pharmacokinetic and metabolism data. In a New Drug Application (NDA) in the United States, three complete sections are devoted to this topic: Section 5, nonclinical pharmacokinetics; Section 6, clinical pharmacokinetics; and Section 8, clinical pharmacology. Pharmacokinetic information is actually used in almost every section of the NDA, or equivalent international documents, as it draws together concepts in pharmacology, toxicology, and clinical practice. It also links nonclinical and clinical experiences, validates (or else questions the validity of) species used in toxicology, and provides a rational basis for therapy based on drug concentration–effect relationships.

In order to incorporate pharmacokinetic concepts during drug discovery and development, pharmacokinetic and drug metabolism disciplines are required to interact on an almost continuous basis with other disciplines. The intent of this chapter is to describe the interactive nature of pharmacokinetics and drug metab-

Regulatory Submissions

olism (PDM) in pharmaceutical R&D. The information in this chapter is adapted from material presented elsewhere (*1*).

Drug Discovery and Development

The process of discovery and development for new drug candidates is summarized in Figure 21.1. The process can be considered as having three components: discovery, nonclinical development, and clinical development. As a compound progresses through these components, it draws on different resources and disciplines. However, unlike other disciplines, pharmacokinetics plays a pivotal role throughout the entire process, from initial discovery to filing the final application for marketing approval, and beyond.

Discovery

During this period, chemical leads are generated, culminating in declaration of a development candidate. Historically, pharmacokinetics has not played a significant role during this period. This has changed. Metabolism expertise now plays an important role in molecule design to reduce, enhance, or redirect metabolic profiles and activity of drug candidates. This is achieved through considerable interaction among drug metabolism sciences, synthetic chemistry, and molecular drug design. This level of interaction continues to increase as in vitro metabolism technology using pure cell lines and enzyme systems increases.

The role of pharmacokinetics during the predevelopment period has also increased. Most organizations now require pharmacokinetic characterization, particularly absorption, elimination, and protein binding characteristics, before advancing a drug candidate into development. Thus, before a drug candidate enters a nonclinical development program, it has been at least partially characterized in terms of its overall pharmacokinetics and metabolism.

Radiosynthesis is generally not initiated until a drug candidate reaches lead compound status. Preparation of radioactively labeled compound earlier would be costly in terms of the uncertain status of the compound and also the preliminary nature of synthetic procedures. Therefore, analytical work during the prelead discovery period is done with "cold" compound. This requires development of sensitive and specific assays for a large number of novel compounds, most of which will likely not survive beyond early nonclinical development.

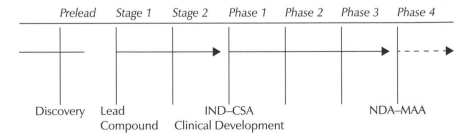

FIGURE 21.1   The drug discovery and development process. IND, Investigational New Drug application; CSA, Clinical Study Application (Europe); NDA, New Drug Application; MAA, Marketing Authorization Application (Europe). Reproduced with permission from reference 1.

On many occasions, pharmacokinetic and systemic availability information is requested to provide an initial screen in order to determine the suitability of compounds for advancement. This is an inefficient approach and is generally a misuse of resources. If a series of compounds is being screened for advancement to lead status, then it is currently more economical to use some form of pharmacologic screen that examines activity than to use a pharmacokinetic screen for which analytical procedures have to be developed and validated for individual compounds. This situation is changing, however, and the rate of change will likely accelerate. The impetus to change is the recent introduction of liquid chromatography (LC) coupled to one or more mass spectrometry (MS) instruments (LC–MS, LC–MS–MS) with atmospheric pressure ionization, and also direct-insertion MS. By focusing on quantitation of molecular ions, and de-emphasizing or completely avoiding LC separation, a rapid and efficient analytical methodology is achieved. This is a quantum leap forward in drug analysis and will greatly increase the efficiency and impact of pharmacokinetic involvement in nonclinical drug discovery, as well as in nonclinical and clinical drug development.

Nonclinical Development

In Figure 21.1, nonclinical development is the period from declaration of lead compound status to submission of the Investigational New Drug (IND) application. The spectrum of studies that may comprise a nonclinical pharmacokinetic and metabolism development program, and their approximate chronological sequence, is shown in Figure 21.2. In this figure, typical studies are described for the rat and dog. While the rat is essentially the obligatory rodent species used, monkey may be substituted for dog as the nonrodent species, depending on the compound class. However, this is generally avoided if possible because of the cost of procuring and maintaining monkeys.

The duration of a nonclinical development program may range from a few months to 2 or 3 years, depending on the compound. In the figure, the term "analytical development" is repeated several times throughout development. This reflects the need to continually modify and revalidate analytical methods as knowledge of compounds increases and as drug disposition studies advance to additional species and biological matrices.

A biopharmaceutical profile of the compound may be prepared at an early stage in nonclinical development in order to provide preliminary documentation on physical properties, stability, protein binding, bioavailability, and pharmacokinetics of a compound. This is typically followed by more extensive bioavailability and pharmacokinetic studies in appropriate species to gain insight as to how the compound behaves in these species, with possible prediction to humans. Radioactively labeled compound is prepared at this stage, so that mass balance, autoradiography, and metabolism studies may be conducted.

Most autoradiography studies, which are done predominantly in the rat and mouse, are conducted after single doses of radiolabeled drug. Some regulatory agencies, such as the Japanese Koseisho, require these studies to be conducted after repeated dosing, using radiolabeled drug for all doses. In some instances both single and repeated doses may be requested. This consumes a considerable quantity of time, and of radiolabeled drug, but may be necessary to adequately charac-

PRELEAD	Stage 1	Stage 2	IND	Phase 1	Phase 2	Phase 3	NDA

Analytical development
Bioavailability
    Biopharmaceutical profile
    Dog bioavailability and pharmacokinetics
    Analytical development
        Rat bioavailability and pharmacokinetics
        Dog multiple-dose pharmacokinetics
        Rat mass balance
        Analytical development
            Dog mass balance
                Autoradiography
                Toxicology dose ranging
                2-week toxicology
                    Rat single-dose toxicokinetics
                    Rat multiple-dose toxicokinetics
                    Dog single-dose toxicokinetics
                    Dog multiple-dose toxicokinetics
                    Rat, dog preliminary metabolism
                        13-week toxicology dog and rat
                      Dog major metabolites
                      Rat major metabolites
                          Regulatory submission documentation
                          Rat tissue distribution
                              Toxicology, 52-week definitive
                                  Placental transfer
                                  Enzyme induction

FIGURE 21.2   Typical nonclinical pharmacokinetic studies. Reproduced with permission from reference 1.

terize disposition and tissue accumulation of radiolabel after single and repeated doses. Although autoradiographic data obtained in animals are considered by many to be important for extrapolation to humans, because physicochemical properties of tissues are similar across mammalian species, the argument has been made that routine autoradiography as part of a general development program should be de-emphasized (2). This argument is based on the fact that autoradiography is based on radiolabel rather than on a known chemical entity (see Chapter 8), and also on the difficulty of using radioactivity concentrations in any organ or tissue to predict drug effect, beneficial or otherwise.

Recent advances in metabolism technology have made it increasingly feasible to carry out early metabolism characterization for drug candidates. Use of metabolism information can save considerable time in compound selection, while knowledge of the rate and pattern of compound degradation, possible creation of active metabolites, and the nature of metabolite disposition are critical to informed and successful compound development.

Toxicokinetics.   An increasingly important component of a nonclinical development program is associated with toxicology. The principal objective of toxicology studies is to establish an appropriate safety margin for a drug candidate in humans.

Toxicokinetic studies, on the other hand, are of more recent interest, so recent in fact that the term toxicokinetics is not accepted by many, and is not understood by most. Toxicokinetics is in its infancy now, but it is maturing and expanding rapidly. The subject of toxicokinetics is described in Chapter 20, which also compares this topic with pharmacokinetics.

Pharmacokinetic–Pharmacodynamic Relationships.   Studies of pharmacokinetic–pharmacodynamic relationships are an important component of clinical development programs. However, they are also playing an important role in nonclinical development. This is particularly true for cardiovascular agents for which measurable pharmacodynamic parameters are available in order to determine concentration–effect relationships. However, pharmacokinetic–pharmacodynamic relationships also need to be investigated wherever possible in other therapeutic areas in order to generate information that is useful for subsequent Phase 1 and Phase 2 studies in humans, and also to obtain accurate prediction of the therapeutic dosage range and margin of safety.

It is evident from Figure 21.2 that nonclinical studies extend beyond the time of IND submission and may extend into Phase 3 of the clinical program or beyond. This occurs particularly when observations made during clinical studies require additional nonclinical work for clarification or resolution of problems.

Overall PDM support for a drug development program is thus a continuum between nonclinical and clinical phases, and such support should not be divided, as has mistakenly been done by some, into rigidly defined nonclinical and clinical components.

Interdisciplinary Interactions.   During nonclinical discovery and development, PDM departments interact with many disciplines. During early discovery, interaction is primarily with the disciplines of chemistry and biology. As a drug candidate moves into the nonclinical development pipeline, interactions with chemistry and biology continue, but they are also expanded to include product development, toxicology, and clinical pharmacology. As described in Chapter 20, toxicokinetic data, and drug monitoring data from longer term toxicology studies, provide critical concentration–effect information to help establish a realistic safety margin and also dosage ranges in initial clinical dose studies.

As will be demonstrated in the next part of this chapter, interactions between PDM and other disciplines continue to evolve during clinical development, culminating in multidisciplinary strategies for successful submission of final regulatory documents for marketing approval.

The spectrum of clinical pharmacokinetic and metabolism studies that may be conducted, and their approximate chronological sequence, are described in Figure

Clinical Development

21.3. Referring to Figure 21.1, the entire clinical development program is conveniently subdivided into Phases 1, 2, and 3.

Phase 1.   The major objective during Phase 1 clinical studies is to examine the safety of a drug candidate in a small population of individuals. These studies are normally conducted in healthy adult volunteers. However, for some drug classes, such as anticancer agents and some cardiovascular compounds, they may be conducted in appropriate patient populations. Initial doses administered at this early stage are well below the anticipated therapeutic dosage range, and subjects or patients are carefully monitored for tolerance. The pharmacokinetic component of these studies is extensive, and rapid analytical turnaround is routinely required consistent with increasing emphasis on concentration-based dose escalation and study termination.

Typical studies involving pharmacokinetics that are conducted during Phase 1 are described in Table 21.1. Initial single rising-dose and repeated-dose tolerance

Phase 1	Phase 2	Phase 3	NDA-MAA
Single-rising-dose tolerance			
Multiple-dose tolerance			
Pilot bioavailability			
Bioavailability, positive control			
Food effect			
Therapeutic drug monitoring			
Single-dose pharmacokinetics			
Multiple-dose pharmacokinetics			
Metabolism			
Mass balance			
Food effect			
Dose proportionality			
Formulation bioavailability			
Pharmacokinetics vs. competitive product			
Renal disease			
Liver disease			
Elderly vs. young			
Formulation bioavailability			
Drug interactions			
Therapeutic drug monitoring			
Integrated data analysis			
Bioavailability and bioequivalence, market image formulation vs. clinical trials formulation			
Clinical pharmacokinetic report			
Regulatory submission documentation			

FIGURE 21.3   Typical clinical pharmacokinetic studies. Reproduced with permission from reference 1.

studies both contain substantial pharmacokinetic components. In addition to tolerance information, these studies provide preliminary information on compound pharmacokinetics after single and repeated doses, accumulation, and any pharmacokinetic changes that may arise from repeated doses. Other studies conducted during Phase 1, or early Phase 2, although concentrating on tolerance, are predominantly pharmacokinetic in nature. They may include absolute systemic availability relative to parental dosage or relative systemic availability relative to another enteral dosage form, dose proportionality, and also relative bioavailability of any reformulated compound that may be used as a positive control during subsequent Phase 2 or 3 efficacy studies.

Ingestion of food is now recognized as a key factor influencing drug absorption (*see* Chapter 7), and a food–drug interaction study is virtually mandatory during Phase 1. As the final "market image" formulation would not yet have been developed, additional food-effect studies will be required later in the program. However, the nature of this interaction should be examined early using a formulation that is hopefully similar or identical to that used in clinical trials in order to guide later dosage regimens.

Thus, whereas the main interest during Phase 1 clinical studies has historically been to examine tolerance and safety, much of the focus is now on pharmacokinetics. The trend of placing increasing emphasis on bioavailability and pharmacokinetic disposition of drug candidates and their metabolites during Phase 1 continues throughout the clinical program, and represents recognition of the important relationship between circulating levels of parent compound or active metabolites(s) and pharmacologic activity. Some excellent examples of collaborations between clinical pharmacologists and scientists in drug metabolism, bioanalysis, and other disciplines to resolve drug development problems have recently been described by Powell (3).

Phase 2.   During Phase 2 a drug candidate is typically tested in small, well-defined patient populations. Although a measure of efficacy may be derived from Phase 1 studies for some drug classes, formal determination of efficacy usually occurs in Phase 2. During this phase many drug candidates fail because of lack of efficacy or unacceptable side effects, and are withdrawn from development. Typical collaborative studies involving pharmacokinetics that are conducted during Phase 2 are described in Table 21.2. During this phase, the pharmacokinetics of a drug candidate is often characterized after single and repeated doses in patients or in healthy individuals, and comparison may be made between the drug candidate and competitive products. In the competitive pharmaceutical environment, a

---

Phase 1 Studies Involving Pharmacokinetics   **TABLE 21.1**

Single-rising-dose tolerance
Multiple-dose tolerance
Bioavailability
Food effect
Bioavailability, positive control

modest pharmacokinetic advantage of one product over another may be critical to its success. Additional food effect studies may also be conducted during Phase 2 in order to refine observations made during Phase 1, and to better define the labeling.

Apart from continuing pharmacokinetic studies, three additional components are introduced during Phase 2. These are mass balance, metabolism, and therapeutic drug monitoring. Mass balance and metabolism studies traditionally use radioactively labeled compound and are generally conducted in volunteers. Mass balance studies are designed to examine recovery of total radioactivity following oral and/or intravenous administration of labeled compound, and may identify drug sequestration or covalent binding to tissue in the event of incomplete mass balance. Metabolism studies are essential at this stage, not only to better understand the pathways and rate of formation of biologically active or inactive metabolites, but also to validate the animal species used in toxicity studies.

If humans generate metabolites that are qualitatively or quantitatively different from those observed in toxicity species, then the question of toxicity species validity arises. The question is invariably raised as to whether a metabolite formed in humans, or formed in humans to a greater extent than in the toxicity species, should be reexamined in that toxicity species. If so, then by what route should the metabolite be administered? If given intravenously, it may give rise to specific toxic effects that might not occur in the case of in vivo generation from parent drug. If given orally, the same argument may apply, but the problem is compounded by uncertain metabolite bioavailability. There are no rules in these circumstances, and compounds must be considered on an individual basis.

One of the most important activities that starts during Phase 2, and continues throughout the remainder of the clinical development program, is that of therapeutic drug monitoring. Careful and accurate plasma sampling and/or urine collection during appropriate segments of efficacy and tolerance protocols can provide a wealth of information regarding dose–concentration–effect relationships for a drug candidate. This is probably one of the most important pharmacokinetic activities and pharmacokinetic–clinical collaborations in a clinical development program. Data from such collaborations may be used to define dose–effect and pharmacokinetic–pharmacodynamic relationships for a drug. These are described briefly later in this chapter.

Phase 3.   The main focus during Phase 3 is on pivotal efficacy trials. These trials, conducted in large patient populations, provide efficacy and tolerance data to

TABLE 21.2	Phase 2 Studies Involving Pharmacokinetics

Therapeutic monitoring, small-scale clinical efficacy, and tolerance trials
Single-dose pharmacokinetics
Multiple-dose pharmacokinetics
Pharmacokinetics vs. competitive product
Food effect
Dose proportionality
Mass balance
Metabolism

support the marketing application. As in the Phase 2 efficacy studies, advantage should be taken during these studies to apply population or integrated data analysis techniques to learn as much as possible regarding the pharmacokinetic–pharmacodynamic characteristics of a drug candidate in a large, heterogeneous patient population. It would probably be unrealistic to attempt to conduct this level of investigation on all participating patients at all clinical investigation sites. However, use of a representative sample of patients from several sites will provide valuable information. Typical studies involving pharmacokinetics that are conducted during Phase 3 are described in Table 21.3.

Apart from the population or mixed effect modeling approach, more traditional studies in special populations still comprise a major part of a Phase 3 development program. Studies in elderly patients, patients with hepatic or renal impairment, and perhaps other special populations are important in order to characterize drug disposition in a broad spectrum of patients. Drug interaction studies are also mandatory. In this case the investigator has to decide how many studies to conduct and how best to conduct them in order to obtain the most useful information for a considerable capital and resource outlay. As many as 6 to 8 interaction studies may be required, with appropriate focus on drugs that may be given concomitantly with the drug candidate and/or drugs that are likely to interact mechanistically and whose interaction may be therapeutically important. Clearly, compounds that are highly protein bound or are extensively metabolized and have narrow therapeutic indices are candidates for these studies. There is a strong case to conduct in vitro studies to "screen" compounds for possible interactions before embarking on lengthy and costly clinical studies. This concept is still developing within the industry, but it is attractive as a potential way to increase efficiency and reduce costs during drug development.

Market-image formulations for a drug candidate are seldom developed before the compound enters Phase 3. Thus, most clinical studies, often including pivotal multicenter efficacy trials, may be conducted with clinical samples that may differ in form or composition from the final market image. Once a market-image formulation is available, bioequivalence studies are required to compare these with the formulations used in clinical trials. Bioinequivalence is seldom encountered in these studies, but if it is it can profoundly disrupt and delay a clinical program. This situation can be avoided if market-image formulations are prepared early during a clinical development program, preferably before the small-scale efficacy

Phase 3 Studies Involving Pharmacokinetics   **TABLE 21.3**

Therapeutic monitoring, pivotal efficacy trials
Renal failure
Hepatic failure
Elderly vs. young
Other special patient populations
Drug interactions
Bioavailability and bioequivalence of market-image formulation vs. clinical trials
     formulations
Regulatory submission documentation

and safety studies in Phase 2. This may usually be impractical, but it is nonetheless a worthwhile goal to have all clinical efficacy studies conducted using the market-image formulation. This might cause stress in some disciplines but would relieve stress in others.

Pharmacokinetics and drug metabolism studies are thus a major component of a Phase 3 clinical development program. Not only are a considerable number of discrete pharmacokinetic studies conducted in order to characterize drug disposition under a variety of conditions and thus guide doses in patient populations, but there is also a growing pharmacokinetic component to the large, multicenter efficacy studies.

Pharmacokinetic–Pharmacodynamic Relationships. During Phase 2 and Phase 3 of a clinical program, there are opportunities, either during therapeutic monitoring of large clinical trials, through the conduct of specific concentration–effect studies, or both, to establish pharmacokinetic–pharmacodynamic relationships. For some therapeutic classes, this is relatively straightforward. For example, cardiovascular agents, antibiotics, and antimicrobial drugs have several measurable pharmacodynamic end points that are reliable, repeatable, objective in nature, and can be validated. Other therapeutic areas are more difficult to study because of the difficulty of identifying measurable and reliable pharmacologic end points. This has been the case with many agents that act on the central nervous system (CNS). In some cases surrogate end points have been used, such as electroencephalograms as indicators of benzodiazepine activity. Attempts are continually being made to establish reliable pharmacologic end points for a variety of drugs, and considerable success has been achieved in several difficult therapeutic areas including anesthetic agents, analgesics, some CNS drugs, anticonvulsants, anticoagulants, corticosteroids, and anticancer drugs (4). Establishing pharmacokinetic–pharmacodynamic relationships has been particularly difficult for CNS drugs because of the subjective nature of pharmacologic end points and the development of tolerance (5). Although several subjective tests are still used, such as neuromotor and motor coordination tests, other more objective tests such as electroencephalography (EEG) have been introduced. For example, Greenblatt et al. (6) have demonstrated a relationship between plasma concentrations and EEG effects for diazepam and midazolam. Kroboth et al. (7) have demonstrated relationships among psychomotor performance, EEG, and plasma levels for alprazolam.

Interactions. During Phase 3, PDM scientists interact extensively with clinical pharmacology and clinical development, as well as with regulatory departments, to provide the bulk of the pivotal data to support drug labeling. Interactions between various disciplines during the multicenter trials are often difficult and in some cases may be costly, but the ultimate value added by an integrated approach is substantial and fully justifies commitment of additional resources to obtain maximum information from these clinical studies.

Regulatory Submissions and Drug Labeling

The PDM contribution to the final marketing submission document was briefly described earlier in this chapter. Suffice it to reflect here on the highly interactive nature of this documentation. Preparation of Sections 5, 6, and 8 of the U.S. NDA, and equivalent sections of international filings, is a multidisciplinary exercise. Section 5 requires integration of pharmacokinetics, metabolism, and toxicology and toxicokinetics information, with input from chemistry and pharmacology. Sections 6 and 8 require close collaboration among pharmacokinetics, clinical pharmacology, clinical development, and product development, with input from other disciplines. Prudent integration of relevant information is necessary to provide meaningful characterization of safety and efficacy to support marketing approval.

Increasing emphasis on drug and metabolite concentration–effect information is also reflected in drug labeling. In the past, clinical pharmacokinetic data were minimal and were largely submerged within clinical pharmacology material. With the change of focus within the U.S. Food and Drug Administration, both the nonclinical and clinical pharmacokinetics sections of the labeling have greatly increased and now often include figures and tables to provide detailed information on drug disposition. For further reading on regulatory submissions in general for pharmaceutical products, the reader is referred to recent reviews for submissions in the United States (8), Japan (9), and Europe (10).

Postsubmission and Postmarketing Studies

The period between submission for marketing and approval is often one of considerable negotiation between a company and regulatory authorities. Issues have to be clarified, claims substantiated, and labeling justified and finally agreed on. Although many disciplines are called upon during this period, pharmacokinetics is frequently involved and is often central to the process. As a result of the shift in emphasis toward concentration–effect and pharmacokinetic–pharmacodynamic relationships, many questions are raised that require a collaborative response or even additional collaborative studies to clearly define drug behavior.

Once marketing approval is obtained, an intensive program may be initiated to support the marketed product, explore new dosage forms and indications, respond to international issues and, as the patent expiration time approaches, defend the product against generic erosion. During this period, pharmacokinetics departments work in close collaboration with clinical departments, marketing, and also product development to provide appropriate support for these different areas. Development of modified dosage forms, sustained release or transdermal preparations, and other alternative drug delivery forms requires extensive pharmacokinetic input to obtain approval of a new product, hopefully without the need for additional clinical efficacy trials.

Conclusions

The disciplines of pharmacokinetics and drug metabolism have evolved rapidly from being minor, almost insignificant contributors to the drug discovery and development process, to the present situation where they play a major, pivotal role in all phases, from early discovery through nonclinical and clinical development. The increasing role of pharmacokinetics in drug discovery and development has resulted from greater scientific awareness of its importance and also advances in analytical and computer technology. Analytical methodology will continue to present challenges as the pharmacologic potency of new molecules increases, driving doses down to microgram levels, and as the proportion of biotechnology-derived molecules entering the development pipeline increases. Major recent advances in mass spectrometric and immunoanalytic technology are leading to increasing pharmacokinetic contributions to both drug discovery and development.

Through its continuous involvement during the entire process of discovery, development, and postmarketing, PDM is unique among disciplines in the pharmaceutical industry. This far-reaching responsibility, involving extensive collaborations with other disciplines, is gratifying for those who "practice the art". However, this role brings with it extensive responsibilities that must be discharged in a rapidly changing research environment. This is the challenge that pharmacokinetics has to meet, both now and in the future.

Summary

Pharmacokinetics and drug metabolism studies play a critical and central role in drug discovery and development.

Three sections of a New Drug Application (NDA) are devoted to nonclinical or clinical pharmacokinetics.

Pharmacokinetic studies are conducted throughout a discovery and development program, ranging from metabolism and disposition studies on early chemical leads, to postapproval market support studies.

Pharmacokinetic–pharmacodynamic studies are important during nonclinical and clinical development to establish definitive drug and/or metabolite concentration–effect relationships.

References

1. Welling, P. G. In *Pharmacokinetics of Drugs*; Welling, P. G.; Balant, L. P., Eds.; Springer-Verlag: Heidelberg, Germany, 1994; pp 3–19.
2. Monro, A. M. *Drug Metab. Dispos.* **1994**, *22*, 341–342.
3. Powell, J. R. In *The Drug Development Process: Increasing Efficiency and Cost Effectiveness*; Welling, P. G.; Lasagna, L.; Banakar, U. V., Eds.; Marcel Dekker: New York, 1996; pp 295–316
4. *Handbook of Pharmacokinetic/Pharmacodynamic Correlation*; Derendorf, H.; Hochhaus, G., Eds.; CRC: Boca Raton, FL, 1995; p 482.
5. Bellantuono, C.; Reggi, N.; Tognoni, G.; Gorattini, S. *Drugs* **1980**, *19*, 195–201.
6. Greenblatt, D. J.; Ehrenberg, B. L.; Gunderman, J.; Scavone, J. M.; Harmatz, J. S.; Shader, R. L. *Clin. Pharmacol. Ther.* **1989**, *45*, 356–362.
7. Kroboth, P. D.; Smith, R. B.; Erb, R. J. *Clin. Pharmacol. Ther.* **1988**, *43*, 270–275.
8. Cochetto, D. M.; Nardi, R. V. *Managing the Clinical Drug Development Process*; Marcel Dekker: New York, 1992; pp 163–193.
9. Currie, W. J. C. In *The Drug Development Process: Increasing Efficiency and Cost Effectiveness*; Welling, P. G.; Lasagna, L.; Banakar, U. V., Eds.; Marcel Dekker: New York, 1996; pp 353–376
10. Marshall, V. M. In *The Drug Development Process: Increasing Efficiency and Cost Effectiveness*; Welling, P. G.; Lasagna, L.; Banakar, U. V., Eds.; Marcel Dekker: New York, 1996; pp 421–442

Integration of Pharmacokinetics into Research and Development

22

he explosive growth in the awareness, technological advances, and impact of pharmacokinetics and drug metabolism (PDM) in pharmaceutical R&D (*see* Chapters 1 and 21) may be compared with similar events in biotechnology. Both disciplines, or families of disciplines, have evolved rapidly to make an increasing impact on drug discovery and development. Biotechnology has evolved in a well-defined environment. It belongs to discovery and, together with molecular modeling and novel mass screening technology, is providing a new driving force in discovery. Thus, biotechnology has a well-defined place in pharmaceutical R&D.

PDM has not been so fortunate. The rapid emergence of pharmacokinetics and metabolism has at once given those concerned with drug discovery and development new opportunities to facilitate the process, but also considerable problems. The question, where does it belong? which has been answered readily for biotechnology (indeed no question really existed), has caused considerable debate and conflict for PDM. Only now that the importance of these disciplines is becoming acknowledged has some resolution been obtained, but this has been fragmentary and traumatic in some cases. The question "where does PDM belong?" may be more correctly stated as "to whom does it belong?". The considerable interactions between PDM and other disciplines, and the impact that these may have on the success of R&D programs, have caused debate on ownership and identity. This chapter addresses this problem, examines the alternative philosophies together with organizational structures, and offers opinion as to the optimal approach. The arguments developed in this chapter have been expressed by this writer elsewhere (*1*).

The sequence of events that occurs during an R&D program was described earlier in Figure 21.1. Although the processes of discovery and development are in a state of continual change to meet evolving scientific and regulatory circumstances, the

The R&D Sequence

343

basic sequence of R&D is relatively constant. Currently, the major focus of R&D managers is to minimize the time interval between discovery and New Drug Application (NDA) submission, and of course costs.

PDM contributions to the entire process are often considered separately under discovery and development. However, from a PDM perspective, a true distinction between these segments of the program does not exist. Indeed, such a distinction probably does not exist for any discipline when one appreciates that the whole process of discovery and development is a dynamic continuum rather than a series of discrete events.

Statement of the Problem

The increasing complexity of drug discovery and development and the rapidly expanding role played by PDM in all sections of the process have together created a problem of some dimension. The resource issue is obvious. Demands on PDM resources have become extremely heavy during all stages of discovery and development. Bearing in mind the high compound attrition rate as new chemical entities progress through the developmental pipeline, careful resource management is required, particularly at the discovery end, bearing in mind that most compounds studied during this stage will ultimately die. A second major issue is that of management of PDM resources to meet, in the most efficient way, the needs of discovery and development programs. Discovery or development teams, that are multidisciplinary in nature and are formed to guide or "champion" a compound or group of compounds through the discovery and development stages, naturally seek resource commitment, and covet "membership" from the various contributing disciplines, including PDM. This frequently leads to competition between discovery and development teams for PDM resources, and also to conflict in resource management. How then should PDM resources be structured and managed to meet the many demands?

The Alternatives

Several alternative approaches to management and allocation of PDM resources have been used across the pharmaceutical industry in attempts to optimize the quality of PDM support across R&D. Which alternative is used by a particular firm or organization may result from individual preferences within management, the lobbying power of particular interests, or influences of other disciplines and structures within an organization. In some horizontally structured organizations, that is, those whose management is divided along therapeutic lines, total resource management may be exercised within a therapeutic area. For example, cardiovascular R&D will have its own team of PDM scientists, committed to and managed by managers of that therapeutic class. The same holds for central nervous system (CNS) R&D, chemotherapy, etc. In these cases, PDM resources are divided and are administered separately within each therapeutic area.

A second alternative, or cluster of alternatives, is to administratively distribute PDM resources vertically across the discovery and development pipeline. This may take the form of separate PDM support for discovery and development, or for nonclinical and clinical development, or for discovery, exploratory development, and full-scale development, or combinations of these. With this alternative, PDM resources are separately administered within segments of the PDM pipeline.

Resources may thus be committed to each major segment of the pipeline, with flexibility of resource distribution across therapeutic areas and projects within a particular segment of discovery and development. A common structure within this alternative is one that separates nonclinical discovery and development from clinical development.

A third alternative is to administer PDM as a single "team" of scientists with allocation of resources vertically across therapeutic areas and horizontally across the R&D pipeline, managed within PDM. This may be called a holistic or monolithic approach in that total PDM resources are centrally administered within PDM, with resource allocation and relocation vertically and horizontally maintained flexible (negotiable) by means of matrix administration within the overall PDM structure.

The three alternative PDM structures are summarized in Table 22.1 and are illustrated in Figure 22.1.

Comparison of the Alternatives

It is instructive to examine the advantages and disadvantages of the three alternative PDM administrative structures, both from the perspective of the quality of support that each can provide to an R&D program and also from that of professional satisfaction and career development of PDM scientists.

Alternative One: Horizontal Administration

With this alternative, separate PDM resources are committed to each therapeutic area. Thus, a group of pharmacokineticists and metabolism scientists is separately allocated and administered to support a single therapeutic area, such as cardiovascular. The main advantage of this approach is that the managers responsible for overall development in that therapeutic area have control over PDM resources associated with that area. This could apply to the entire spectrum of activity within an area, from discovery to NDA documentation, or it may be subdivided vertically within that area to give a subclass of this alternative, that is, subdivision of cardiovascular support into nonclinical and clinical. Whether subdivided or not, the therapeutic area would have a team of PDM resources committed to and administered within the therapeutic area, thus ensuring PDM support. The therapeutic area manager or managers would administer PDM activities to ensure appropriate allocation of resources between projects. The PDM scientists involved would have intimate knowledge of a therapeutic area and of drug candidates and chemical classes

Alternative Administrative Structures for Pharmacokinetics and Drug Metabolism (PDM)

TABLE 22.1

1.	Horizontal:	PDM activities are administered within therapeutic areas.
2.	Vertical:	PDM activities are administered within specific segments of the R&D pipeline.
3.	Monolithic:	PDM activities are administered as a single unit, with resource allocation vertically across therapeutic areas and horizontally across the R&D pipeline by internal matrix administration.

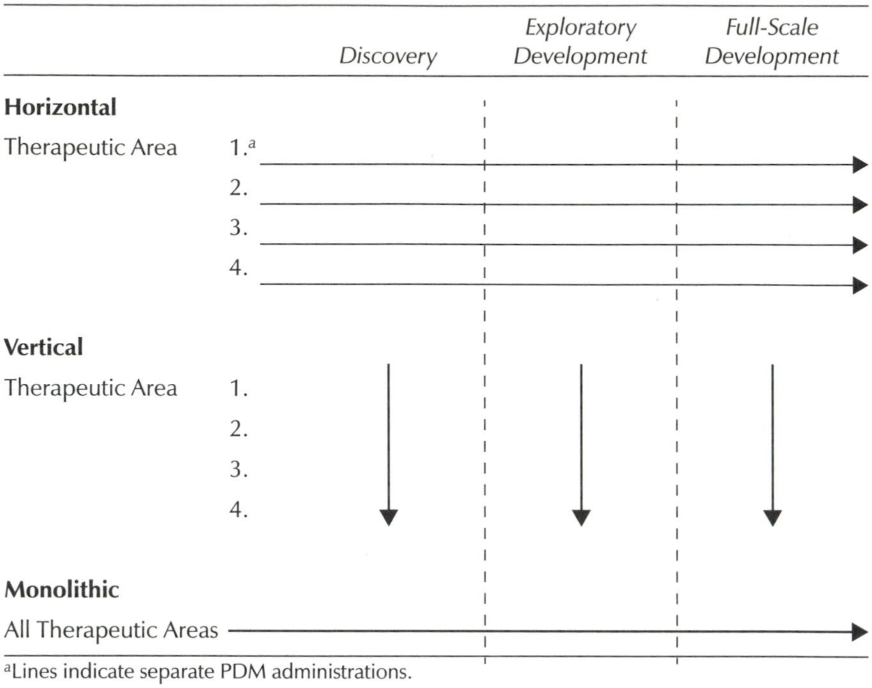

	Discovery	Exploratory Development	Full-Scale Development
Horizontal			
Therapeutic Area 1.[a]			
2.			
3.			
4.			
Vertical			
Therapeutic Area 1.			
2.			
3.			
4.			
Monolithic			
All Therapeutic Areas			

[a]Lines indicate separate PDM administrations.

FIGURE 22.1   Diagrammatic representation of the horizontal, vertical, and monolithic approaches to PDM administration in R&D.

associated with that area. There would be close links, "family", between PDM scientists and other disciplines supporting each area, and there would be administrative resource commitment necessary for effective team support.

The disadvantages of this alternative, whether subdivided along the discovery and development pipeline or not, reside in the fragmentation of PDM resources, loss of critical mass, and flexibility in PDM support from one therapeutic area to another. There is also high probability of minimal exchange of information between the PDM groups associated with different therapeutic areas unless particular efforts are made to ensure that this occurs. The advantages and disadvantages of the horizontal alternative are summarized in Table 22.2.

Alternative Two: Vertical Administration

With this alternative, separate PDM resources are administratively committed to different segments of the discovery and development pipeline. There may be two or three distinct PDM groups, each supporting a different phase of R&D, with compounds being handed from one group to another as they move through the pipeline.

The type of PDM activity may be technically dissimilar from one segment of the pipeline to another. As indicated in Chapter 21, the discovery phase requires extensive analytical and in vitro and in vivo metabolism in addition to pharmacokinetic input. Early development requires extensive toxicokinetic, metabolism, and pharmacokinetic support, whereas full-scale development requires input in the areas of clinical pharmacokinetics, pharmacodynamics, drug interactions, bioequivalence, and pharmacokinetics in disease states and other patient populations.

The advantages of the vertical approach reside in the commitment of PDM resources to particular phases in the discovery and development pipeline. Thus, the manager of a drug discovery program would administratively manage PDM resources that are committed to that program and would have the flexibility of allocating PDM resources among the therapeutic areas within discovery. This structure would enable PDM scientists to focus on a particular segment of activity, with little or no dilution of effort across the large number of PDM activities that occur as a compound moves through the R&D pipeline. The popular practice among companies of separating PDM resources among discovery, nonclinical development, and clinical development provides opportunity for PDM scientists to focus on these particular areas, and the relevant managers to concentrate their assigned PDM resources on the activities associated with these areas. With this approach, PDM scientists have the opportunity to specialize in particular technologies in R&D, such as on pharmacokinetics in animals, metabolism, and toxicokinetics on the one hand, and clinical pharmacokinetics, dose response, and drug interactions in humans on the other.

The disadvantages of this alternative are generally quite different from those associated with Alternative One, although there are some similarities. The major problems faced are those of lack of ownership in and commitment to a particular compound or therapeutic class, restricted exchange of information, reduced efficacy, possible waste of resources, and also territory, communication, and accountability problems. There may also be loss of professional satisfaction on the part of PDM scientists as a result of activities being restricted to one segment of the overall pipeline. Lack of ownership must also occur if a compound passes through various hands during its transition through the pipeline, and waste of resources may result from inadequate or inefficient transfer of knowledge from one group of

Advantages and Disadvantages of the Horizontal Alternative for PDM Administration in R&D

TABLE 22.2

Advantages	Disadvantages
Control of PDM resources within a therapeutic area	Fragmentation of overall PDM resources
Close familiarity of PDM scientists with a particular therapeutic area	Lack of critical mass
Close links between PDM scientists and other disciplines supporting each therapeutic area	Lack of flexibility for PDM support across therapeutic areas
Administrative commitment for effective therapeutic team support	Minimal information exchange between PDM scientists supporting different therapeutic areas

PDM scientists to another. A scientist who is intimately familiar with certain aspects of a compound may or may not pass all this information along, and not all nuances of a compound's characteristics may be readily available. Territorial concerns and lack of accountability are also potential problems in this situation, and excellent inter- and intradepartmental teamwork is necessary in order to maintain a "team spirit" across the different PDM "departments". Maintenance of accountability by the separate PDM groups can be a significant problem. Nobody owns or is prepared to assume total responsibility for an overall PDM program to support a particular compound if it passes from hand to hand through the pipeline.

Professional satisfaction may or may not suffer in the case of Alternative Two, depending on individual circumstances. A PDM scientist administratively associated with full-scale clinical development may find sufficient professional satisfaction in being associated with the exciting, if harrowing, final stage of development leading to the NDA. On the other hand, he or she may become frustrated by lack of control of the design and conduct of pharmacokinetic studies during this period because of their multifactorial nature and also the multidepartmental inputs that are inevitable at this stage. PDM scientists working in early development, on the other hand, may obtain great satisfaction from the high level of scientific activity involved during this phase and yet may be frustrated at having to pass a compound on to another PDM administration just when it is "really showing promise".

A final and important disadvantage of this approach lies in the preparation of the NDA document. As indicated in Chapter 21, this is a highly integrated document that combines large bodies of information from nonclinical and clinical PDM, together with information from many other departments, to produce an integrated document for marketing authorization submission. Preparation and integration of PDM components into a meaningful and reader-friendly format require close collaboration among all aspects of PDM activity, ranging from the discovery phase to studies on market-image formulations. At this point, close collaboration among scientists from all sections of PDM, and between them and other nonclinical and clinical disciplines is essential. With Alternative Two, in which PDM support is fragmented along the R&D pipeline, this type of collaboration and integration of data is more difficult than with Alternative One. The advantages and disadvantages of Alternative Two are summarized in Table 22.3.

Alternative Three: Monolithic Administration

The third alternative, the monolithic or holistic approach, differs from the other two alternatives in that all PDM activities, from discovery to the NDA submission, are coordinated under a single administration with horizontal and vertical support being managed by the PDM internal organization, generally in the form of a matrix structure. Close collaboration is assured between PDM scientists and therapeutic teams by a team membership approach, and different therapeutic areas are supported by different groups within the PDM organization. As a compound progresses from the discovery phase through the development pipeline, it is handled by appropriate groups within PDM, or it may continue with the same group depending on the appropriate support structure for that compound or therapeutic class.

The advantages of this alternative are severalfold, and they include focused PDM management, economy of resources for operating and capital budgets, little or no duplication of effort, ownership and accountability in a particular compound or therapeutic class, efficient technology transfer, lack of territorial problems, and a high critical mass for resource allocation flexibility. With this approach, a single body of PDM scientists is responsible for supporting an entire spectrum of R&D activity while champions for therapeutic areas or development phases do, and should, exist within the PDM organization. In addition, the third alternative does not promote "elitism", which often occurs in a fragmented organization. PDM management responsibility would be to equitably support projects across R&D, both vertically among therapeutic areas and horizontally along the R&D pipeline. The ability to plan across a broad spectrum of activities enables PDM management to apply resources fairly and to muster resources to focus on particular areas when necessary. The large critical mass generates a high degree of flexibility and permits ready mobilization of PDM expertise to meet ever-changing demands, unencumbered by administrative barriers. Economy of management of capital and operating resources is possible because technology transfer is minimal and in

Advantages and Disadvantages of the Vertical Alternative for PDM Administration in R&D   **TABLE 22.3**

Advantages	Disadvantages
PDM resources administratively committed to particular phases of the discovery and development pipeline	Fragmentation of PDM resources
Flexibility of resource allocation among therapeutic areas within a particular phase of the pipeline	Loss of critical mass
PDM scientists are able to focus on a particular segment of discovery or development	Loss of flexibility for PDM support along the R&D pipeline
No dilution of PDM activity as compounds move along the R&D pipeline	Lack of commitment to therapeutic areas
PDM scientists can specialize in particular aspects of R&D	Potential waste of resources
	Difficult and possibly inefficient information and technology transfer between separate PDM groups as compounds move through the pipeline
	Lack of accountability
	Difficult communication
	Lack of professional satisfaction
	Problems of integrating data and concepts in final regulatory documents

many cases unnecessary, with no duplication of effort. As a single department is responsible for all PDM activities, accountability and ownership are mandatory. There is nowhere to pass the buck!

These advantages naturally require a high degree of flexible and responsive management within PDM. This quality is a must, and no organization, no matter how simple or complex, can function efficiently without it.

Through a combination of line and matrix interactions, which are common and necessary in this type of multifaceted organization, rapid sharing or transfer of information technology is possible, unencumbered by obstacles that might occur with the other alternatives. These factors, together with continued ownership of projects within the department, represent considerable advantages.

The disadvantages of this approach are more difficult to identify. However, if the main advantages of Alternatives One and Two are those of commitment and close familiarity with a particular therapeutic area, or discovery or development phase, then lack of such commitment is a potential disadvantage with the monolithic PDM structure This may not necessarily be the case, however, because in the latter structure the extent of commitment should be maintained at an appropriate level and exercised through the different sections within PDM. The only other possible disadvantage would be the need for PDM to efficiently manage its own resources in order to provide effective support for a large spectrum of activities. In order to meet this need, PDM management would need to exercise a total and unbiased commitment to the whole R&D pipeline.

The professional advantages to PDM scientists working in a monolithic administrative environment are manyfold. If all PDM activities, across disciplines, across therapeutic areas, and throughout the discovery and development phases, are handled segmentally, under a single administration, there is considerable opportunity for scientists to be exposed to the whole spectrum of PDM activities. Active participation by any one scientist in all areas may not occur. However, a high degree of involvement within PDM disciplines to support particular compounds, and a broad spectrum of exposure with this alternative, nonetheless provides a rich and professionally stimulating environment. Together with this comes the concept of pride of ownership and responsibility. The buck stops here! The advantages and disadvantage of this third alternative are given in Table 22.4.

Conclusions

Three alternative approaches for the structure of overall PDM support for drug discovery and development have been debated here in an attempt to tease out the most important factors distinguishing each, and to compare their advantages and disadvantages.

The arguments do not apply to all R&D situations, nor to some particular types of expertise. It would be illogical to fragment such technologies as autoradiography, immunoanalysis, metabolism, mass spectrometry, and other highly specialized activities. These activities should be and generally are centralized within a PDM organization. Similarly, the debate also does not necessarily apply to small or highly specialized R&D organizations in which the PDM contribution may be minimal, or may not begin to contribute until potential new chemical entities enter development, possibly at a different organization.

The arguments are thus limited to large organizations—the major pharmaceutical houses—which have comprehensive discovery and development programs. Within these types of organizations, the three alternative PDM structures described here are currently being used. There are other alternatives, but the three considered here represent the great majority of cases. Of these, Alternative Two may be the most common, often expressed as separation of discovery, nonclinical support, and clinical support, followed in decreasing frequency by Alternatives One and Three.

It is difficult to compare the alternatives objectively because every organization has its own internal requirements, influences, and interests. An alternative that works well for one organization may not work for another. As with all intraorganizational situations, most schemes will work given good management, good people, and, of course, a productive and successful environment. Also, any structural approach must ultimately work to the benefit of the entire organization. An organization that is good for a PDM scientist may or may not be satisfactory for interacting colleagues in other departments.

In the current pharmaceutical discovery and development environment of continual challenge and change, together with resource limitations with little foreseeable relief in sight, organizational structures are bound to involve compromises and will be subject to stress and challenges. The advantages and disadvantages of the three alternatives must therefore be considered in the light of these factors.

Given the arguments for the three alternative PDM structures, one can draw certain conclusions. Alternatives One and Two may be more satisfactory from a

Advantages and Disadvantages of the Monolithic Approach for PDM Administration in R&D

TABLE 22.4

Advantages	Disadvantages
PDM activities coordinated under a single administration	Possible lack of commitment to particular therapeutic or developmental areas
High critical mass for resource allocation	
Flexibility	
Horizontal and vertical support managed by internal matrix structure	
Economy of resources, both operating and capital	
Little or no duplication of effort	
Ownership	
Accountability	
Efficient information and technology transfer	
No elitism	
High degree of professional satisfaction among PDM scientists	

purely "local" perspective, that is, from the perspective of a manager of a therapeutic area or of a segment of an R&D pipeline. These managers will have control of appropriate PDM resources to utilize them to their best advantage. PDM scientists employed within these areas will also become expert in, but not necessarily limited to, these particular areas. On the other hand, these alternatives suffer from the potential problem of barriers to transfer of knowledge and information, compound familiarity, and lack of critical mass for PDM scientists necessary to maintain essential flexibility. Overall PDM support program will be under the direction of several managers, and this may give rise to problems. Certainly this is unlikely to be optimal for PDM scientists, even if a high degree of specialization is sought by a PDM scientist. Fragmented PDM organizations will not be conducive to good collaboration or to good information and technology exchange between different groups. The overall potential for top-quality PDM support for R&D may not be realized.

Alternative Three, on the other hand, will undoubtedly deprive a therapeutic area or pipeline segment manager of complete control over PDM support for the area of his or her particular interest. However, from a more global perspective, PDM administrative control by therapeutic area or development segment would be replaced by healthy competition based on priorities as agreed by R&D upper management, and thus an equitable distribution of PDM resources across therapeutic areas and across the R&D pipeline would be obtained. Of course, if there were unlimited resources there would be no issue. But this is rarely the case. PDM departments across the industry are invariably understaffed, and there is currently much debate regarding shortcomings in PDM resources to meet regulatory demands, and what spectrum of PDM support is necessary to adequately support marketing applications. Resolution of these questions will take time. Meanwhile, PDM departments labor for the most part with inadequate resources.

In addressing the problems of the ideal PDM structure to support R&D, this writer has come from the perspectives both of a PDM scientist and an administrator. Thus, a certain bias is inevitable. However, an attempt has been made to fairly represent and analyze the alternatives that are most commonly employed in the pharmaceutical industrial environment, and to critically assess the impact that the alternatives may have on the efficiency with which PDM may support drug discovery and development, on the professional satisfaction of PDM scientists, and on PDM as a professional discipline now central to the drug R&D process.

Although the monolithic alternative has clear advantages over the others, each of the alternatives can be made to work provided the program has good management and a collegiate atmosphere can be established through close collaboration. It's all a matter of teamwork.

Summary

Several PDM administrative structures are employed by industry to provide PDM support to drug discovery and development.

The three major structures are based on therapeutic areas (horizontal administration), different segments of the R&D pipeline (vertical administration), and a single PDM unit (monolithic or holistic administration).

Each alternative has advantages and disadvantages in terms of flexibility, critical technical mass, collaboration, and project commitment.

On balance, the advantages of the monolithic or holistic approach outweigh those of the other alternatives in terms of flexibility and high-quality project support, collaboration, and professional satisfaction.

Reference

1. Welling, P. G. In *Pharmacokinetics, Regulatory, Industrial, Academic Perspectives*, 2nd ed.; Welling, P. G.; Tse, F. L. S., Eds.; Marcel Dekker: New York, 1995; pp 261–280.

Appendixes

APPENDIX I

Computer Methods and Software for Pharmacokinetic Data Analysis

he approach to pharmacokinetic instruction in this book is deliberately based on graphical data analysis. This is without doubt the best way to learn pharmacokinetic principles, and to understand the meaning of, and relationships among, pharmacokinetic parameters.

However, the complex nature of pharmacokinetic analysis, whether it be based on a simple one-compartment model or a multicompartment model with Michaelis–Menten kinetics, requires the use of computer programs to generate values of pharmacokinetic parameters, establish the correctness of the kinetic model, and determine the goodness of fit of the data provided by the model.

Computers used for pharmacokinetic analysis have evolved from mainframe systems to the current wide variety of user-friendly software packages for personal computers (PCs). A wide variety of PC software programs is available, and it may be difficult for the potential user to select the best package for a particular task. A review on currently available PC software for pharmacokinetic analysis was recently published by Gex-Fabry and Balant (1). This review summarizes key features of computer-based pharmacokinetic analysis, and also provides an excellent description of more than 40 software packages currently available from sources in the United States, Europe, Japan, and New Zealand. In the review, key features in computer-aided data interpretation are summarized under the headings of data entry, model specification, error model definition, parameter estimation, judging quality of parameter estimates, measuring goodness of fit, and printing and plotting results.

For data entry, such programs as SIPHAR, which includes a spreadsheet-like system, and the programs MINSQ and MKMODEL are highlighted. For pharmacokinetic model specification, several programs are identified, including PH/EDSIM, SAAM, MCAB, ScoP, and TOPFIT. The program TOPFIT includes 16 different models with Michaelis–Menten kinetics, and it is also recommended for use in pharmacodynamic model analysis. The PC software package PCNONLIN benefits from over 20 years of mainframe and microcomputer environments and includes an extensive library of about 20 predefined 1–3 compartment models.

For population kinetics a variety of programs is available, in particular NON-MEM and USC*PACK and the Japanese equivalents MLG and MLT12. Most of the more recently introduced programs incorporate a numerical integration component that is essential for pharmacokinetic models such as those containing Michaelis–Menten kinetics, the differential equations for which cannot be solved analytically.

Elaboration of the advantages of particular programs for specific data analysis is beyond the scope of this book. Suffice it to say that the wide variety of programs available differ in their method of data entry, number of predefined models, ability to develop more complex models, sophistication of algorithms, and quality of display and printout. These factors should be taken into account, together with the need and capabilities of the user, in selecting an appropriate program for data analysis.

The systematic description of available software in the cited review, which includes a short description of each software package together with source contact information, provides an excellent and current reference for software selection.

APPENDIX II
Worked Answers to Problem Sets

$$\text{Hepatic clearance} = \frac{\text{Dose}}{\text{AUC}^{0 \to \infty}}$$

$$\frac{100 \times 1000 \ \mu g}{200 \ \mu g \cdot \text{min/mL}} = 500 \ \text{mL/min}$$

$$E_H = \frac{500 \ \text{mL/min}}{1200 \ \text{mL/min}} = 0.42$$

Maximum availability $= 1 - 0.42 = 58\%$

Problem 5.2

$$\frac{40\%}{58\%} \simeq 69\%$$

Problem 8.1

$C_0 = 16.6 \ \mu g/mL = \text{Dose}/V_d$

Dose $= 250 \ \text{mg} = 250 \times 1000 \ \mu g$

$$V_d = \frac{250 \times 1000 \ \mu g}{16.6 \ \mu g/mL} = 15{,}060 \ \text{mL} \simeq 15 \ \text{L}$$

Problem 8.2   80% of 250 mg = 200 mg in body

$0.7 \times 16.6 \, \mu g/mL = 11.62 \, \mu g/mL$

$11.62 \, \mu g/mL \times 3000 \, mL = 34.86 \, mg$

$200 - 34.86 = 165.14 \, mg$ free drug

Free concentration $= 16.6 - 11.62 = 4.98 \, \mu g/mL$

$$V_d = \frac{165.14 \times 1000 \, \mu g}{4.98 \, \mu g/mL \times 1000} = 33.16 \, L$$

Problem 8.3

$$C_F = \frac{A_T}{V_F + 3\gamma + (V_F - 3)\varepsilon} = \frac{100 \, mg}{42 + 3\left(\dfrac{90}{10}\right) + (39)0} = 1.45 \, \mu g/mL$$

$C_B = C_F \cdot \gamma = 1.45 \times 9 = 13.05 \, \mu g/mL$

$C_T = C_B + C_F = 13.05 + 1.45 = 14.5 \, \mu g/mL$

$A_F = C_F \cdot V_F = 1.45 \, \mu g/mL \times 42,000 \, mL = 60.9 \, mg$

$$V_{app} = \frac{A_T}{C_T} = \frac{100 \times 1000 \, \mu g}{14.5 \, \mu g/mL} = 6896.6 \, mL$$

Problem 8.4 $\gamma = 90\%$ in plasma

$\varepsilon = 80\%$ in tissue

$$C_F = \frac{A_T}{V_F + 3\gamma + (V_F - 3)\varepsilon} = \frac{100 \, mg}{42 + 3\left(\dfrac{90}{10}\right) + 39\left(\dfrac{80}{20}\right)} = 0.44 \, \mu g/mL$$

$C_B = 0.44 \times \gamma = 3.96 \, \mu g/mL$

$C_T = 4.4 \, \mu g/mL$

$A_F = C_F \cdot V_F = 0.44 \times 42,000 = 18,480 \, \mu g = 18.5 \, mg$

$$V_{app} = \frac{A_T}{C_T} = \frac{100 \, mg \times 1000}{4.4 \, \mu g/mL} = 22,727 \, mL = 22.7 \, L$$

Problem 8.5

$$C_F = \frac{100}{42 + 3\left(\dfrac{60}{40}\right) + 39(4)} = 0.49 \, \mu g/mL$$

$C_B = 0.49 \times 1.5 = 0.74$

$C_T = 0.49 + 0.74 = 1.23 \, \mu g/mL$

$A_F = 0.49 \times 42{,}000 = 20{,}580 \, \mu g = 20.6 \, mg$

$$V_{app} = \frac{A_T}{C_T} = \frac{100 \times 1000 \, \mu g}{1.23 \, \mu g/mL} = 81.3 \, L$$

Problem 11.1

$$\text{Renal clearance} = \frac{\text{Amount cleared}}{\text{Plasma concentration}} = \frac{3 \, mg/min}{10 \, \mu g/mL} = \frac{3000 \, \mu g/min}{10 \, \mu g/mL}$$

$$= 300 \, mL/min$$

Problem 11.2

Plasma clearance $= 2 \times$ renal clearance $= 600 \, mL/min$

Problem 11.3

$Cl_p = V_d \cdot k_{el} = 600 \, mL/min$

$$k_{el} = \frac{600 \, mL/min}{42{,}000 \, mL} = 0.014 \, min^{-1} = 0.86 \, h^{-1}$$

$$t_{1/2} = \frac{0.693}{0.86} = 0.81 \, h$$

Problem 11.4

No. If a drug is cleared by methods that include active tubular secretion, clearance is not corrected for protein binding.

Problem 12.1

$$Cl_{cr} = \frac{(140 - 40)(50)}{(72)(1.6)} = \frac{5000}{115.2} = 43 \, mL/min$$

Problem 12.2

$$Cl_{cr} = 0.85 \frac{(140 - 50)(45)}{(72)(2.4)} = \frac{3442.5}{172.8} = 20 \, mL/min$$

Problem 12.3

From equation 12.6

$k = 0 + X \cdot Cl_{cr}$

$0.35 = X(130), \therefore X = 2.7 \times 10^{-3}$

(i) $k = 0 + 2.7 \times 10^{-3} \, (40) = 0.108 \, h^{-1}, t_{1/2} = 6.4 \, h$

(ii) $k = 0 + 2.7 \times 10^{-3} \, (10) = 0.027, t_{1/2} = 26 \, h$

Problem 12.4

Altered dose every 8 h:

$\qquad$ Uremic dose = 50(0.027/0.35) = 4 mg every 8 h

Altered dosage interval:

$\qquad$ Uremic dosage interval = 8(26/2) = 104 h

Regimen = 50 mg every 104 h

Problem 13.1

$C_A = 10\ \mu g/mL$

$C_B = 20\ \mu g/mL$

$$V_d = \frac{D}{C} = \frac{D}{10}\ \text{or}\ \frac{D}{20}$$

$t_{1/2}A = 2t_{1/2}B$

$k_{el}A = 1/2\ k_{el}B$

$$Cl_pA = \frac{D}{10} \times 1/2\ k_{el}B$$

$$Cl_pB = \frac{D}{20} \times k_{el}B$$

$$\therefore Cl_pA = Cl_pB$$

Problem 13.2

$C_0 = 25.0\ \mu g/mL$

$C_{t=2} = 12.5\ \mu g/mL$

$t_{1/2} = 2\ h$

$$Cl_p = V_d \cdot k_{el} = \frac{250 \times 1000\ \mu g}{25\ \mu g/mL} \times \frac{0.693}{2} = 3{,}465\ mL \cdot h^{-1} = 57.8\ mL/min$$

$$\approx 58\ mL/min$$

Problem 13.3

25% cleared by kidneys = 14.5 mL/min

$$\frac{14.5}{58} = 0.25$$

Problem 13.4

$t_{1/2} = 4\ h$

$k_{el} = 0.173\ h^{-1}$

$C_t = C_0 e^{-k_{el}t}$

$4.0 = C_0 e^{-0.173t} = C_0 \cdot 0.84$

Therefore, $C_0 = 4.8 \; \mu g/mL = \dfrac{D}{V_d}$

$V_d = \dfrac{100 \times 1000 \; \mu g}{4.8 \; \mu g/mL} = 20.8 \; L$

Problem 13.5

$D = 200 \; mg$

$C_{t=10} = 3.28 \; \mu g/mL$

$C_{t=16.93} = 1.64 \; \mu g/mL$

(i)   Therefore, $t_{1/2} = 6.93 \; h$

(ii) $k_{el} = 0.1 \; h^{-1}$

(iii) $3.28 = C_0 e^{-0.1 \times 10}$, $C_0 = 8.92$

$V_d = \dfrac{200 \times 1000 \; \mu g}{8.92 \; \mu g/mL} = 22.4 \; L$

(iv) $Cl_p = V_d \cdot k_{el} = \dfrac{22.4 \times 1000 \times 0.1}{60 \; min} = 37.3 \; mL/min$

Problem 13.6

$AUC^{0 \to \infty}$ by trapezoid $= 175.33 \; \mu g \cdot h/mL$

$k_{el} = 0.5 \; h^{-1}$

$AUC^{0 \to \infty} = 175.33 + \dfrac{13.5}{0.5} = 202.3 \; \mu g \cdot h/mL$

Problem 13.7

From the data in the table:

$k_{el} = 0.14 \; h^{-1}$

$t_{1/2} = 4.95 \; h$

$V_d = 12 \; L$

$Cl_p = \dfrac{12{,}000 \; mL \times 0.14}{60 \; min} = 28 \; mL/min$

$Cl_r = Cl_p = 28 \; mL/min$

$C_0 = \dfrac{D}{V_d} = \dfrac{250 \times 1000}{12 \times 1000} = 20.83 \; \mu g/mL$

At $t = 1$ h,

$$C_t = C_0 e^{-k_{el}t} = 20.83 e^{-0.14 \times 1} = 18.11 \ \mu g/mL$$

$$\mathrm{AUC}^{0 \rightarrow \infty} = \frac{D}{V_d \cdot k_{el}} = \frac{250 \times 1000}{28 \ \mathrm{mL/min}} = 8928.6 \ \mu g \cdot \mathrm{min/mL} = 148.8 \ \mu g \cdot \mathrm{h/mL}$$

Problem 13.8

$k_0 = 0.5$ mg/min

$t_{1/2} = 4.5$ h

$k_{el} = 0.154 \ \mathrm{h}^{-1} = 2.57 \times 10^{-3} \ \mathrm{min}^{-1}$

(i) $C_{ss} = \dfrac{0.5 \times 1000 \ \mu g/min}{5000 \times 2.57 \times 10^{-3} \ \mathrm{mL/min}} = 38.9 \ \mu g/mL \simeq 39 \ \mu g/mL$

(ii) $C_t = C_{ss} \left(1 - e^{-k_{el}t}\right)$

$0.95 = 1\left(1 - e^{-0.154t}\right) = 1 - e^{-0.154t}$

$-0.154t = \ln(0.05)$

$+0.154t = +2.996$

$t = 19.45$ h

$0.99 = \left(1 - e^{-0.154t}\right)$

then, as above, $t = 29.9$ h

Problem 13.9

$t_{1/2} = 1.5$ h, $k_{el} = 0.462 \ \mathrm{h}^{-1} = 7.7 \times 10^{-3} \ \mathrm{min}^{-1}$

(i) $C_{ss} = \dfrac{0.5 \times 1000 \ \mu g/min}{5000 \times 7.7 \times 10^{-3} \ \mathrm{mL/min}} = 12.99 \mu g/mL$

(ii) $0.95 = 1 - e^{-0.462t}$

$-0.462t = \ln(0.05), \ t = 6.48$ h

$0.99 = 1 - e^{-0.462t}$

$-0.462t = \ln(0.01)$

$+0.462t = +4.605, \ t = 9.96 \ \mathrm{h} \simeq 10$ h

Problem 13.10

In problem 13.8, $C_{ss} = 39 \ \mu g/mL$

$$C_{ss} = \frac{k_0}{V \cdot k_{el}} = 39 = \frac{k_0}{5000 \times 7.7 \times 10^{-3} \ \mathrm{min}}$$

$k_0 = 39 \ \mu g/mL \times 5000 \ mL \times 7.7 \times 10^{-3} \ \mathrm{min}^{-1} = 1501.5 \ \mu g/min = 1.5$ mg/min

$C_t = C_{ss}e^{-k_{el}t}$

At $t = 1$ h: $C_1 = 39e^{-0.462 \times 1} = 24.57$ µg/mL

At $t = 6$ h: $C_6 = 39e^{-0.462 \times 6} = 2.44$ µg/mL

At $t = 12$ h: $C_{12} = 39e^{-0.462 \times 12} = 0.15$ µg/mL

Problem 13.11

$V_d = 5000$ mL

$C_{ss} = 39$ µg/mL

$$\text{Initial bolus dose} = \frac{39 \times 5000}{1000} = 195 \text{ mg}$$

Problem 13.12

$D = 500$ mg, $V_d = 12$ L, $t_{1/2}(k_a) = 0.25$ h

$k_a = 2.77$ h^{-1}, $t_{1/2} = 12$ h, $k_{el} = 0.058$ h^{-1}

(i) $\text{AUC} = \dfrac{FD}{V \cdot k_{el}} = \dfrac{0.8 \times 500}{12 \times 0.058} = 574.7$ µg·h/mL

(ii) $C_{max} = \dfrac{FD}{V} \left(\dfrac{k_a}{k_{el}} \right)^{\frac{k_{el}}{k_{el} - k_a}} = \dfrac{0.8 \times 500}{12} \left(\dfrac{2.77}{0.058} \right)^{\frac{0.058}{0.058 - 2.77}}$

$\qquad = 33.3(47.76)^{-0.021} = 33.3 \times 0.92 = 30.6$ µg/mL

$t_{max} = \dfrac{1}{k_a - k_{el}} \ln\left(\dfrac{k_a}{k_{el}} \right) = \dfrac{1}{2.77 - 0.058} \ln\left(\dfrac{2.77}{0.058} \right)$

$\qquad = (0.369)(3.87) = 1.43$ h

Problem 14.1

$t_{1/2} = 18$ h, $k_{el} = 0.039$ h^{-1}

(i) $\text{AUC} = \dfrac{0.8 \times 500}{12 \times 0.039} = 854.7$ µg·h mL

(ii) $C_{max} = (33.3)\left(\dfrac{2.77}{0.039} \right)^{\frac{0.039}{0.039 - 2.77}} = 31.3$ µg/mL

(iii) $t_{max} = (0.37)(4.26) = 1.58$ h

Problem 14.2

$\text{Ratio } \dfrac{A}{B} = \dfrac{210}{130} \times \dfrac{150}{250} = 0.97 = 97\%$

Problem 14.3

Problem 14.4 $D = 400$ mg, $F = 50\%$

$k_a = 2.0$ h^{-1}

$t_{1/2} = 3.46$ h

$k_{el} = 0.2$ h^{-1} = (a)

(b) $\dfrac{A_u^\infty k_a}{k_a - k_{el}}$ = Intercept

$A_u^\infty = \dfrac{400 \times 0.5}{2} = 100$ mg

Intercept $= \dfrac{100 \times 2}{2.0 - 0.2} = 111.1$ mg

Problem 14.5   (i) $k_{el} = 0.2$ h^{-1}

(ii) $t_{1/2} = 3.5$ h

(iii) $k_a = 1.5$ h^{-1}

(iv) $t_{1/2} = 0.46$ h

(v) AUC$^{0\to\infty}$ $= 105.8$ μg·h/mL $= \dfrac{0.8 \times 500}{V \times 0.2}$

$V = 18.9$ L

(vi) $V_d \cdot k_{el} = Cl_p = \dfrac{18,900 \times 0.2}{60} = 63$ mL/min

Problem 14.6 400 mg $\times 0.75 = 300$ mg $= A_u^\infty$

$$\ln(A_u^\infty - A_u) = \ln\left(A_u^\infty \cdot \frac{k_a}{k_a - k_{el}}\right) - k_{el}t$$

t	$\ln(A_u^\infty - A_u)$
0	5.85
0.25	5.80
0.5	5.75
1.0	5.65
2.0	5.45
4.0	5.05
8.0	4.25
12.0	3.45
24.0	1.05

$k_{el} = 0.2 \text{ h}^{-1}$

$$\frac{k_e}{k_{el}} = \frac{A_u^\infty}{D}$$

$$k_e = \frac{k_{el} \cdot A_u^\infty}{FD} = \frac{0.2 \times 300}{0.8 \times 500} = 0.15 \text{ h}^{-1}$$

$k_m = k_{el} - k_e = 0.2 - 0.15 = 0.05 \text{ h}^{-1}$

Problem 14.7

$$Cl_r = V_d \cdot k_e = \frac{18,900 \times 0.15}{60} = 47.3 \text{ mL/min}$$

Problem 15.1

$D = 200 \text{ mg}, \tau = 6 \text{ h}, V_d = 42 \text{ L}, t_{1/2} = 6 \text{ h}$

(i) $\quad C_{(max)} = \dfrac{D}{V_d} = \dfrac{200}{42} = 4.76 \text{ µg/mL}$

(ii) $\quad C_{\infty(max)} = \dfrac{D}{V(1 - e^{-k_{el}\tau})} = \dfrac{4.76}{1 - e^{-(0.693/6)6}} = 9.52 \text{ µg/mL}$

(iii) $\quad C_{\infty(min)} = \dfrac{De^{-k_{el}\tau}}{V(1 - e^{-k_{el}\tau})} = 9.52e^{-0.693} = 4.76 \text{ µg/mL}$

(iv) $\quad \overline{C}_\infty = \dfrac{D}{V \cdot k_{el} \cdot \tau} = \dfrac{4.76}{0.693} = 6.87 \text{ µg/mL}$

Problem 15.2

(i) $\quad C_{\infty(max)} = \dfrac{300}{42(1 - e^{-0.116 \times 4})} = \dfrac{7.14}{1 - 0.63} = 19.24 \text{ µg/mL}$

(ii) $\quad C_{\infty(min)} = 19.24e^{-0.116 \times 4} = 19.24 \times 0.63 = 12.1 \text{ µg/mL}$

(iii) $\quad \overline{C}_\infty = \dfrac{D}{V \cdot k_{el} \cdot \tau} = \dfrac{7.14}{0.116 \times 4} = 15.4 \text{ µg/mL}$

Problem 15.3

(i) $\quad C_{\infty(max)} = \dfrac{50}{5(1 - e^{-(0.693/8)8})} = 20.0 \text{ µg/mL}$

(ii) $\quad \varepsilon = \tau/t_{1/2} = 1$

$$\frac{C_{n(max)}}{C_{\infty(max)}} = 1 - 2^{-n\varepsilon}$$

$$0.95 = 1 - 2^{-n\varepsilon}$$

$$0.05 = 2^{-n\varepsilon}$$

$$\ln 0.05 = -n(\ln 2)$$

$$-2.996 = -n(0.693)$$

$$n = 4.32 \text{ doses}$$

(iii) $4.5 \times 8 = 36.0$ h

Problem 15.4

(i) $C_{\infty(max)} = \dfrac{D}{V(1 - e^{-0.035 \times 8})} = \dfrac{50}{5(0.24)} = 41.7$ µg/mL

(ii) $n\varepsilon = 4.32$

$\varepsilon = \tau/t_{1/2} = 8/20 = 0.4$

$n = \dfrac{4.32}{0.4} = 10.8$ doses

(iii) $4.5 \times 20 = 90$ h   or $10.8 \times 8 = 86.4$ h

Problem 15.5

$$\overline{C}_{\infty} = 14.43 = \frac{D}{5 \times 0.035 \times 8}$$

$$\therefore D = 20.2 \text{ mg}$$

Problem 15.6

$D = 50$ mg

$$\overline{C}_{\infty} = 14.43 \text{ µg/mL} = \frac{50}{5 \times 0.035 \times \tau}$$

$$\tau = 19.8 \approx 20 \text{ h}$$

Problem 15.7

Loading dose $= V_d \cdot C_{\infty(max)}$

$$C_{\infty(max)} = \frac{D}{V(1 - e^{-0.035 \times 20})} = \frac{50}{5(0.5)} = 20 \text{ µg/mL}$$

Loading dose $= 20$ µg/mL $\times 5000$ mL $= 100$ mg

$D = 250$ mg, $k_a = 0.8$ h^{-1}, $k_{el} = 0.08$ h^{-1} **Problem 15.8**

(i) $T_{max} = \dfrac{1}{k_a - k_{el}} \ln\left(\dfrac{k_a}{k_{el}}\right) = 1.39(2.3) = 3.2$ h

(ii) $T_{max(ss)} = 1.39 \ln\left(\dfrac{0.378}{0.080}\right) = 2.16$ h

$D = 10$ mg, $Cl_p = 50$ mL/min, $V_d = 5$ L **Problem 16.1**

$Cl_p = V_d \cdot k_{el} = 5000$ mL $\cdot k_{el} = 50$ mL/min

$k_{el} = 0.01$ min$^{-1} = 0.6$ h^{-1}

(i) $k_e = 0.75 \times k_{el} = 0.45$ h^{-1}

(ii) $k_m = 0.15$ h^{-1}

(iii) $k_{me} = 0.2$ h^{-1}

$D = 100$ mg, 10% not absorbed **Problem 16.2**

$F = 0.9$

$k_{el} = 0.3$ h^{-1}

Intercept $= \dfrac{A_u^\infty \cdot k_a}{k_a - k_{el}} = 75 = \dfrac{60 k_a}{k_a - 0.3}$

(i) $\dfrac{k_e}{k_{el}} = \dfrac{A_u^\infty}{FD}$

$k_e = \dfrac{0.3 \times 60}{0.9 \times 100} = 0.2$ h^{-1}

(ii) $k_m = k_{el} - k_e = 0.1$ h^{-1}

(iii) $k_a = 1.5$ h^{-1}

$t_{1/2} = 4$ h, $k_{el} = 0.17$ h^{-1} **Problem 16.3**

6 h **Problem 16.4**

6 h **Problem 16.5**

Problem 17.1 $D = 200$ mg, $\beta = 0.1$ h^{-1}

(i) $V_1 = \dfrac{D}{A + B} = \dfrac{D}{C_0} = \dfrac{200 \text{ mg}}{40 \times 10^{-3} \text{ mg/mL}} = 5{,}000$ mL $= 5$ L

$V_{d\beta} = V_1(k_{el}\beta) = 5\left(\dfrac{0.3}{0.1}\right) = 15$ L

(ii) $V_{d(extrap)} = 5\left(\dfrac{1.2 - 0.1}{0.4 - 0.1}\right) = 18.3$ L

(iii) $V_{dss} = V_1\left(1 + \dfrac{k_{12}}{k_{21}}\right) = 5\left(1 + \dfrac{0.6}{0.4}\right) = 12.5$ L

(iv) $Cl_p = V_1 \cdot k_{el} = 5000 \times \dfrac{0.3}{60 \text{ min}} = 25$ mL/min

(v) $Cl_r - V_1 \cdot k_e = 5000$ mL $\times 2.75 \times 10^{-3}$ min$^{-1} = 13.75$ mL/min

Problem 17.2 $\beta = k_{21} = 0.20$
$t_{1/2}(\beta) = 3.5$ h

Problem 17.3 $Cl_p = \dfrac{D}{\dfrac{A}{\alpha} + \dfrac{B}{\beta}} = \dfrac{250 \times 1000 \text{ μg}}{\dfrac{25 \text{ μg/mL}}{3.5 \text{ h}^{-1}} + \dfrac{7.5 \text{ μg/mL}}{0.6 \text{ h}^{-1}}} = 12{,}727$ mL/h $= 212$ mL/min

Problem 17.4 $t_{1/2}(\beta) = 1.16 \simeq 1.2$ h

Problem 17.5   Triphasic

Problem 17.6 $t_{1/2}(\beta) = 2.17$ h

Problem 17.7 $\text{AUC} = \dfrac{A}{\alpha} + \dfrac{B}{\beta} + \dfrac{C}{\gamma} = 49.17 + 406.25 - 60.97 = 394.5$ μg·h/mL

Problem 17.8 $Cl_p = \dfrac{0.8 \times 100 \times 1000 \text{ μg}}{394.5 \text{ μg·h/mL}} = 202.8$ mL/h $= 3.38$ mL/min

(i)   Renal clearance = hepatic clearance = 1.69 mL/min

Problem 17.9

Problem 18.1

(i)   Kidney extraction ratio = $E = \dfrac{C_i - C_0}{C_i} = \dfrac{10 - 5}{10} = 0.5$

(ii) Renal clearance = 1200 mL/min $\times$ 0.5 = 600 mL/min

Problem 18.2

$$\frac{10 - 0}{10} = 1$$

Problem 18.3

$Q = 450$ mL/min

$C_m = C_0 \cdot R_m = 4\ \mu g/mL \times 1.8 = 7.2\ \mu g/mL$

Problem 18.4

Concentration in tissue = $1.8 \times 0.2 \times 4\ \mu g/mL = 1.4\ \mu g/mL$

Problem 18.5

Hepatic extraction = $0.4 = \dfrac{10 - C_0}{10}$

(i)   Hepatic clearance = $E_H \times$ hepatic blood flow

    Blood flow = 1500 mL/min

    Plasma flow = 840 mL/min

(ii) $C_0 = 6.0\ \mu g/mL$

(iii) $Cl_H = 0.4 \times 840$ mL/min = 336 mL/min

    $840/1500 = 0.56$

    $6.0/0.56 = 10.7\ \mu g/mL$

Problem 19.1

$V_M = 5\ \mu g/mL \cdot h$

$K_M = 10\ \mu g/mL$

(i) $k' = \dfrac{5}{10} = 0.5\ h^{-1}$

(ii) $t_{1/2} = 1.39$ h

$C_0 = 8\ \mu g/mL$

Problem 19.2

Intercept $= 13.5\ \mu g/mL = C_0^\star$

$t_{1/2} = 1.5\ h,\ k' = 0.462\ h^{-1}$

$$K_m = \frac{C_0}{\ln\!\left(\dfrac{C_0^\star}{C_0}\right)} = \frac{8}{\ln\!\left(\dfrac{13.5}{8}\right)} = 15.3\ \mu g/mL$$

$$V_M = 2.3\ \frac{(0.462)(15.3)}{2.3} = 7.07\,\mu g/mL{\cdot}h$$

Problem 19.3

$6 \times 0.025 = 0.15$

$0.018 \times 4\ h = 0.072$

At 10 p.m., concentration $= 0.15 - 0.072 = 0.078$ mg%

At 11 p.m., concentration $= 0.078 - 0.018 = 0.06$ mg%

Problem 19.4

At 11 p.m., blood alcohol $= 0.1$ mg%

At 10 p.m., blood alcohol $= 0.118$ mg%

$X(0.025) - 4(0.018) = 0.118$

$X = 7.6$ oz

Problem 19.5

(i) C_{max} from tablet $= 15\ \mu g/mL$, C_{max} from suspension $= 25\ \mu g/mL$

$$AUC^\infty\!\left(\frac{\text{fast}}{\text{slow}}\right) = \frac{C_0^2/2V_M(\text{fast})}{C_0^2/2V_M(\text{slow})} = \frac{625}{225} = 2.8$$

(ii)   Desired AUC $= 225\ \mu g{\cdot}h/mL$

100-mg fast-release dose yields an AUC of 625 $\mu g{\cdot}h/mL$

In order to achieve an AUC of 225 $\mu g{\cdot}h/mL$, the fast-release dose would have to be reduced by:

$$\frac{100 \times 225}{625} = 36\ mg$$

APPENDIX III
Nomenclature

A	(1) Amount of drug in the body (numerical subscript indicates compartment); (2) intercept of slope of blood level curve at $t = 0$ in two-compartment model; (3) surface area of membrane
AAG	α_1-Acid glycoprotein
A_0	Amount of drug in the body at zero time
A_F	Amount of free drug in the body
$A_{n(max)}$	Maximum amount of drug in the body after n doses
$A_{n(min)}$	Minimum amount of drug in the body after n doses
$A_{n(t)}$	Amount of drug in the body at any time during a dosage interval after n doses
A_{ss}	Amount of drug in the body at steady state during zero-order input
A_T	Amount of total drug in the body
A_u	Cumulative urinary recovery of unchanged drug
A_u^∞	Total urinary recovery of unchanged drug
$A_{1(2)}$	Amount of drug in first or second compartment of the two-compartment model
AUC	Area under drug-concentration curve
$AUC^{0 \to t}$	Area under drug-concentration curve from zero to time t
$AUC^{0 \to \infty}$	Area under drug-concentration curve from zero to infinite time
$AUC^{0 \to \tau}$	Area under drug-concentration curve during a multiple dosage interval, at steady state
$AUMC^{0 \to \infty}$	Area under the moment curve from zero to infinite time
α	Fast composite rate constant for the two-compartment open model
B	Intercept of slope of blood level curve at $t = 0$ in two-compartment model
BW	Body weight
β	(1) Slow composite rate constant for the two-compartment open model; (2) fraction of drug bound in plasma

373

C	Drug concentration (subscript indicates fluid, tissue, or compartment, bound or unbound)
C'	Intercept of k_a slope of drug plasma concentration curve at $t = 0$ in two-compartment model
C_i	Drug concentration in afferent blood
C_o	Drug concentration in efferent blood
C_0	Initial drug concentration
C_{max}	Peak or maximum drug concentration
$C_{n(max)}$	Maximum drug concentration after n doses
$C_{n(min)}$	Minimum drug concentration after n doses
$C_{n(t)}$	Drug concentration at any time during a dosing interval after n doses
$C_{\infty(max)}$	Maximum drug concentration after repeated doses, at steady state
$C_{\infty(min)}$	Minimum drug concentration after repeated doses, at steady state
$\overline{C}_{\infty(t)}$	Drug concentration at any time during a dosing interval, at steady state
C_{∞}	Average drug concentration during a repeated dose interval, at steady state
C_{ss}	Drug concentration at steady state during zero-order drug input
C_t	Drug concentration at time t after a single dose
Cl_{cr}	Creatinine clearance
Cl_H	Hepatic clearance
Cl_{int}	Intrinsic clearance of unbound or protein-free drug
Cl_p	Plasma clearance
Cl_r	Renal clearance
Cl_T	Drug clearance by particular tissue T
Cl_{int}'	Intrinsic hepatic clearance of unbound drug
ΔC	Change in drug concentration during time interval t
CRFA	Cumulative relative fraction absorbed
CSA	Clinical Study Application (Europe)
CSX	Clinical Study Exemption (U.K.)
D	Dose (subscripts n and u indicate normal and uremic states, respectively)
E	Extraction ratio of any organ
E_h	Hepatic extraction ratio
ESR	Erythrocyte sedimentation rate
ε	(1) Ratio of bound to free drug in extravascular fluids; (2) the ratio $\tau{:}t_{1/2}$ during repeated drug doses
F	Fraction of administered dose that is available to the systemic circulation
f	Fraction of drug that is cleared unchanged in urine
f_B	Free fraction of drug in blood
F_g	Fraction of administered dose that survives initial passage through the GI tract
F_h	Fraction of administered dose that survives initial passage through the liver
f_M	Free fraction of drug in muscle

F_l — Fraction of administered dose that survives initial passage through the lung

F_T — Free drug fraction in tissue T

f' — Fraction of systemically available drug that is cleared as a metabolite

IND — Investigational New Drug (U.S.)

γ — Ratio of bound to free drug in plasma

J — Direction and rate of passive transport of a substance across a biological membrane

K — Drug–protein equilibrium dissociation constant

K_m — Michaelis constant

K_p — Equilibrium distribution ratio

k — First-order rate constant (numerical subscript denotes compartment: 1, first (central) compartment; 2, second (peripheral) compartment)

k_a — First-order rate constant for appearance of drug into the systemic circulation

k_{ab} — Intrinsic first-order rate constant for drug absorption

k_d — First-order rate constant for drug degradation at the absorption site

k_e — First-order rate constant for urinary excretion of unchanged drug

k_{el} — First-order rate constant for drug elimination by all routes

k_m — First-order rate constant for drug metabolism

k_{me} — First-order rate constant for urinary excretion of a metabolite

k_n — First-order elimination rate constant for a drug under condition of normal renal function

k_{nr} — First-order rate constant for drug elimination by nonrenal mechanisms

k_0 — Zero-order rate constant for drug absorption

k_u — First-order elimination rate constant for a drug under condition of uremia

k_{12} — First-order rate constant for transfer of drug from the first compartment to the second compartment in the two-compartment open model

k_{21} — First-order rate constant for transfer of drug from the second compartment to the first compartment in the two-compartment open model

λ — A first-order rate constant

M — Amount of metabolite in the body

MAA — Marketing Authorization Application (Europe)

M_u — Cumulative urinary recovery of metabolite

M_u^∞ — Total urinary recovery of metabolite

n — Number of binding sites per protein molecule

nP — Total concentration of receptors

NDA — New Drug Application (U.S.)

P — Permeability of the membrane to a substance

P — Protein concentration

P_f — Concentration of free receptors on protein

$P_f C$ — Concentration of occupied receptors

pK_a — Negative logarithm of the dissociation constant for acid

pK_w — Negative logarithm of the dissociation constant for water

Q	Blood flow rate (subscript denotes a particular organ)
R	(1) Residual obtained by drug level curve-stripping; (2) accumulation factor during repeated drug dose; (3) partition coefficient for drug disposition into organ (subscript) at equilibrium
r	Number of moles of drug bound to a mole of protein
R_T	Partition coefficient of total drug between tissue T and blood, at equilibrium
S_{cr}	Serum creatinine
T	(1) Duration of zero-order drug input; (2) quantity of drug in tissue compartment
t	Time elapsed since a single drug dose, or since the last dose
t_0	Lag time. The time interval between drug administration and the first appearance of measurable drug in the circulation
$t_{1/2}$	Drug elimination half-life
$t_{1/2(abs)}$	Half-life of drug absorption
$t_{1/2n}$	Drug elimination half-life under condition of normal renal function
$t_{1/2u}$	Drug elimination half-life under condition of uremia
T_{max}	Time of peak or maximum drug concentration
$T_{\infty(max)}$	Time of maximum drug concentration during repeated doses, at steady state
t'	Time elapsed since end of zero-order drug release
τ	Dosage interval (subscripts n and u indicate normal and uremic states, respectively)
U	Concentration of drug in urine
V	Volume (subscript indicates fluid, tissue, or compartment)
V'	Urine flow rate
V_{app}	Apparent drug distribution volume in the body
V_{dss}	Overall distribution volume of a drug that obeys two-compartment model kinetics
$V_{d(area)}$, $V_d\beta$	Overall distribution volume of a drug that obeys two-compartment model kinetics, calculated by the area method
$V_{d(extrap)}$	Overall distribution volume of a drug that obeys two-compartment model kinetics, calculated by the extrapolation method
V_m, V_{max}	Maximum rate of elimination in a saturable system
X	Amount of drug remaining to be absorbed

APPENDIX IV
Glossary

Achlorhydria   diminished secretion of hydrochloric acid into the stomach.

Active transport   membrane transport that uses cellular energy for drug movement, with or against a concentration gradient.

Amorphic forms   polymorphs that have no crystal structure.

Autoinduction   induction by a drug of its own metabolism.

Blood–brain barrier   defense structure that minimizes access of water-soluble substances to the brain.

Bile   fluid produced in the liver that enters the GI tract at the duodenum; constituents include bile salts, bilirubin, phospholipids, and salts.

Bioavailability   extent to which administered drug enters the systemic circulation.

Buccal administration   administration of drug into the buccal cavity, for absorption via the oral mucosa.

Catenary chain kinetics   sequence of kinetic steps in the form A→B→C.

Channels   pores within membranes that facilitate transport of small polar molecules—water channels, ion channels.

Chirality   property of a molecule possessing one or more asymmetric carbon atoms.

Celiac disease   inflammatory condition of the small intestine, associated with ingestion of gluten.

Cerebrospinal fluid   fluid volume of approximately 120 mL that surrounds and protects the central nervous system (CNS).

Cotransporters and exchangers   membrane cotransport mechanisms, driven by sodium or hydrogen ion gradients.

Crohn's disease   inflammatory condition of unknown etiology, affecting the small and large intestine.

Diffusion rate-limited   drug movement that is dependent on the rate at which drug molecules can diffuse across a membrane.

Enantiomer   one possible form of a compound that is asymmetric about a single carbon atom.

Endocytosis   receptor-mediated endocytosis occurring when a substrate binds to a specific membrane receptor before being internalized.

Enteral administration   drug administration via the GI tract.

Enterohepatic circulation   continuous recirculation of substances between the liver and intestine by means of the bile.

Enzyme induction   increased activity of drug-metabolizing enzyme(s).

Enzyme inhibition   decreased activity of drug-metabolizing enzyme(s).

Excipients   inert binders, diluents, disintegrants, etc., included in a drug formulation.

Extrinsic proteins   proteins attached to the cytoplasmic face of a cell membrane by electrostatic forces.

Facilitated diffusion   mechanism proposed to explain membrane transport of water-soluble substances, that is, the glucose transporter.

First-pass effect   clearance of a drug by GI or hepatic metabolism or biliary excretion during the absorption process, before drug reaches the systemic circulation.

Genetic polymorphism   genetically based polymorphism in expression of metabolism isozymes.

Glomerulus   series of capillary loops enclosed within Bowman's capsule in the kidney

Glycocalyx   sulfated mucoprotein lining the villi and microvilli of the intestine.

Hepatic clearance   volume of blood cleared of drug by the liver per unit time.

Hepatic extraction ratio   ratio of hepatic clearance to hepatic blood flow.

Horizontal administration   administration of PDM resources to support drug discovery and development by therapeutic area.

Inhalation   administration of drug into the lungs, for local or systemic activity.

Intact hepatocyte hypothesis   hypothesis stating that the diseased liver is considered to comprise variable ratios of normal hepatocytes and hepatocytes that are essentially nonfunctional due to a pathological condition.

Intact nephron hypothesis   hypothesis stating that the diseased kidney is considered to comprise variable ratios of normal nephrons and nephrons that are essentially nonfunctional due to a pathological condition.

Interdigestive migrating motility complex   pattern of GI contractions in the fasting state thought to be responsible for clearance of nondisintigrating dosage forms from the stomach.

Intraarterial administration   administration of drug directly into an artery.

Intranasal administration administration of drug into the nasal cavity, for local or systemic activity.

Intrathecal administration administration of drug directly into cerebrospinal fluid.

Intramuscular administration administration of drug into muscle tissue.

Intravenous administration administration of drug into the venous circulation.

Intrinsic proteins globular proteins held within the membrane biolayer by hydrophobic interactions, and at the membrane surface by electrostatic interactions.

In vitro–in vivo correlation correlation between in vitro dissolution rate and in vivo bioavailability.

Isoform a group of enzymes that catalyze the same reaction (isoform: same or equal forms); also, isoenzyme, isozyme.

Market image marketed formulation of a drug.

Membrane transport mechanisms by which substances cross membranes.

Microsomes liver fraction obtained from sequential centrifugation of liver homogenate, associated with drug metabolism.

Monolithic administration administration of PDM resources as a single unit to support drug discovery and development.

Nephron series of differentiated tubules in the kidney associated with excretion, electrolyte balance, and homeostasis.

Pinocytosis nonspecific process by which substrate is taken up by invagination of cell membrane.

Parenteral administration drug administration via any route other than the GI tract for systemic activity.

Passive-facilitated diffusion membrane crossing driven by a concentration gradient.

Passive transfer mechanism of membrane transport that depends on the concentration gradient of a substance across the membrane.

Peptide transporter substance(s), as yet unidentified, that promotes absorption of di-and tripeptides by a carrier-mediated process.

Perfusion rate-limited drug movement that is dependent on how fast a particular tissue is perfused by blood.

Phase I reactions first phase of drug metabolism in which molecules are chemically activated by oxidation, reduction, and/or hydrolysis.

Phase II reactions second phase of drug metabolism, involving conjugation.

Plasma clear supernatant recovered after red blood cells have been precipitated, obtained from a blood sample in the presence of an anticoagulant.

Plasma clearance volume of plasma cleared of drug per unit time as a result of all elimination pathways.

Plasma half-life   time taken for drug concentration in plasma to be reduced by one-half.

Polymorphs   compounds that can form crystals with different molecular arrangements.

Pumps   proteins capable of transporting ions across membranes against an electrochemical gradient.

Rectal administration   administration of drug into the rectum for local or systemic activity.

Renal clearance   volume of plasma cleared of drug per unit time as a result of urinary excretion.

Serum   clear supernatant recovered after red blood cells have been allowed to clot, obtained from a blood sample in the absence of an anticoagulant.

Sick cell hypothesis   hypothesis relating liver disease to changes in hepatic cell and other liver functions.

Solvates   crystals that incorporate one or more solvent molecules during preparation.

Solvation   incorporation of one or more solvent molecules into the crystalline structure of a substance.

Space of Disse   outer space of the hepatic sinusoid.

Splanchnic circulation   that portion of the systemic circulation that bathes the abdominal area.

Systemic availability   amount of administered drug that reaches the systemic circulation.

Systemically acting drugs   drugs that must enter the systemic circulation to exert their therapeutic effect.

Thekes   Greek word meaning sheath or box; derivation of intrathecal administration.

Transdermal administration   administration of drug by means of occluded devices through the skin for systemic activity.

Unstirred layer   series of fluid layers, each progressively more stirred, extending from the intestinal epithelial cell surface to the bulk phase of the intestinal lumen.

Vaginal administration   administration of drug via the vaginal route, for local or systemic activity.

Vertical administration   administration of PDM resources to support drug discovery and development by development phase.

Whole body autoradiography   technology to measure radioactivity distribution in a whole animal.

Index

A

Absorption
  clinical factors and interactions affecting
      drug absorption, 95–117
  differences between pharmacokinetics
      and toxicokinetics, 327–328
  drug–food interactions, 105–116
  effect on drug pharmacokinetics, 3–9
  influencing factors, 63–76
  pharmacokinetic characteristics, 87–88
  pharmacokinetics and toxicokinetics,
      325–326
Absorption lag time, dosage forms, 236–237
Absorption mechanisms, interactions of
      drugs, 102–105
Absorption models, transdermal delivery, 32
Absorption rate constant, equations, 255
Accumulation phase, drug concentrations,
      249*f*
Accumulation rate, drug, 250–252
Accumulation, first-order elimination, 311–
      315
Achlorhydria, effect of disease on drug
      absorption, 96
Acid secretion, effect on enteral
      administration, 48–49
Acidic drugs, ionization constants, 14*t*
Acquired immunodeficiency syndrome
  effect of food on drug absorption, 108–110
  types of liver disease, 164

Acronyms and abbreviations, list, 373–376
Active metabolites, types, 260*t*
Activity, delayed onset, controlled-release
      forms, 85
Acyclovir, bioavailability in dog, human,
      and monkey, 56*t*
Administration rate, advantage of
      intravenous dosage route, 22
β-Adrenergic antagonists, transdermal
      systems, 35–36
Adrenocorticotropic hormone, intranasal
      administration, 27–28
Aerosol, nasal deposition, 27
Age
  effect on drug metabolism, 156–159
  factor influencing first-pass effect, 71
Albumin, binding to plasma proteins, 124
Alcohol dehydrogenase, first-pass
      metabolism in stomach, 64
Alkaline fluid, bile, effect on enteral
      administration, 50
Alternative administrative structures,
      pharmacokinetics and drug
      metabolism, 345*t*
Alternatives, drug R & D, 344–345
Alveolar macrophages, removal of
      particulate material from respiratory
      tract, 30
Ambenonium chloride, effect of food on
      absorption, 107*f*
Aminorex, bioavailability, 59*f*

Aminosalicylic acid, drug delivery, 65

Amorphism, improvement of drug absorption and stability, 76–78

Angiotensin converting enzyme, carrier-mediated absorption, 104

Angiotensin-converting enzyme inhibitors, transport mechanisms, 52

Animal models
inhalation administration, 31
intranasal administration, 28
oral drug absorption, 52–60

Anti-inflammatory agents, nonsteroidal, 90–91

Antimicrobial agents, effect of liver impairment on drug metabolism, 170

Antiporters, membrane transport, 17

Arachnoid villi, routes to CNS, 137–139

Area under drug-concentration curve, influence of saturable kinetics, 318–321

Area under curve, drug-concentration, 319*f*–320

Area under drug-plasma curve, derivation, 278–279

Areas under drug blood-concentration curves, oral and intravenous doses, 68–69

Ascites, effects on pharmacokinetics, 165

Aspirin, enteric-coated tablets and capsules, 83*f*

Autoradiography studies, role of pharmacokinetics, 333–335

Availability, systemic, possible reduction, controlled-release forms, 85*f*

Azo reduction, Phase I metabolism reactions, 150–154

Azone, absorption enhancers, 92

B

Bacterial enzymes, metabolism of drugs, 64–65*f*

Basic drugs, ionization constants, 14*t*

Beer, first-order elimination kinetics, 312

Bile salts and acids, enterohepatic cycling, 69–70

Bile salts, absorption enhancers, 90–91

Bile
approximate pH values, 44*f*–45
effect on enteral administration, 49–50

Biliary excretion, schematic of enterohepatic circulation, 70*f*

Binding characteristics, intravascular and extravascular, 130*t*

Bioavailability assessment, first-order elimination, 311–315

Bioavailability, influencing factors, 63–76

Biodegradable microsphere, use in vaginal drug absorption, 37–38

Biotransformation step, drug to metabolite, 260

Blood
component of physiological model, 4
concentration, versus time profiles, 286*f*, 314*f*
flow rate
model, drug pathway to organ, 300–305
percent in some organs and tissues, 308*t*
rate-limited transport, 300–305
levels, two-compartment model kinetics, 274*f*
perfusion rates, tissue effects, 122*t*
general physiological pharmacokinetic model, 301*f*, 302*f*
model of drug transport, 305*f*

Blood–brain barrier, protection of central nervous system, 11

Bloodstream, sites of drug metabolism, 145–147

Body fluids
distribution of drug, 272*f*
volumes, 120*t*

Body tissues, distribution of drug, 272*f*

Body volume, percent in some organs and tissues, 308*t*

Bolus intravenous injection
one-compartment open model, 202–206
quantity of drug in body, equation, 246
two-compartment model, 272–273

Bolus IV injection, typical drug profile, 207*t*

Bolus-loading dose, administration, 217

Bone, distribution of drug, 272*f*

Bone-marrow concentrations, drug model simulations, 307*f*

Bound drug, percent in body, 131*f*, 132*f*, 133*f*, 134*f*, 135*f*

Bowman's capsule, structure and function, 175–178

Brain
distribution of drug, 272*f*
sites of drug metabolism, 145–147

Buccal administration, drug routes, 24–25

Buccal cavity, drug administration, 19–40

C

Calcium channel blocking agents, effect of liver impairment on drug metabolism, 170

Capillary endothelial cells, pathways of CSF flow, 136–139

Capillary membranes, drug pathway to organ, 300–305

Capillary permeability, drug distribution, 121–122

Capsules, dosage forms, 79–80f

Cardiovascular agents, effects of liver disease on pharmacokinetics, 169–170

Cardiovascular drugs, those available in controlled-release form, 84f

Carrier-mediated absorption, interactions of drugs with shared absorption mechanisms, 104

Celiac disease, effect on drug absorption, 97

Cell membrane, structure of, 11–12

Cell permeation enhancers, absorption enhancers, 92–93

Central nervous system
drugs, those available in controlled-release form, 84f
effect of liver impairment on drug metabolism, 170
infections, intrathecal administration, 20
intraventricular administration, 21
penetration of drug, 135–139
pharmacodynamic response in liver disease, 165

Cephalexin, plot of mean wall permeability, 53f

Cephalic phase, gastric acid secretion, 48

Cerebrospinal fluid, drug penetration, 135–139

Channels, membrane transport, 15

Chemical factors, improvement of drug stability and absorption, 75–77f

Chemical structure, bile excretion of drug, 70

Cholate, effect on rectal insulin absorption, 91f

Cholestyramine, prolonged elimination rate of digitoxin, 71

Cilited cells, removal of particulate material from respiratory tract, 30

Cirrhosis, types of liver disease, 164

Clearance
effect of liver disease, 166–169
renal clearance, 178–184

Clinical development, role of pharmacokinetics, 335–340

Clinical factors, effect on drug absorption, 95–117

Clinical study application, drug discovery and development process, 332f

Clonidine, transdermal systems, 35

Cloning, transport proteins, 14

Coated tablets, drug dosage forms, 82

Cocaine, intranasal administration, 26

Cockcroft and Gault, method of creatinine clearance, 189–190

Colon, drug delivery, 65

Compartment modeling, approach to pharmacokinetics, 4,5f

Computer methods, pharmacokinetic data analysis, 357–359

Concentration relationships, effect on distribution, 120–121

Conjugation reaction, Phase II mechanism, 143–149

Continuous ambulatory peritoneal dialysis, end-stage renal disease, 195

Controlled drug release
advantages, 83–85
disadvantages, 85–86

Controlled-release dosage forms
pharmacokinetics, 88–89
drug dosage forms, 82–93

Controlled-release products, bacterial metabolism of drugs, 64

Controlled-release, unsuited, drugs, 87–88

Cotransporters, membrane transport, 16–17

Coughing, removal of particulate material from respiratory tract, 30

Creatinine clearance, methods of measuring, 188–191

Crohn's disease, effect on drug absorption, 97

Crystalline compounds, improvement of drug absorption and stability, 76–78

Cumulative urinary excretion, unchanged drug, 237–241

Cytochromes P-450, drug metabolism, 152–154

D

Danazol, food effect on drug absorption, 113–114

Data analysis, pharmacokinetic instruction, 357–359

Definitions, list, 377–380

Delayed drug absorption, interactions that cause, 110–112

Delivery enhancers, transdermal drug delivery, 32–34

Desmopressin, intranasal administration, 27–28

Development, drug, role of pharmacokinetics, 331

Dialyzable, drugs, 195t

Didanosine, effect of food on drug absorption, 108–110

Diffusion
effects on drug distribution, 122–124
model of polar solutes across plasma membranes, 16*f*

Digitoxin
administered, half-life, 251
disposition, physiological model in rat, 6*f*
effects of liver disease on pharmacokinetics, 169–170
prolonged elimination rate of drug, 71

Direct action, drug–drug interaction affecting absorption, 101–104

Discovery, drug, role of pharmacokinetics, 331

Diseases, gastrointestinal tract, effect on drug absorption, 95–100

Disintegration, tablet-form drug dosage, 80*f*

Dissolution process, schematic, 77*f*

Dissolution, polymorphism effect, 77

Dissolution, tablet-form drug dosage, 80*f*

Distribution volume, influence on drug half-life, 183*t*

Distribution
effect of liver disease, 166–169
effect on drug pharmacokinetics, 3–9
improvement through physiological models, 298–310
pharmacokinetics and toxicokinetics, 325–326

Diuretic drugs, those available in controlled-release form, 84*f*

Dog, animal models for oral drug absorption, 57–59

Dosage adjustment, methods, 191–194

Dosage design and adjustment, first-order elimination, 311–315

Dosage forms, improvement of drug absorption and stability, 78–80*f*

Dose, systemic circulation, 224*f*

Dose dumping, controlled-release forms, 85

Dose size, advantage of intravenous dosage route, 22

Dose-response relationships, first-order elimination, 311–315

Doxycycline, mean serum concentrations 103*f*

Drug absorption
increased by food, 112*t*–114
direct effect of food, 106
effect of pH, 43–45
improvement through physiological models, 298–310
interactions causing reduction, 106–110

Drug absorption—*Continued*
physicochemical and formulation factors, 75–93
plots, 232–234
multiple dosing, 253*t*

Drug accumulation
rate, 250–252
repeated doses, 245–250, 252–253

Drug blood-concentration curve, area under, 68–69

Drug concentration
bloodstream, maximum, 230–231
first compartment, 288–292
intravenous bolus injection plot, 204*f*

Drug discovery, role of pharmacokinetics, 331

Drug distribution
into an elimination organ, calculation, 301–305
pharmacokinetics, 119–143

Drug elimination
influence of saturable kinetics, 318–321
prediction by creatinine clearance, 190–191
rate, organ clearance, calculation, 299–300
renal impairment, 187–197

Drug input
first-order, two-compartment model, 285–288
one-compartment open model, 201–221

Drug metabolism
pharmaceutical R&D, 343–353
pharmacokinetics, 145–161

Drug profile
fast and slow intravenous injection, 22*f*
multiple intravenous doses, 254*f*

Drug release characteristics, oral controlled-release dosage forms, 89*f*

Drug transport, biological membranes, 11–17

Drug-concentration curve
area, 234–236
drug from absorption site, 224–225

Drug-concentration profiles, interpretation to obtain parameter estimates, 273–276

Drug–drug interactions, effect on drug absorption, 100

Drug–food interactions, absorption, 105–116

Drugs
binding to tissues, 125–128
those available in controlled-release form, 84*f*
transdermally delivered, 34–37

Duodenal secretions, effect on enteral administration, 50
Duodenum, approximate pH values, 44f–45

E

Elimination
general physiological pharmacokinetic model, 301f, 302f
metabolite pharmacokinetics, 259–261
one-compartment open model, 201–221, 223–243
Elimination half-life
common drugs, 213t, 219t
effect of liver disease, 166–169
first-order rate constant, 205
Elimination kinetics, useful expressions, 313–315
Elimination rate, relationship with clearance, 182–184
Elimination rate constant
equations, 255, 261
various drugs, 192t
Enantiomers, effect on drug metabolism, 159
End-stage kidney disease, hemodialysis, 194–196
Endocytosis, membrane transport, 17
Enteral administration
dosage routes, 19–40
factors influencing absorption and bioavailability, 63–76
Enteral routes, drug administration, 43–60
Enterohepatic circulation, effect on absorption and bioavailability, 69–71
Enzyme
bile, effect on enteral administration, 50
gut wall metabolic, 64
metabolizing gut lumen, 64
Enzyme induction, effect on drug metabolism, 159–160
Enzyme inhibition, pharmacological effect, 160
Epithelium, difference between small intestine and stomach, 46
Equilibrium, volume of distribution, 279–282
Erythrocyte sedimentation rate, factors in celiac and Crohn's disease, 99–100
Erythromycin
effect of pH, 44–45
food interactions, 111–112
Esophagus, drug administration, 19–40
17β-Estradiol, sublingual administration, 25f

Estradiol
mean serum concentrations, 36f
transdermal systems, 35
Estrone, sublingual administration, 25f
Ethambutol, mean serum concentrations before and after antrectomy and gastroduodenostomy, 98f
Ethanol
first-order elimination kinetics, 312
first-pass metabolism in stomach, 64
Exchangers, membrane transport, 16–17
Excipients, drug dosage forms, 81–82
Excretion
effect on drug pharmacokinetics, 3–9
general physiological pharmacokinetic model, 301f, 302f
improvement through physiological models, 298–310
pharmacokinetics and toxicokinetics, 325–326
renal, differences between pharmacokinetics and toxicokinetics, 327–328
Expression cloning, transport proteins, 14
Extravascular drug binding, plasma protein, 128–135
Extrinsic proteins, cell membrane, 12
Eye, application of a drug solution, 225

F

Facilitated diffusion
membrane transport, 15
model of polar solutes across plasma membranes, 16f
Fast first-order rate, drug absorption, 255f
Fat
distribution of drug, 272f
general physiological pharmacokinetic model, 301f, 302f
Fatty acids, medium-chain, absorption enhancers, 91–92
Film autoradiography, distribution of labeled drugs, 140–141
First compartment, drug concentration, 283f, 288–292
First-order absorption
case, 253–256
one-compartment open model, 223–243
First-order drug input, two-compartment model, 285–288
First-order elimination
descriptions, 311–322
zero-order drug input, 213–219
First-order kinetics, multiple dose, 245–256

First-pass effect
  differences between pharmacokinetics
    and toxicokinetics, 327–328
  hepatic metabolism, 66–67*f*
  influence of saturable kinetics, 318–321
First-pass metabolism
  drug administration, 19–40
  increased potential, controlled-release
    forms, 85
  pharmacokinetic characteristics, 87–88
  quantitative aspects, 68–69
  rectal administration, 39
First-pass pulmonary metabolism, effect on
  inhalation administration, 30
First residuals, method of obtaining, 290*f*
Folds of kerkring, influence on surface area
  of small intestine, 47*f*
Follicle stimulating hormone, sublingual
  administration, 25*f*
Food, influence on the gastrointestinal tract,
  105–110
Formation, metabolite pharmacokinetics,
  259–261
Formulation factors
  effect on drug absorption, 75–93
  improvement of drug absorption and
    stability, 78–80*f*
Free drug, percent in body, 131*f*, 132*f*, 133*f*,
  134*f*, 135*f*
Frequency of drug dosing, general equation,
  251–252
Fuzzy-coat, effect on drug absorption, 46–47

G

Gastric acid secretion, phases, 48
Gastric juice, effect on enteral
  administration, 49–50
Gastric phase, gastric acid secretion, 48–49
Gastrointestinal anatomy, differences
  among species, 52–53
Gastrointestinal blood flow, relation to drug
  absorption, 50–51
Gastrointestinal disease, influence on drug
  absorption, 95–100
Gastrointestinal drug, those available in
  controlled-release form, 84*f*
Gastrointestinal epithelial lining mem-
  branes, absorption and secretion, 11
Gastrointestinal lumen, site of drug
  degradation, 63
Gastrointestinal microflora, metabolism
  foreign compounds, 65*f*
Gastrointestinal pH, effect on enteral
  administration, 43–45

Gastrointestinal physiology
  differences among species, 52–53
  effect on enteral administration, 45–48
Gastrointestinal secretion, effect on enteral
  administration, 48–49
Gastrointestinal tract
  effect on liver function, 164
  enteral routes of drug administration,
    43–60
  influence of food, 105–110
  physiology, 43–45
  residence time with oral controlled-
    release products, 86
  residence time, 218
  sites of drug metabolism, 145–147
Glial connective tissue, routes to CNS, 137–
  139
Glomerular filtration rate, renal clearance,
  178–184
Glomerulus, renal excretion, 177–178
Glossary, list, 377–380
Glycerides, medium chain, absorption
  enhancers, 91–92
Glycocalyx, effect on drug absorption, 46–47
P-Glycoprotein, membrane transport, 17
Graphical estimation of parameters,
  absorption and elimination, 226–232
Griseofulvin
  increased absorption from particle size
    reduction, 78
  peak blood levels after administration
    to humans and minipigs, 57*f*
Growth hormone releasing factor,
  intranasal administration, 27–28
Gut
  absorption via splanchnic circulation, 67*f*
  general physiological pharmacokinetic
    model, 301*f*, 302*f*
  metabolism, 64–65
Gut lumen, drug metabolism, 64
Gut wall
  drug metabolism, 64
  site of drug degradation, 63

H

Half-life
  general equation, 251–252
  pharmacokinetic characteristics, 87–88
  relationship with clearance, 182–184
Health care, cost containment, 85
Heart, general physiological
  pharmacokinetic model, 301*f*, 302*f*
Hemodialysis, end-stage kidney disease,
  194–196

Hepatic arteries, blood supply to liver, 66
Hepatic blood flow, hepatic extraction ratio, 68
Hepatic clearance
  hepatic extraction ratio, 68
  mathematical description, 181–182
Hepatic extraction ratio
  drugs with high, intermediate, and low values, 69f
  oral availability, 68
Hepatic function, changes, 164–165
Hepatic metabolism
  drug administration, 19–40
  first-pass effect, 66–67f
Hepatic vein, absorption via the splanchnic circulation, 67f
Hepatitis, types of liver disease, 164
Horizontal administration, drug R&D program, 347
Horizontal alternative for PDM, advantages and disadvantages, 347t
Hormones, those available in controlled-release form, 84f
Human immunodeficiency virus, effect of food on drug absorption, 108–110
Hydrogen ion gradients, cotransport drive, 16

I

In vitro–in vivo correlations, drug dosage forms, 80–81
Indirect action, drug–drug interactions affecting absorption, 100–104
Indomethacin, transdermal delivery, 36
Inferior vena cava, absorption via the splanchnic circulation, 67f
Inhalation, drug route, 28–31
Insulin bioavailability, medium-chain fatty acid and glycerides, 91–92
Insulin
  intranasal administration, 28
  use of, absorption enhancers, 90–91f
Intact hepatocyte theories, liver failure, 165
Intact nephron hypothesis, chronic renal failure, 187–188
Interactions, drug, differences between pharmacokinetics and toxicokinetics, 327–328
Interdisciplinary interactions, nonclinical development, 335
Interferon, intranasal administration, 27–28
Interstitial fluid, drug pathway to organ, 300–305

Intestinal diseases, apparent effects on drug absorption, 98t
Intestinal drug absorption enhancers, list, 90t
Intestinal infections, effect on drug absorption, 100
Intestinal phase, gastric acid secretion, 49
Intestinal surgery, effects on drug absorption, 98t
Intraarterial administration, drug routes, 20
Intracellular fluid, general physiological pharmacokinetic model, 301f, 302f
Intracellular fluid, model of drug transport, 305f
Intramuscular administration, drug routes, 23–24
Intranasal administration, drug routes, 25–28
Intrathecal administration, drug routes, 20–23
Intravascular drug binding, plasma protein, 128–135
Intravenous administration
  common drugs, 213t, 219f
  drug routes, 21–23
  kinetic scheme, 261f
  one-compartment open model, 201–221
  systemic availability, 65–66
  two-compartment open model, 271–295
  urinary recovery of unchanged drug, 211f
Intrinsic absorption rate constant, drug absorption, 224
Intrinsic proteins, cell membrane, 12
Investigation of new drug application, drug discovery and development process, 332f
Ion channels, membrane transport, 15
Ionization constants, acidic and basic drugs, 14t
Isoxicam, population approach to pharmacokinetic data, 7f

K

Kidney
  general physiological pharmacokinetic model, 301f, 302f
  relationship to liver for drug elimination, 66
  sites of drug metabolism, 145–147
  structure and function, 175–178
Kinetic parameters, graphical estimation, 276–278
Kinetics
  multiple-dose, 245–256
  urinary excretion, 210–211

L

β-Lactam antibiotics, transport
    mechanisms, 52
Large intestine
    approximate pH values, 44*f*–45
    diseases, effect on drug absorption, 100
    drug absorption, 47–48
Lineweaver–Burk expression, estimates of
    V_m and K_m from plasma-level data, 326
Lipid content, membrane, 12*f*
Lipid film, separation of intracellular and
    extracellular fluids, 11–12
Lipid-soluble molecules, intramuscular
    injection, 23
Liposomes, absorption enhancers, 92
Liquor, first-order elimination kinetics, 312
Liver
    absorption via the splanchnic circulation,
      67*f*
    distribution of drug, 272*f*
    first-pass metabolism, 320–321
    general physiological pharmacokinetic
      model, 301*f*, 302*f*
    primary site for drug metabolism, 66–67*f*
    site of drug degradation, 63
Liver disease
    effect on drug metabolism, 163–172
    factor influencing first-pass effect, 71
Loading dose
    intravenous bolus, quantity of drug, 217*f*
    steady-state levels, 252–253
Loo–Riegelman, one-compartment model
    approach, 232–233*t*
Lower rectum, drug administration, 19–40
Lung
    drug metabolism, 65–66
    site of drug degradation, 63
Luteinizing hormone releasing hormone
    intranasal administration, 27–28
    sublingual administration, 25*f*

M

Macaque, animal models for oral drug
    absorption, 54–57
Marketing authorization application, drug
    discovery and development process,
    332*f*
Maximum drug concentration in plasma,
    calculations, 230–231*t*
Maximum rate of elimination, saturable
    reaction, 313–318
Membrane permeabilities, range, 12–14

Membrane transport
    antiporters, 17
    cotransporters and exchangers, 16–17
    endocytosis, 17
    mechanisms, 14–15
    P-glycoprotein, 17
    pinocytosis, 17
    pumps, 16
    simple or passive membrane transfer, 12–
      14
    water layer, unstirred, 14–15
Membrane-limited transport, description,
    305–306
Membrane-modulated system, transdermal
    drug delivery, 33*f*–34
Metabolism
    clearance, mathematical description,
      181–182
    differences between pharmacokinetics
      and toxicokinetics, 327–328
    effect on drug pharmacokinetics, 3–9
    gut, 64–65
    improvement through physiological
      models, 298–310
    model for loss of drug from plasma, 181*f*
    pharmacokinetics and toxicokinetics,
      325–326
Metabolite concentrations in plasma,
    analysis, 262–265
Metabolite pharmacokinetics, administered
    drug, 259–270
Metabolites, active, type, 260*t*
Metered-dose inhalers, drug delivery to lung,
    31
Methacycline, mean serum concentrations
    103*f*
Method of residuals, drug concentration
    data analysis, 289*t*
Methotrexate
    drug model simulations, 307*f*
    mean levels, whole brain and spinal cord
      region, 21*t*
Methyldopa, bioavailability in dog, human,
    and monkey, 56*t*
Metoclopramide, drug–drug interactions
    affecting absorption, 100–101
Micelles, mixed, absorption enhancers, 92
Michaelis–Menten elimination,
    estimates from plasma-level data, 316–318
    saturable reaction, 313–318
Michaelis–Menten kinetics, physiological
    pharmacokinetic models, 305–306
Microcapsules, use in vaginal drug
    absorption, 37–38

Microchannel array detection, distribution of labeled drugs, 140–141

Microflora, gastrointestinal, metabolism of foreign compounds, 65*f*

Microscopic rate constants, equations, 291

Microvilli, influence on surface area of small intestine, 47*f*

Milk, effect on drug absorption, 109–110

Minerals, those available in controlled-release form, 84*f*

Mitomycin C, transdermal delivery, 36–37

Mixed-effect modeling, approach to pharmacokinetics, 7–9

Model-independent pharmacokinetics, approach to pharmacokinetics, 4–6

Molecular weight, bile excretion of drug, 70

Monkey, animal models for oral drug absorption, 54–57

Monolithic administration, drug R&D, 348–350

Motility factors, effect on enteral administration, 45–48

Mouth, approximate pH values, 44*f*–45

Mucosal surface of the epithelium, difference between small intestine and stomach, 46

Multiple dosing, drug accumulation, 253*t*

Multiple-dose kinetics, drug therapy, 245–256

Multiwire proportional counting, distribution of labeled drugs, 140–141

Muscle
  distribution of drug, 272*f*
  general physiological pharmacokinetic model, 301*f*, 302*f*
  sites of drug metabolism, 145–147

N

Nadolol, bioavailability in dog, human, and monkey, 56*t*

Nanoparticles, absorption enhancers, 93

Narcotic analgesics, effect of liver impairment on drug metabolism, 170

Nasal cavity, anatomy and physiology, 26*f*

Nebulized spray, nasal deposition, 27

Nephron, structure and function, 175–178

New drug application
  drug discovery and development process, 332*f*
  regulatory submissions, 331–332

Nicotine, transdermal systems, 35

Nitro reduction, Phase I metabolism reactions 150–154

Nitroglycerin
  controlled-release dosage forms, 87–88
  transdermal systems, 34–35

Nomenclature, list, 373–376

Nonclinical development, role of pharmacokinetics, 333–335

Nondialyzable, drugs, 195*t*

Nonlinear pharmacokinetics, descriptions, 311–322

NONMEN, pharmacokinetic profile of central tendency, 8*f*

Nonsteroidal anti-inflammatory agents, absorption enhancers, 90–91

O

One-compartment kinetics, multiple dose, 245–256

One-compartment open model
  elimination and absorption, 223–243
  intravenous dosage, 201–221
  typical compartment models, 5*f*

Oral administration
  plasma penicillin levels, 76*f*
  two-compartment open model, 271–295

Oral availability, hepatic extraction ratio, 68

Oral bioavailability, unchanged drug, 63

Oral controlled-release dosage forms
  drug release characteristics, 89*f*
  list, 88*f*

Oral drug absorption, animal models, 52–60

Organ clearance, physiological pharmacokinetic models, 299–300

Oxidative drug metabolism, mechanisms, 152

Oxytetracycline, mean serum concentrations 103*f*

Oxytocin, intranasal administration, 27–28

P

Paclitaxel, first-pass metabolism, 321

Pancreatic juice, effect on enteral administration, 50

Parent drug, resolving the pharmacokinetics of metabolite, 265–268

Parenteral administration, dosage routes, 19–40

Partial gastrectomy, effect of disease on drug absorption, 96

Particle size
  effect on dissolution rate of solid particles, 78
  effect on inhalation administration, 30

Passive absorption, interactions of drugs that share common absorption mechanisms, 102–104

Passive membrane transfer, membrane transport, 12–14

Pencillin G sodium, use of water-soluble salts, 75

Pencillin V, use of water-soluble salts, 75

Peptides, intranasal administration, 27–28

Perfusion rates, tissue effects, 122t

Perfusion, effects on drug distribution, 122–124

Perfusion-limited model, drug pathway to organ, 300–305

Perfusion-rate-limited drug, equilibration in tissue, 123t

Peritoneal dialysis, end-stage renal disease, 195

Permeabilities, range, 12–14

Permeability, muscle capillary to water-soluble molecules, 121t

Pharmacodynamics, components of drug discovery, 325–330

Pharmacokinetic characteristics, drugs unsuited for controlled release, 87–88

Pharmacokinetic model, physiological, description, 297–300

Pharmacokinetic parameters, list, 267t

Pharmacokinetic-pharmacodynamic relationships, nonclinical development, 335

Pharmacokinetics
    components of drug discovery, 325–330
    controlled-release dosage forms, 88–89
    pharmaceutical R&D, 343–353

Phase 1, clinical development, 336–337

Phase 2, clinical development, 337–338t

Phase 3, clinical development, 338–340

Phase I and II Mechanisms, mechanisms of drug metabolism, 147–156

Phenobarbital, intramuscular injection, 23

Phosphor imaging, distribution of labeled drugs, 140–141

Physical factors, improvement of drug absorption and stability, 76–78

Physicochemical factors, effect on drug absorption, 75–93

Physiological feedback, differences between pharmacokinetics and toxicokinetics, 327–328

Physiological modeling, approach to pharmocokinetics, 4, 5f, 6f

Physiological pharmacokinetic models, descriptions, 297–310

Pig, animal models for oral drug absorption, 56–57

Pinocytosis, membrane transport, 17

Plasma, drug model simulations, 307f

Plasma clearance, AUC of drug concentration curve, 235

Plasma clearance
    derivation, 278–279
    differentiation from renal clearance, 179–182
    expression with two-compartment model, 281

Plasma drug concentration, plot versus time, 226f

Plasma membranes, structural model, 13f

Plasma profiles, release rates, 235t

Plasma proteins, drugs at therapeutic drug concentrations, 125t

Polymorphism, improvement of drug absorption and stability, 76–78

Population pharmacokinetics, approach to data interpretation, 6–9

Portal circulation, blood supply to liver, 66

Portal system, absorption via the splanchnic circulation, 67f

Postmarketing studies, role of pharmacokinetics in drug discovery, 341

Postsubmission, role of pharmacokinetics in drug discovery, 341

Presystemic clearance
    differences between pharmacokinetics and toxicokinetics, 327–328
    effect of liver disease, 168

Problem sets, worked answers, 359–372

Profiles, drug, multiple intravenous doses, 254f

Progesterone, effect of food on drug absorption, 114

Propantheline, drug–drug interactions affecting absorption, 100–101

Propranolol, factors in celiac and Chrohn's disease, 99–100

Protein binding
    differences between pharmacokinetics and toxicokinetics, 327–328
    drug distribution, 125–127

Protein content, membrane, 12f

Pulmonary metabolism, effect on inhalation administration, 30

Pumps, membrane transport, 16

Pyridostigmine bromide, those available in controlled-release form, 84f

Q

Quantitative aspects, first-pass metabolism, 68–69

Quinidine, mean serum concentrations before and after antrectomy and gastroduodenostomy, 98f

R

Rabbit, animal models for oral drug absorption, 54
Radiolabeled drug
    determination of distribution, 139–141
    role of pharmacokinetics, 333–335
Rapid decline of drug concentrations, drug profile, 273–292
Rapid intravenous injection, two-compartment open model, 273f
Rate constants, various drugs 192f
Rate of administration, advantage of intravenous dosage route, 22
Rate of drug loss, equations, 260–261
Rate of transport, determination, 12
Rate-limited transport, blood flow, 300–305
Reabsorption, schematic of enterohepatic circulation, 70f
Rectal administration, drug route, 38–39
Rectum, drug absorption, 47–48
Regulatory submissions, pharmacokinetic–pharmacodynamic relationships, 340–341
Regulatory submissions, role of pharmacokinetics, 331
Release rates, plasma profiles, 235t
Renal clearance, derivation, 278–279
Renal excretion
    differences between pharmacokinetics and toxicokinetics, 327–328
    pharmacokinetics, 175–185
Renal function
    effects on pharmacokinetics, 165
    methods of measuring, 188–190
Renal impairment
    acute renal failure, 187–197
    chronic renal failure, 187–197
    drug elimination, 187–197
    prediction by creatinine clearance, 190–191
Renal tubular secretion, drugs that undergo, 180t
Repeated doses, drug accumulation, 245–253
Research and development, integration of pharmacokinetics, 342–353
Residual, method of obtaining, 290f
Respiratory agents, those available in controlled-release form, 84f
Respiratory tract, inhalation drug delivery, 29–30f

Rhesus, animal models for oral drug absorption, 54–57
Rings, use in vaginal drug absorption, 37–38

S

Salicylate levels
    administration of enteric-coated tablets to human, dog, and rabbit, 55f
    enteric-coated tablets and capsules, 83f
Sampling approach, limit of population pharmacokinetics, 9
Saturable kinetics, influence on drug concentration, 318–321
Saturable process, first-order elimination, 311–315
Scatchard plot, concentration and binding, 126–128f
Scopolamine, transdermal drug delivery, 31, 34
Second residuals, method of obtaining, 290f
Second-compartment, drug concentrations, 283f
Sequencing, transport proteins, 14
Serum level, administration of enteric-coated tablets to human, dog, and rabbit, 55f
Sex, effect on drug metabolism, 156–159
Sick cell hypothesis, liver failure, 165
Sigma-minus plot
    construction, 211
    urinary excretion, 240f
Simple membrane transfer, membrane transport, 12–14
Simple one-compartment kinetics, multiple dose, 245–256
Skin, transdermal drug delivery, 31–32
Slow decline of drug concentrations, drug profile, 273–292
Slow first-order rate, drug absorption, 255f
Small bowel, resection, effect of disease on drug absorption, 96
Small intestine
    approximate pH values, 44f–45
    disease, effect on drug absorption, 97–100
    drug absorption, 45–48
Sodium bicarbonate, tetracycline HCl absorption, 102f
Sodium ion gradients, cotransport drive, 16
Sodium-glucose cotransporter, description, 16–17
Software, pharmacokinetic data analysis, 357–359
Solid solutions, dosage forms, 79

Solubility
  differences between pharmacokinetics
    and toxicokinetics, 326–327
  diffusion, simple or passive membrane
    transfer, 12–14
  pharmacokinetic characteristics, 87–88
Solutions, dosage forms, 78–79
Solvation, influence on drug dissolution
    rate, 77–78
Species, effect on drug metabolism, 156–159
Splanchnic circulation
  absorption via the splanchnic circulation,
    67f
  drug absorption, 50–51
  drug administration,, 19–40
Spleen, sites of drug metabolism, 145–147
Stability, differences between pharmaco-
    kinetics and toxicokinetics, 326–327
Statement of the problem, drug R&D, 344–
    345
Steady state
  drug concentration, 249f
  levels, loading dose, 252–253
  two-compartment model, 280
Steric factors, effect on drug metabolism,
    156–159
Steroids, intranasal administration, 28
Stomach
  approximate pH values, 44f–45
  effect of disease on drug absorption, 96
  emptying rate, drug–drug interactions
    affecting absorption, 100–101
Sublingual administration, drug routes, 24–25
Sulfisoxazole, mean serum concentrations
    before and after antrectomy and
    gastroduodenostomy, 98f
Surfactants, absorption enhancers, 90–91
Surgery, stomach, effect of disease on drug
    absorption, 96–97t
Suspensions, dosage forms, 79
Systemic circulation, dose, 224f

T

Tablets, dosage forms, 79–80f, 82
Taxol, first-pass metabolism, 321
Terms, list, 377–380
Testosterone, blood concentrations in rats
    following nasal, intravenous, and
    intraduodenal administration, 29
Tetracycline HCl absorption, sodium
    bicarbonate, 102f
Tetracycline
  controlled-release antimicrobial agent, 83
  mean serum concentrations 103f

Theophylline
  cumulative absorption plots, 86f
  food interactions, 111–112
  transdermal delivery, 36
Therapeutic index, narrow, 86–87
Thyrotropin releasing hormone, intranasal
    administration, 27–28
Time
  intravenous bolus injection plot, 204f
  plot versus plasma drug concentration,
    226f
  versus blood concentration profiles, 314f
Time course, pharmacokinetic
    characteristics, 87–88
Time course, zero-order infusion, 215f
Time dependency, amount of metabolite in
    body, 263f
Time of maximum drug concentration,
    calculations, 230–231t
Time profile, blood concentration, 286f
Tissue
  component of physiological model, 4
  binding to drugs, 125–128
  distribution kinetics, 282–283
Toxicity, local limiting factor in intranasal
    drug delivery, 27
Toxicodynamics, components of drug
    discovery, 325–330
Toxicokinetics
  components of drug discovery, 325–330
  nonclinical development, 335
Transdermal administration, drug route,
    31–37
Transfer, one-compartment open model,
    201–221
Transport mechanisms
  GI tract absorption efficiency, 51–52
  membrane, 14
Trapezoidal rule, area under the plasma
    curve following bolus intravenous
    injection, 207–210
Tumor, general physiological
    pharmacokinetic model, 301f, 302f
Two-compartment model kinetics, model-
    independent parameters, 292–294
Two-compartment open model
  intravenous or oral administration, 271–
    295
  typical compartment models, 5f

U

Ultrafine particles, absorption enhancers, 93
Unchanged drug, cumulative urinary
    excretion, 237–241

Upper rectum, drug administration, 19–40
Urinary excretion rate, administration of
    enteric-coated tablets to human, dog,
    and rabbit, 55f
Urinary excretion
    analysis, 262–265
    kinetics, 210–211
    model for loss of drug from plasma, 181f
    sigma-minus plot, 240f
Urine
    drug excretion vs. time, 285f
    measurement of metabolites, 264–265
    quantity of metabolite recovered, 265t

V

Vaginal administration, drug route, 37–38
Valproic acid, bioavailability, 59f
Vascular perfusion, intramuscular injection,
    23
Vasopressin, intranasal administration,
    27–28
Velocity, effect on inhalation
    administration, 30
Vertical administration, drug R&D, 346–
    348, 349t
Villi, influence on surface area of small
    intestine, 47f
Vitamins, those available in controlled-
    release form, 84f
Volume of distribution, equilibrium, 279–
    282

W

Wagner–Nelson, one-compartment model
    approach, 232–233t
Wall permeability, cephalexin perfusion, 53f
Water channels, membrane transport, 15
Water layer, unstirred, membrane transport,
    14–15
Water-soluble compounds, absorption,
    disadvantage for controlled release, 87
Water-soluble molecules, intramuscular
    injection, 23
Water-soluble salts, effect on stability and
    solubility of weak acids and bases, 75–
    77f
Well-perfused organ, physiological flow
    model, 299f
Whole-body autoradiography,
    determination of drug distribution,
    139–141

Y

Yogurt, effect on drug absorption, 109–110

Z

Zero-order drug input, first-order
    elimination, 213–219
Zero-order infusion, time course, 215f

Acquisition: Cheryl Shanks
Copyediting: Jay C. Cherniak
Production: Amie Jackowski

Cover design: Amy O'Donnell
Text design and composition: Betsy Kulamer, Washington, DC
Printing: Maple Press, York, PA

PROPERTY OF
SENECA COLLEGE
LEARNING COMMONS
SENECA@YORK